University of Plymouth Library

Subject to status this item may be renewed
via your Voyager account

http://voyager.plymouth.ac.uk

Exeter tel: (01392) 475049
Exmouth tel: (01395) 255331
Plymouth tel: (01752) 232323

CHEMICAL PESTICIDES

MODE OF ACTION AND TOXICOLOGY

Jørgen Stenersen

CRC PRESS

Boca Raton   London   New York   Washington, D.C.

Library of Congress Cataloging-in-Publication Data

Stenersen, Jørgen.
    Chemical pesticides: mode of action and toxicology / Jørgen Stenersen.
        p. cm.
    Includes bibliographical references and index.
    ISBN 0-7484-0910-6
    1.   Pesticides--Toxicology.   I. Title.

    RA1270.P4S74 2004
    615.9'02--dc22

                                                                        2004043568

Visit the CRC Press Web site at www.crcpress.com

©2004 by Jørgen Stenersen

No claim to original U.S. Government works
International Standard Book Number 0-748-40910-6
Library of Congress Card Number 2004043568
Printed in the United States of America 2 3 4 5 6 7 8 9 0
Printed on acid-free paper

Dedication

To Eira, my grandchild

Without your inspiration, references from 2002 and 2003 would have been missing and the book already outdated! Although much wisdom can be extracted from Tomlin's The Pesticide Manual, *even more is found in* The Norwegian Folk Tales *and Astrid Lindgren's* Pippi Langstrømpe, *which you for long periods gave me an excuse to concentrate upon.*

About the Author

Jørgen H.V. Stenersen, Dr. Philos., is a professor in ecotoxicology at the Biological Institute, University of Oslo. He graduated as Cand. Real. in biochemistry in 1964 (University of Oslo) and subsequently worked at the Norwegian Plant Protection Institute on research related to possible side effects of pesticides. His first interests were the mechanisms behind insect resistance to insecticides, with emphasis on DDT resistance in stable flies. He was also engaged in studies of the extent of DDT contamination of soil, fauna, and humans as a result of DDT usage in orchards.

During a one-year stay at the Agricultural Research Laboratory in London, Ontario, Canada, he became interested in the effects of pesticides on earthworms at the biochemical level. This became his research focus for some years. A stay at the Biochemical Institute of the University of Stockholm led to research in the comparative biochemistry of biotransformation enzymes (glutathione transferases).

In 1985, Dr. Stenersen was a senior lecturer in ecotoxicology at the University of Oslo, becoming a professor in 1994. Since then he has been devoted to the education of environmental toxicologists. It is his opinion that the basic knowledge of health-oriented toxicologists and ecotoxicologists should be the same, or at least overlap so they can compete in the same job market and "speak the same language." He is a member of the Norwegian Committee for Approval of EUROTOX Registered Toxicologists and is responsible for the master's and Ph.D. studies of toxicology at the Biological Institute, University of Oslo.

The mode of action of pesticides, development of resistance, and their side effects are central topics in this pursuit. The present book is an enlarged revision of an earlier book written in Norwegian (*Kjemiske plantevernmidler*, 1988, Yrkeslitteratur as, Oslo, pp. 218), and is suited to introductory courses in general toxicology. Dr. Stenersen's current research interest is the application of biochemical methods in terrestrial ecotoxicology.

Acknowledgments

I thank in particular Professor John Ormerod at the Department of Biology, University of Oslo; Senior Scientist Avi Ring at the Norwegian Defense Research Institute; and Senior Scientist Christian Thorstensen at the Norwegian Plant Protection Institute for reviewing the many chapters, correcting, and proposing rewriting. Thanks also go to Baard Johannessen for providing materials to the chapter about interaction. I also thank my students that had to hear and learn more about pesticides than they actually needed for their exams and future jobs.

I also thank Catherine Russel at Taylor & Francis for her difficult job of correcting my English.

Contents

chapter one

Introduction

1.1 Motivation

The mode of action of pesticides is extremely fascinating because the subject covers so many fields of biology and chemistry and has many practical implications.

All disciplines of biology have developed greatly since 1,1,1-tri-chloro-di-(4-chlorophenyl)ethane — better known as DDT — and the other synthetic pesticides were introduced just after the Second World War. At that time, the knowledge of the normal biochemical and physiological processes in organisms was not sufficiently clarified to make it possible for us to understand properly either the mode of action of the pesticides at the target site or their uptake, distribution, and degradation in the ambient environment. The development of resistance of various pests to pesticides should have been possible to predict at that time, even before the use of these pesticides had expanded so much, but how rapidly or to what degree resistance would develop and what biochemical mechanisms where behind the development had to be a matter of experience and research.

We now know how nerve impulses are transmitted, how plants synthesize amino acids, and how fungi invade plant tissue. The textbooks in the various biological disciplines have become enormous, but in spite of this, they do not tell us where and why pesticides interfere with the normal processes. Other toxicants are mentioned occasionally, but only when they have been tools for the exploration of the normal processes. The intention of this book is therefore to try to collate some of the knowledge in the respective biological sciences and to explain the points at which the pesticides have an effect. While reading this book students are encouraged to consult textbooks in biochemistry, nerve physiology, plant biochemistry, and so forth, in order to get a more comprehensive explanation of the normal processes disturbed by pesticides. To understand the toxicology of pesticides, it is first necessary to learn organic chemistry, biochemistry, almost all disciplines of plant and animal physiology at the cellular or organismic level, and ecology, as well as the applied sciences within agriculture. This is, of course, impossible, but these disciplines will for many students be much

1

more interesting when put into the context of an applied science, e.g., pesticide science. At the least, myself and many of my students have become motivated to go back to learn more of organic chemistry and the biological sciences when confronted with the pesticides or other groups of toxicants, curious about why they are toxic or not for different organisms.

1.2   Pesticides and opinion

From 1962 to about 1975, there was hot debate about pesticides. Everyone had an opinion about them. Knowledge of chemistry, agriculture, toxicology, and so on, was not necessary. The debate was a precursor to the conflicts in the 1970s about environmental issues. In those days words like *pollutants, environmental contamination, biocides, pesticides, DDT, mercury,* etc., were synonymous. People were putting together all negative properties of synthetic compounds: they were all called biocides, were persistent with a tendency to bioaccumulate, and were all regarded as carcinogens. The scientific and technological establishment had, of course, difficulties in meeting this avalanche of opinion. Toxicology was, and still is, a much less developed science than, for instance, the science of making bridges, or other fields where hazard and risk assessments must play an essential role.

Pesticides are toxic substances applied on plants that are going to be food. It is therefore not difficult to understand the great focus from the public on these substances. The legislation and control of their use were not very much developed, and at the same time, the need for pest control agents was very high, but the growing urban population was a little removed from this reality. We must, of course, not forget the very high and unchallenged optimism of the first decade after the Second World War. DDT and the newer persistent pesticides should solve all problems of controlling insect-borne diseases as well as preventing food loss due to insect pests. The use of pesticides was indiscriminate and the remonstrance was few. Rachel Carson's book *Silent Spring* (1962) was an important warning and should be read (with caution) today.

Today the legal systems for approving pesticides are much more demanding than they were during the first optimistic years. This is illustrated by the situation in Norway. In 1965 there was only one person in a part-time job that collected data about the toxicology of pesticides in order to advise official authorities about approval and safe use. Today there are at least a dozen toxicologists as well as a few agriculturists to do the same job. The safety and agricultural advantages of all pesticides have to be scrutinized before approval. A committee of independent experts now advises the agricultural minister.

1.2.1   The fly in the soup

Pesticide residues in food were and still are much discussed, and many people are still convinced that vegetables not treated with pesticides are without poisons, taste better, have more vitamins, etc., than pesticide-treated products. Their views are not shared by more rational and scientific minds. However, it is very important for all kinds of narrow- or broad-minded experts to be aware of the important psychological factors that determine the quality of food. For instance, I do not like to eat soup if a blowfly has drowned in it. Even if it is properly removed, and even if nutrition experts convince me that the fly has increased the nutritional value of the soup by adding vitamins and proteins, I will not have the soup. Our comprehension of food quality and a pleasant environment is based upon feelings and sensations, and not on knowledge of chemical structures and dose–response extrapolations or other exact data. Food that has not been in contact with synthetic chemicals feels better, and a landscape without visible or invisible garbage is much more pleasant. We "experts" must accept that people, ourselves included, want food with neither blowflies nor synthetic chemicals, and prefer landscapes without synthetic and artificial elements. Such nonscientific emotions are important driving forces for exploring the possibilities of using nonchemical methods to combat pests.

1.2.2   Low-tech food production

There are some more political reasons to be against the use of pesticides.

Most of the pesticides have been developed in the large-scale chemical industry with a few dominating concerns. It has not been a task for the small backyard industries because the amount of work behind each substance is too great. In addition, the necessary lobbying activity and the whole marketing process are outside the sphere of the small businessman or Peter Smart-like inventor.

The development and production of pesticides may be classified as high-tech, and the use of such products is not always conforming to the changing ideas about a better and truer way of life. Moreover, the political and economic dependence of powerful multinational companies, impossible to control by democratic institutions, may be dangerous in the end. To grow (own) vegetables without the use of synthetic poisonous chemicals that have been produced by such multinational, powerful companies definitely feels better and safer. Such views were vocalized in the 1970s, but today they may appear a little outdated because so many other products used in our everyday life (mobile telephones, computers, polymers, etc.) are extremely high-tech. Nevertheless, organic and biodynamic farming and husbandry, without pesticides and antibiotics, are increasingly popular. Furthermore, the public does not receive the use of transgenic plants with enthusiasm.

1.2.3   Conclusion

The public awareness of the possible negative side effects of pesticide production and use has definitely led to a greater requirement for responsibility from the chemical industry, greater prudence on the part of farmers, and stricter legislation. However, the pesticides have certainly improved our lives by being versatile tools in food production and in the combat of insect-borne diseases.

1.3   A great market

1.3.1   The number of chemicals used as pesticides

The Pesticide Manual from 1979 (C. Worthing, 6th edition, British Crop Protection Council) presents 543 active ingredients. Approximately 100 of these are organophosphorus insecticides and 25 are carbamates used against insects. The issue of *The Pesticide Manual* from 2000 (T. Tomlin, 12th edition, British Crop Protection Council, 49 Downing St., Farnham, Surrey GU9 7PH, U.K., www.bcpc.org) describes 812 pesticides and lists 598 that are superseded. Today's 890 synthetical chemicals are approved as pesticides throughout the world and the number of marketed products is estimated to be 20,700. Organophosphorus insecticides are still the biggest group of insecticides with, according to *The Pesticide Manual*, about 67 active ingredients on the market, but the pyrethroids are increasing in importance, with 41 active ingredients. The steroid demethylation inhibitors (DMIs) constitute the main group of fungicides (31). Photosynthesis inhibitors (triazines 16, ureas 17, and other minor groups) and the auxin-mimicking aryloxyalkanoic acids (20) are still very popular as herbicides, but many extremely potent inhibitors of amino acid synthesis (e.g., the sulfonylureas (27)) are becoming more important.

It is very interesting to study lists of pesticides for sale in 1945 or earlier. Lead arsenate, mercury salts, and some organic mercury compounds, zinc arsenate, cyanide salts, nicotine, nitrocresol, and sodium chlorate were sold with few restrictions. Very few of these early pesticides are now regarded as safe. The world had a very strong need for safe and efficient pesticides like DDT. This fantastic new substance started to appear on the lists of approved pesticides under various names (Gesarol, Boxol S, pentachlorodiphenylethane, etc.) at that time.

The herbicide 2,4-D got a similar status as the first real efficient herbicide that made mechanization in agriculture possible. "The discovery of 2,4-D as an herbicide during World War 2 precipitated the greatest single advance in the science of weed control and the most significant in agriculture" (cited in Peterson, 1967).

1.3.2   Amounts of pesticides produced

Successful pesticides are produced in massive quantities. It has been estimated that between 1943 and 1974 the world production of DDT alone reached 2.8×10^9 kg (Woodwell et al., 1971). DDT was the first efficient synthetic pesticide and had all the good properties for an insecticide any person at that time could imagine. It is extremely stable, and only one treatment may suffice for good control of insect pests. It was cheap to produce and had (and still has) a low human toxicity, but is extremely active toward almost all insects. As a tool in antimalarial campaigns, it was extremely efficient. By the end of World War II it was used to combat insect-transmitted diseases and agricultural and household pests like flies and bedbugs. The production reached the maximum in 1963 with 8.13×10^7 kg in the U.S. alone. Bans and restrictions of DDT usage have since reduced the production volume of this first and efficient modern pesticide. Today an international treaty has been signed to restrict its use to very few applications in vector control.

DDT is therefore not very important as a commercial product anymore. There are no patent protections. Because of environmental problems, its usefulness is limited. Furthermore, insect resistance to DDT would in any case have restricted its usefulness.

However, other pesticides are now an integral part of agriculture throughout the world and account for approximately 4.5% of total farm production costs in the U.S. Pesticide use in the U.S. averaged over 0.544×10^9 kg of active ingredients in 1997, exceeding a price of $11.9 billion, whereas the world pesticide consumption in 1995 has been estimated to be 2.6×10^9 kg of active ingredient. The newer superactive pesticides, including herbicides such as glufosinate and glyphosate and insecticides such as the synthetic pyrethroids, can be used at very low volumes.

When measured in dollars, the herbicides dominate the market as shown by the table:

Sales	
Herbicides	47.6%
Insecticides	29.4%
Fungicides	17.4%
Others	5.5%

Herbicides are applied to 92 to 97% of acreage planted with corn, cotton, soybeans, and citrus; three quarters of vegetable acreage; and two thirds of the acreage planted with apples and other fruit.

The Nordic countries have fewer insect pests in agriculture and very few human or veterinary diseases that are transmitted by insects. There are restrictions against the use of aircraft for insecticide spraying in forestry and agriculture. Insecticides are therefore, by volume, much less significant than the herbicides.

Eighty-seven percent of the global pesticide use is in agriculture, and Europe, the U.S., and Japan constitute the biggest market, especially for herbicides, whereas insecticides dominate Asia, Africa, and Latin America. Figure 1.1 shows the approximate amounts of active ingredients in the various regions of the world.

The global chemical–pesticide market of about $31 billion is increasing 1 to 2% per year; the cost for developing a single new pesticide was estimated to be about $80 million in 1999, and the demand for much toxicological research on each single new substance is the most important reason for this

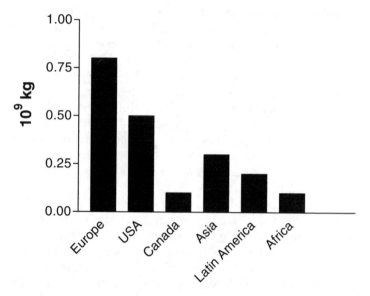

Figure 1.1 Mass of active ingredients from pesticides in different regions of the world. (From data in Board on Agriculture and Natural Resources and Board on Environmental Studies and Toxicology, C.o.L.S. 2000. The Future Role of Pesticides in U.S. Agriculture/Committee on the Future Role of Pesticides in U.S. Agriculture. 301 pp.)

high cost. It is, of course, much cheaper to develop a new pesticide when the mode of action is known. Therefore, it is not surprising that new organophosphorus insecticides and herbicidal urea derivatives are marketed every year. The pyrethroids constitute a new group of similar reputation. The exact mode of action was long not understood, but at Rothamstead Experimental Station and other institutes, basic studies of structure–activity relationships were carried out, making it possible to develop more active compounds.

1.3.3 Marketing

Very few multinational agrochemical companies dominate the market. Due to vertical and horizontal integration, the number of companies becomes fewer every year. For instance, the Swiss companies CIBA and Geigy amalgamated to become CIBA-Geigy, which amalgamated with Sandoz to form Novartis, which merged with AstraZeneca to form Syngenta. AgroEvo has merged with Rhône-Poulenc to form Aventis. The new era of biotechnology that has just started will speed up this process. Companies will try to get hold of the seed market for transgenic crops made resistant to insect pests and diseases or made tolerant to herbicides. It is worth mentioning that many countries like India, Brazil, China, and South Africa have great producers of pesticides. Often they take up the production of older pesticides without patent protection and produce pesticides that for various reasons are no longer approved in the U.S. or Europe. An example is the very toxic organophosphorus insecticide monocrotophos, which was cancelled in the U.S. in 1988 but is produced and used in Asia.

1.3.4 Dirty dozens

The profit rate for a product will decrease over the years because new competing compounds are developed, because resistance may restrict its usefulness, and because new data about ecotoxicological or human health-related toxicity appear. Many organizations engaged in environmental problems try to speed up the process and promote agricultural production without use of pesticides. It is very popular to set up lists of dirty dozens, i.e., compounds that are regarded as very hazardous for health or the environment. Very often, these substances have already been superseded and do not have any patent protection. For instance, the following list produced by the Pesticide Action Network was taken from http://www.pan-uk.org/briefing/SIDA_FIL/Chap1.html at the time of this writing. The year of marketing or patenting has been added. All substances are older than 30 years. Many of them are already in the list of superseded pesticides according to *The Pesticide Manual* (1994 or later) and are therefore of less interest today.

Dirty-Dozen List Found on the Internet

Substance	Year of Marketing/Patenting
Aldicarb	1965
Aldrin	1948
Amitrol	1955
Binapacryl	1960
Camphechlor	1947
Chlordane	1945
Chlordimeform	1966
Chlorobenzilate	1952
Chlorpropham	1951
DBCP[1]	1955
DDT	1942
Dieldrin	1948
Dinoseb	1945
EDB[2]	1946
Endrin	1951
Ethylene oxide	1935
Fluoroacetamide	1955
Heptachlor	1951
Hexachlorbenzene	1945
Hexachlorocyclohexane (mixed isomers)	1940
Isobenzan	1957
Lindane	1942
Mercury compounds	?
Methamidophos	1970

Dirty-Dozen List Found on the Internet (continued)

Substance	Year of Marketing/Patenting
Mirex	1955
Monochlorophos	1965
Paraquat	1958
Parathion	1946
Parathion-methyl	1949
Pentachlorphenol	1936
Phosphamidon	1946
Propham	1946
2,4,5-T	1944

Source: http://www.pan-uk.org/briefing/SIDA_FIL/Chap1.html.

[1] 1,2 dibromo-3-chloropropane

[2] ethylenedibromide

The public concern and the pressure groups may speed up the change to better and safer pesticides. The agrochemical companies' shift toward the development of reduced-risk pesticides is encouraged. More efficient approval procedures are an instrument that may be used to speed up the change. From 1993, the U.S. Environmental Protection Agency (EPA) began a program of expedited review of what could be classified as reduced-risk pesticides. Expedited reviews can reduce the time to registration by more than half.

It may be of interest to study the criteria established by the EPA for reduced-risk pesticides because they are important guiding principles in the development of new pesticides:

The pesticide:

- Must have a reduced impact on human health and very low mammalian toxicity
- May have toxicity lower than alternatives
- May displace chemicals that pose potential human health concerns or reduce exposures to mixers, loaders, applicators, and reentry workers
- May reduce effects on nontarget organisms (such as honey bees, birds, and fish)
- May exhibit a lower potential for contamination of groundwater
- May lower or entail fewer applications than alternatives
- May have lower pest resistance potential (have a new mode of action)
- May have a high compatibility with integrated pest management
- Has increased efficacy

About 20 such reduced-risk pesticides are now registered in the U.S., comprising herbicides (5), insecticides (8), fungicides (5), 1 bird repellent, and 1 plant activator. Their mode of action is based on more new principles. Most of them are listed below, together with the year of registration.

Pesticide	Mode of Action	Year of Registration
Herbicides		
Imazapic	ALS inhibitor (1997)[1]	1997
Imazamox	ALS inhibitor (1997)	1997
Carfentrazone	Inhibits protoporphyrinogen oxidase, giving membrane disruption (1996)	1996
Diflufenzopyr	Inhibits the auxin transport mechanism (1999)	1999
Dimethenamide-P	Cell division inhibitor	1999
Insecticides		
Diflubenzuron	Chitin synthesis inhibitor	1998
Hexaflumuron	Chitin synthesis inhibitor	1994
Pymetrozine	Feeding arrestant	1999
Tebufenozide	Binding (agonistically) to the ecdysone-binding site	1992
Pyriproxyfen	Inhibits the embryogenesis	1998
Spinosad	Activates the nicotinic acetylcholine receptor	1997
Fungicides		
Azoxystrobin	Blocks electron transfer between cytochrome b and cytochrome C_1 in the mitochondria	1997
Cyprodinil	Inhibits synthesis of methionine	1994
Fludioxonil	May inhibit phosphorylation of glucose	1993
Metalaxyl-M	Inhibits synthesis of ribosomal RNA in fungi	1996

[1] ALS: Acetolactate synthase

1.4   Nomenclature, definitions, and terminology

1.4.1   Toxicology, ecotoxicology, and environmental toxicology

The Greek word τοξιχον (toxicon) was used for poisonous liquids in which arrowheads were dipped. The word *toxicology*, derived from this word, has been used as the name of the science within human medicine that describes the effect of poisons on humans. The definition includes uptake, excretion, and metabolism of poisons (toxicokinetics), as well as the symptoms and how they develop (toxicodynamics). We can say that the toxicodynamics tell us what the toxicants do to the organisms; and toxicokinetics, what the organism does with the substance. Toxicology also includes the legislation enforced to protect the environment and human health, and the risk assessments necessary for this purpose. Today a toxicologist is not exclusively working with the species *Homo sapiens* or model organisms like rats, but all kinds of organisms.

The term *ecotoxicology* is defined as "the science occupied with the action of chemicals and physical agents on organisms, populations, and societies within defined ecosystems. It includes transfer of substances and interactions with the environment" (e.g., Hodgson et al., 1998). Ecotoxicology is sometimes used synonymously with *environmental toxicology*; however, the latter also encompasses the effects of environmental chemicals and other agents on humans. Because the basic chemical and physical processes behind the interaction between biomolecules and chemicals are independent of the type of organism, it is not necessary to have a too rigid division between the various branches of toxicology.

1.4.2   Pesticides, biocides, common names, chemical names, and trade names

Pesticides are chemicals specifically developed and produced for use in the control of agricultural and public health pests, to increase production of food and fiber, and to facilitate modern agricultural methods. Antibiotics to control microorganisms are not included. They are usually classified according to the type of pest (fungicides, algicides, herbicides, insecticides, nematicides, and molluscicides) they are used to control. When the word *pesticide* is used without modification, it implies a material synthesized by humans. *Plant pesticide* is a substance produced naturally by plants that defends against insects and pathogenic microbes — and the genetic material required for production.

The term *biocide* is not used much in the scientific literature. It may be used for a substance that is toxic and kills several different life-forms. Mercury salts (Hg^{++}) may be called biocides because they are toxic for microorganisms, animals, and many other organisms, whereas DDT is not a biocide because of its specificity toward organisms with a nervous system (animals).

The word is also sometimes used as a collective term for substances intentionally developed for use against harmful organisms. In a directive from the European Community (EU Biocidal Products Directive 98/8/EC), we find the following definition:

> The new Directive describes biocides as chemical preparations containing one or more active substances that are intended to control harmful organisms by either chemical or biological, but by implication, not physical means. The classification of biocides is broken down into four main groups — disinfectants and general biocides, preservatives, pest control and other biocides and these are further broken down into 23 separate categories.

Pesticides have one or more *standard name(s)* and one or more *chemical name(s)*. The different companies make products with registered *trade names*. They should be different from the standard names, but also have to be approved. The chemical industry also frequently uses a code number for its products. In Germany, for instance, old farmers still know parathion by the number E-605, which was used by Bayer Chemie before a standard name and a trade name were given to O,O'-diethyl paranitrophenyl phosphorothioate. The chemical name is often very complicated and even difficult to interpret for a chemist. The chemical formula, however, is often much simpler and may tell something about the property of the compound even to a person with moderate knowledge of chemistry.

One or more national standardization organizations and the International Organization of Standardization approve standard names. The chemical names are either according to the rules of the International Union of Pure and Applied Chemistry or according to Chemical Abstracts. The so-called Chemical Abstracts Services Registry Number (CAS-RN) is a number that makes it easy to find the product or chemical in databases from Chemical Abstracts. The standard names are regarded as ordinary nouns, but the pesticide products are sold under a trade name that is treated as a proper name with a capital initial letter. We use the various names of a fungicide as an example:

Common Names

British Standards Institution	Captan
International Organization for Standardization (French spelling)	Captan
Japanese Ministry for Agriculture, Forestry and Fishery	Captan
South Africa	Captan
Norsk språkråd (Norwegian standard)	Kaptan

Chemical Names

Chemical Abstracts (CA)
  3a,4,7,7a-tetrahydro-2-[(trichloromethyl)thio]-1*H*-isoindole-1,3(2*H*)-dione
International Union of Pure and Applied Chemistry (IUPAC)
  N-(trichloromethylthio)cyclohex-4-ene-1,2-dicarboximide

Trade Names

Captan, Captec, Merpan, Orthocide, Phytocape, etc.
  (as many as 38 different trade names and chemical names have been recorded for this substance alone)

Chemical Abstracts Services Registry Number (CAS-RN)

133-06-2

Various Codes

SR 406, ENT 26538

Chemical Structure

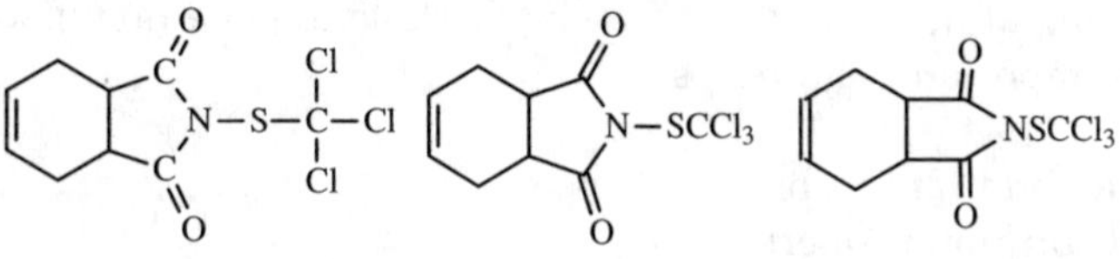

1.4.3   Chemical structures are versatile

The chemical structures are the most versatile way, even for nonchemists, to define a chemical. The chemical structure hides or, better, displays all the properties of the compound. The chemical structures may also be written in several ways. It is therefore not a waste of time to learn some examples of chemical formulas for the more important groups of pesticides.

There are some conventions about how the structures are depicted, but in this book, the structure is drawn to make clear the important points. For instance, the structure for atrazine should be written with this orientation, with the number 1 ring — nitrogen — upward.

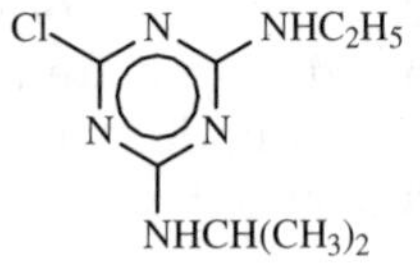

$$Cl \quad N \quad NHC_2H_5$$

$$NHCH(CH_3)_2$$

It is easier to remember and to see the symmetry when written in this direction:

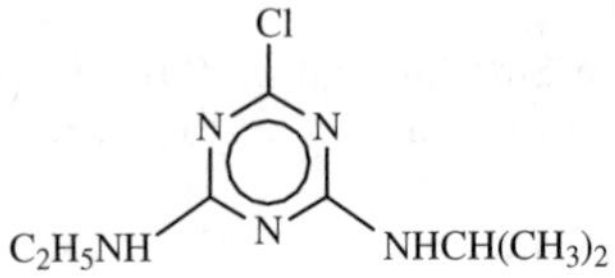

$$Cl$$

$$C_2H_5NH \quad N \quad NHCH(CH_3)_2$$

Remember that the same structural elements may be written quite differently. Carboxyl groups (organic acids) may be drawn in two ways, or in the anionic form, without the hydrogen:

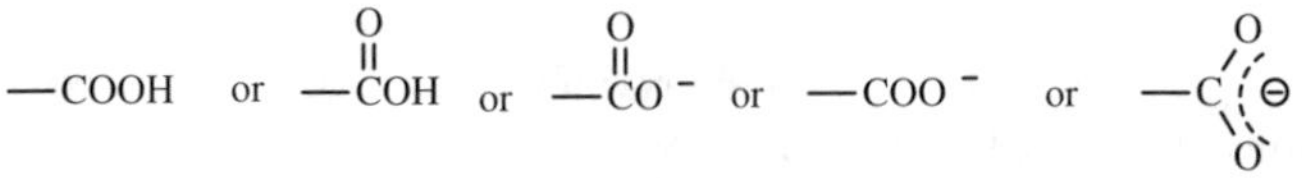

$$-COOH \quad or \quad -\overset{O}{\overset{\|}{C}}OH \quad or \quad -\overset{O}{\overset{\|}{C}}O^- \quad or \quad -COO^- \quad or \quad -C\overset{O}{\underset{O}{\diagdown}}\ominus$$

A methylene bridge can be written in at least three different ways:

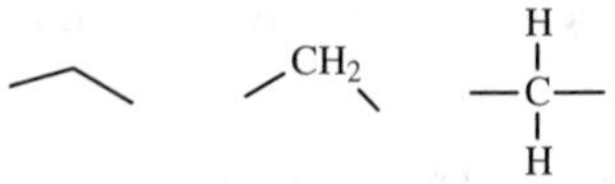

$$\diagup\diagdown \qquad \diagup^{CH_2}\diagdown \qquad -\overset{H}{\underset{H}{\overset{|}{C}}}-$$

Remember also that the paper in this book is flat, but molecules are three-dimensional and their true shape cannot easily be drawn in two dimensions.

By looking at the formula, it may be possible to get a qualified opinion about such important features as:

•   Water and fat solubility
•   Soil sorption property

- Stability toward oxidation, UV light, biotransformation, etc.
- Classification and mode of action
- Stereoisomeri — carbon (or phosphorus) atoms that are connected to four different groups will give stereoisomeric compounds that are biologically different from each other
- Composition and possible xenobiotic character — what elements does the compound contain besides carbon and hydrogen (sulfur, halogen, nitrogen, some odd metals, silicium, etc).

This can be done without much theoretical knowledge of chemistry. Unfortunately, the current knowledge in toxicology is not sufficient to make it possible to deduce all the properties of a chemical just by looking at the structure, but a lot can be said, or at least presumed.

Helpful reading

There is an extensive literature cited section at the end of the book. The following books are useful as general texts.

Biochemistry and cell biology

Alberts, B., Bray, D., Johnson, A., Lewis, J., Raff, M., Roberts, K., and Walter P. 1998. *Essential Cell Biology: An Introduction to the Molecular Biology of the Cell*. Garland Pub., New York. 630 pp.

Alberts, B., Johnson, A., Lewis, J., Raff, M., Roberts, K., and Walter, P. 2002. *Molecular Biology of the Cell*. Garland Science, Taylor & Francis Group, London. 1463 pp.

Nelson, D.L. and Cox, M.M. 2000. *Lehninger Principles of Biochemistry*. Worth Publishers, New York. 1150 pp.

General toxicology

Hayes, A.W. 2001. *Principles and Methods of Toxicology*, Vol. XIX. Taylor & Francis, Philadelphia. 1887 s. pp.

Hodgson, O., Mailman, R.B., Chambers, J.E., and Dow, R.E. 1998. *Dictionary of Toxicology*. MacMillan, New York. 504 pp.

Klaassen, C., Ed. 2001. *Cassarett and Doull's Toxicology. The Basic Science of Poisons*. McGraw-Hill, New York. 1236 pp.

Timbrell, J. 2000. *Principles of Biochemical Toxicology*. Taylor & Francis, London. 394 pp.

Insect biochemistry, plant physiology, and neurophysiology

Breidbach, O. and Kutsch, W. 1995. *The Nervous Systems of Invertebrates: An Evolutionary and Comparative Approach*. Birkhäuser Verlag, Basel, Switzerland. 448 pp.

Levitan, I.K. and Kaczmarek, L.K. 2002. *The Neuron Cell and Molecular Biology*. Oxford University Press, Oxford. 603 pp.

Rockstein, M. 1978. *Biochemistry of Insects*. Academic Press, New York. 649 pp.

Taitz, L. and E. Zeiger. 1998. *Plant Physiology*. Sinauer Associates, Inc., Sunderland, MA.

Pesticides

Bovey, R.W. and Young, A.L. 1980. *The Science of 2,4,5-T and Associated Phenoxy Herbicides*. John Wiley & Sons, New York. 462 pp.

Casida, J.E. and Quistad, G.B. 1998. Golden age of insecticide research: past, present, or future? *Annu. Rev. Entomol.*, 43, 1–16.

Devine, M., Duke, S.O., and Fedke, C. 1993. *Physiology of Herbicide Action*. Prentice Hall, New York. 441 pp.

Fedke, C. 1982. *Biochemistry and Physiology of Herbicide Action*. Springer-Verlag, Heidelberg, Germany. 202 pp.

Gressel, J. 2002. *Molecular Biology of Weed Control*, Vol. XVI. Taylor & Francis, London. 504 pp.

Köller, W. 1992. *Target Sites of Fungicide Action*. CRC Press, Boca Raton, FL. 328 pp.

Schrader, G. 1951. *Die Entwicklung neuer Insektizide auf Grundlage organischer Fluor- und Phosphor-Verbindungen*. Verlag Chemie, Weinheim, Germany. 92 pp.

Schrader, G. 1963. *Die Entwicklung neuer insectizider Phosphrsäure-Ester*. Verlag Chemie GMBH, Weinheim/Bergstr., Germany.

Tomlin, C., Ed. 1994. *The Pesticide Manual: Incorporating the Agrochemicals Handbook*. British Crop Protection Council, Farnham, Surrey.

Tomlin, C., Ed. 2000. *The Pesticide Manual: A World Compendium*, 12th ed. British Crop Protection Council, Farnham, Surrey. 1250 pp.

West, T.F. and Campbell, G.A. 1950. *DDT and Newer Persistent Insecticides*. Chapman & Hall Ltd., London. 632 pp.

Wilkinson, C.F. 1976. *Insecticide Biochemistry and Physiology*, Vol. XXII. Plenum Press, New York. 768 pp.

Worthing, C., Ed. 1979. *The Pesticide Manual: A World Compendium*, 6th ed. British Crop Protection Council, Croydon. 655 pp.

The current Web address of the British Crop Protection Council is www.bcpcorg. It is useful for ordering the current issue of *The Pesticide Manual* and for updating the knowledge of pesticides.

Side effects of pesticides

Board on Agriculture and Natural Resources and Board on Environmental Studies and Toxicology, C.o.L.S. 2000. *The Future Role of Pesticides in U.S. Agriculture/ Committee on the Future Role of Pesticides in U.S. Agriculture*. 301 pp.

Carson, R. 1962. *Silent Spring*. The Riverside Press, Boston, MA. 368 pp.

Ecobichon, D.J. 2001. Toxic effects of pesticides. In *Cassarett and Doull's Toxicology. The Basic Science of Poisons*, Klaassen, C., Ed. McGraw-Hill, New York. pp. 763–810.

Emden, H.P.D. 1996. *Beyond Silent Spring*. Chapman & Hall, London. 322 pp.

Graham, J. and Wienere, B. 1995. *Risk versus Risk*. Harvard University Press, Cambridge, MA. 337 pp.

Mellanby, K. 1970. *Pesticides and Pollution*. Collins, London. 221 pp.

Mineau, P. 1991. *Cholinesterase-Inhibiting Insecticides. Their Impact on Wildlife and the Environment*. Elsevier, Amsterdam. 348 pp.

Richardson, M. 1996. *Environmental Xenobiotics*. Taylor & Francis, London. 492 pp.

Walker, C.H., Hopkin, S.P., Sibly, R.M., and Peakall, D.B. 1996. *Principles of Ecotoxicology*. Taylor & Francis, London. 321 pp.

chapter two

Why is a toxicant poisonous?

Theophrastus Bombastus von Hohenheim, better known in history as Paracelsus, who was born in the Swiss village of Einsiedeln in 1493 and died in 1541, taught us that the severity of a poison was related to the dose (see Strathern, 2000). His citation "All substances are poisons; there is none which is not a poison. The right dose differentiates a poison from a remedy" is found in the first chapter of almost all textbooks of toxicology or pharmacology. However, the molecular theory was formulated more than 300 years later, and the law of mass action not until after the middle of the 19th century. Real rational toxicology and pharmacology are dependent on these laws, and hence could not develop properly before they were known.

Paracelsus' idea that all substances are poisons is, of course, correct; even water, air, and sugar are poisons in sufficient amounts, but by looking at the chemical structures of typical poisons, and trying to sort out the reactions they tend to be involved in, we can roughly put them into seven categories. By using the molecular theory, the law of mass action, and our knowledge of the nature of the chemical processes in organisms, we can condense biochemical toxicology to three sentences, and about seven types of reactions:

1. Toxic molecules react with biomolecules according to the common laws of chemistry and physics, so that normal processes are disturbed.
2. The symptoms increase in severity with increasing concentration of the toxicant at the site of reaction.
3. This concentration increases with increasing dose.

2.1 Seven routes to death

The chemist may prefer to classify toxicants according to their chemical structure, the doctor according to the organ they harm, the environmentalist according to their stability in the environment, and so forth. The biochemist may use a different classification, and we will approach the toxicology of pesticides from the biochemist's perspective. Because of point 1 above, and because the cells in all organisms are very similar, it is possible to classify toxicants into roughly seven categories according to the type of biomolecule they react with. Toxicants in the same category do not need to be chemically related, and one substance may act through several mechanisms. The following simple classification is based on the more comprehensive texts of Ecobichon (2001) and Gregus and Klaassen (2001).

2.1.1   Enzyme inhibitors

The toxicant may react with an enzyme or a transport protein and inhibit its normal function. Enzymes may be inhibited by a compound that has a similar, but not identical structure as the true substrate; instead of being processed, it blocks the enzyme. Typical toxicants of this kind are the carbamates and the organophosphorus insecticides that inhibit the enzyme acetyl cholinesterase. Some extremely efficient herbicides that inhibit enzymes important for amino acid synthesis in plants, e.g., glyphosate and glufosinate, are other good examples in this category.

Enzyme inhibitors may or may not be very selective, and their effects depend on the importance of the enzyme in different organisms. Plants lack a nervous system and acetylcholinesterase does not play an important role in other processes, whereas essential amino acids are not produced in animals. Glyphosate and other inhibitors of amino acid synthesis are therefore much less toxic in animals than in plants, and the opposite is true for the organophosphorus and carbamate insecticides.

Sulfhydryl groups are often found in the active site of enzymes. Substances such as the Hg^{++} ion have a very strong affinity to sulfur and will therefore inhibit most enzymes with such groups, although the mercury ion does not resemble the substrate. In this case, the selectivity is low.

2.1.2   Disturbance of the chemical signal systems

Organisms use chemicals to transmit messages at all levels of organization, and there are a variety of substances that interfere with the normal functioning of these systems. Toxicants, which disturb signal systems, are very often extremely potent, and often more selective than the other categories of poisons. These toxicants may act by imitating the true signal substances, and thus transmit a signal too strongly, too long lasting, or at a wrong time. Such poisons are called *agonists*. A typical agonist is nicotine, which gives signals similar to acetylcholine in the nervous system, but is not eliminated by acetylcholinesterase after having given the signal. Other quite different agonists are the herbicide 2,4-D and other aryloxyalkanoic acids that mimic the plant hormone auxin. They are used as herbicides. An *antagonist* blocks the receptor site for the true signal substance. A typical antagonist is succinylcholin, which blocks the contact between the nerve and the muscle fibers by reacting with the acetylcholine receptor, preventing acetylcholine from transmitting the signal. Some agonists act at intracellular signal systems. One of the strongest man-made toxicants, 2,3,7,8-tetrachlorodibenzodioxin, or dioxin, is a good example. It activates the so-called *Ah* receptor in vertebrates, inducing several enzymes such as CYP1A1 (see p. 181). Organisms use a complicated chemical system for communication between individuals of the same species. These substances are called *pheromones*. Good examples are the complicated system of chemicals produced by bark beetles in order to attract other individuals to the same tree so that they can kill them and make them suitable as substrates. Man-made analogues of these pheromones placed in traps are examples of poisons of this category. The *kairomons* are chemical signals released by individuals of one species in order to attract or deter individuals of another. The plants' scents released to attract pollinators are good examples.

Signals given unintentionally by prey or a parasite host, which attract the praying or parasitizing animal, are important. A good example is CO_2 released by humans, which attracts mosquitoes. The mosquito repellent blocks the receptors in the scent organ of mosquitoes.

2.1.3 Toxicants that generate very reactive molecules that destroy cellular components

Most redox reactions involve exchange of two electrons. However, quite a few substances can be oxidized or reduced by one-electron transfer, and reactive intermediates can be formed. Oxygen is very often involved in such reactions. The classical example of a free radical-producing poison is the herbicide *paraquat*, which steals an electron from the electron transport chain in mitochondria or chloroplasts and delivers it to molecular oxygen. The superoxide anion produced may react with hydrogen superoxide in a reaction called the Fenton reaction, producing hydroxyl radicals. This radical is extremely aggressive, attacking the first molecule it meets, no matter what it is. A chain reaction is started and many biomolecules can be destroyed by just one hydroxyl radical. Because one paraquat molecule can produce many superoxide anions, it is not difficult to understand that this substance is toxic. *Copper* acts in a similar way because the cupric ion (Cu^{++}) can take up one electron to make the cuprous cation (Cu^+) and give this electron to oxygen, producing the superoxide anion ($O_2^{.-}$).

Free radical producers are seldom selective poisons. They work as an avalanche that destroys membranes, nucleic acids, and other cell structures. Fortunately, the organisms have a strong defense system developed during some billion years of aerobic life.

2.1.4 Weak organic bases or acids that degrade the pH gradients across membranes

Substances may be toxic because they dissolve in the mitochondrial membrane of the cell and are able to pick up an H^+ ion at the more acid outside, before delivering it at the more alkaline inside. The pH difference is very important for the energy production in mitochondria and chloroplasts, and this can be seriously disturbed. Substances like *ammonia, phenols,* and *acetic acid* owe their toxicity to this mechanism. Selectivity is obtained through different protective mechanisms. In plants, ammonia is detoxified by glutamine formation, whereas mammals make urea in the ornithine cycle. Acetic acid is metabolized through the citric acid cycle, whereas phenols can be conjugated to sulfate or glucuronic acid. Phenols are usually very toxic to invertebrates, and many plants use phenols as defense substances.

2.1.5   Toxicants that dissolve in lipophilic membranes and disturb their physical structure

Lipophilic substances with low reactivity may dissolve in the cell membranes and change their physical characteristics. *Alcohols, petrol, aromatics, chlorinated hydrocarbons,* and many other substances show this kind of toxicity. Other, quite unrelated organic solvents like *toluene* give very similar toxic effects. Lipophilic substances may have additional mechanisms for their toxicity. Examples are *hexane,* which is metabolized to 2,5-*hexandion,* a nerve poison, and *methanol,* which is very toxic to primates.

2.1.6   Toxicants that disturb the electrolytic or osmotic balance or the pH

Sodium chloride and other salts are essential but may upset the ionic balance and osmotic pressure if consumed in too high doses. Babies, small birds, and small mammals are very sensitive. Too much or too little in the water will kill aquatic organisms.

2.1.7   Strong electrophiles, alkalis, acids, oxidants, or reductants that destroy tissue, DNA, or proteins

Caustic substances like strong acids, strong alkalis, bromine, chlorine gas, etc., are toxic because they dissolve and destroy tissue. Many accidents happen because of carelessness with such substances, but in ecotoxicology they are perhaps not so important. More interest is focused on *electrophilic* substances that may react with DNA and induce cancer. Such substances are very often formed by transformation of harmless substances within the body. Their production, occurrence, and protection mechanisms will be described in some detail later.

2.2   How to measure toxicity

2.2.1   Endpoints

In order to measure toxicity, it is important to know what to look for. We must have an *endpoint* for the test. An endpoint can be very precise and easy to monitor, such as death, or more sophisticated, for instance, lower learning ability or higher risk for contracting a disease. Some endpoints are all-or-none endpoints. At a particular dose some individuals will then get the symptoms specified in the definition of the endpoint and others do not. Tumors or death are such all-or-none endpoints. Such endpoints are often called *stochastic*, whereas endpoints that all individuals reach, to varying but dose-dependent degrees, are called deterministic endpoints. Intoxication by alcohol is a good example. We use the term *response* for the stochastic all-or-none endpoints and the term *effect* for gradual endpoints.

2.2.1.1 *Endpoints in ecotoxicology and pest control*
The fundamental endpoints for nonhuman organisms are:

- Death
- Reduced reproduction
- Reduced growth
- Behavioral change

These endpoints are, of course, connected.

Reduced reproduction is probably the most important endpoint in ecotoxicological risk assessments, whereas in pest control, death or changes in behavior are the most important. We simply want to kill the pest or make it run away. Toxicity tests are often based on what we call surrogate endpoints. We measure the level of an enzyme and how its activity is increased (e.g., CYP1A1) or reduced (acetylcholinesterase), how a toxicant reduces the light of a phosphorescent bacterium, or how much a bacterium mutates. Such endpoints are not always intuitively relevant to human health or environmental quality, but much research is done in order to find easy and relevant endpoints other than the fundamental ones.

2.2.1.2 *Endpoints in human toxicology*
In human toxicology, we have a lot more sophisticated endpoints related to our well-being and health. At the moment, cancer is the most feared effect of chemicals, and tests that can reveal a chemical's carcinogenicity are always carried out for new pesticides. Other tests that may reveal possible effects on reproduction and on the fetus are important. Endpoints such as immunodeficiency, reduced intelligence, or other detrimental neurological effects will play an important role in the future. The problem is that almost all endpoints in human toxicology are surrogate endpoints, and elaborate and dubious extrapolations must be done. The new techniques under development that make it possible to determine the expression of thousands of genes by a simple test will very soon be used in toxicological research, but the interpretation problems will be formidable.

2.2.2   Dose and effect

The law of mass action tells us that the amount of reaction products and the velocity of a chemical reaction increase with the concentrations of the reactants. This means that there is always a positive relation between dose and the degree of poisoning. A greater dose gives a greater concentration of the toxicant around the biomolecules and therefore more serious symptoms because more biomolecules react with the toxicant and at a higher speed. This simple and fundamental law of mass action is one of the reasons why a chemist does not believe in homeopathy. It is also the reason why Paracelsus (1493–1541) was right when saying "All substances are poisons; there is none which is not a poison. The right dose differentiates a poison from a remedy" (Strathern, 2000). The connection between dose or concentration of the toxicant and the severity of the symptoms is fundamental in toxicology. By using the law of mass action, we get the following equilibrium and mathematical expression:

$$B + T \xrightleftharpoons{K} BT$$

$$K = \frac{C_B \cdot C_T}{C_{BT}} \ \text{or} \ \ C_B = K \cdot \frac{C}{C_T + K} \ \text{if } C = C_B + C_{BT}$$

The target biomolecule (B) at the concentration C_B reacts with the toxicant (T) at the concentration C_T to give the destroyed biomolecule (BT) at the concentration C_{BT}. The reaction may be reversible, as indicated by the double arrow. C is the total concentration of the biomolecule and K is the equilibrium constant. If the onset speed of the symptoms is proportional with the disappearance rate of the biomolecules ($-dC_B/dt$), we get this simple mathematical expression telling us that the higher the concentration of the toxicant is, the faster C_B will decrease and the symptoms appear:

$$-\frac{dC_B}{dt} = k_{+1} \cdot C_B \cdot C_T$$

k_{+1} is the velocity constant for the reaction.

These simple formulae illustrate that higher concentrations of a toxicant give a lower amount of the biomolecule and thus stronger symptoms. The onset of symptoms may start when C_B is under a certain threshold or C_{BT} is above a threshold.

The real situation is more complicated. The toxicant may react with many different types of biomolecules. It may be detoxified or need to be transformed to other molecules before reacting with the target biomolecule.

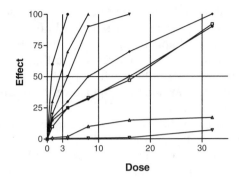

Figure 2.1 A hypothetical example of the effects on eight individuals of a toxicant at different doses.

2.2.3   Dose and response

The sensitivity of the individuals in a group is different due to genetic heterogenicity as well as difference in sex, age, earlier exposure, etc. Therefore, if the effect of a toxicant is plotted against the dose, every individual will get a curve that is more or less different from those of other individuals. In Figure 2.1, some effect on eight individuals is shown. The difference is exaggerated in order to elucidate the points.

Figure 2.1 illustrates a hypothetical example. The effect may be any measurable symptom that has a graded severity. Three individuals seem to be very sensitive, whereas one or two are almost resistant. This figure leads us to a very important concept called *response*. Response (r) is defined as the number of individuals getting symptoms higher than a defined threshold. If we decide that the symptom threshold should be 50, we observe that at doses 3, 10, 20, and 30 the response will be 2, 4, 6, and 6, respectively. When determining the response, we just count how many individuals have the required or higher symptoms.

The relative response (p) is the number of responding individuals divided by the total number given a certain dose. At the marked dose levels in Figure 2.1, the relative responses are 0.25, 0.5, 0.75, and 0.75, respectively. These numbers may be multiplied by 100 to give the percent response.

We very often measure all-or-none symptoms (dead or alive, with tumor or without tumor, numbers of fetus with injury or normal ones) in toxicology. Such symptoms are not gradual. We then have to expose groups of individuals with different doses (D) and determine the number of responding individuals (r) and the relative number (p).

If we have many groups with a high number of individuals and then plot the relative response against the dose, we very often get an oblique

S-shaped graph, with an inflection point at 50% response. The graph may be made symmetrical by plotting *log dose* instead of dose. Furthermore, the S-shaped graphs can be changed into straight lines by transforming the responses to *probit response*. We then presuppose that the sensitivity of the organisms has a normal distribution, which predicts that most individuals have average sensitivity, a few are very robust, very few are almost resistant, and some have high sensitivity.

The log transformation of dose or concentration is easily done with a pocket calculator. Using the formula for the inverse normal distribution in the data sheet *Excel*, one can easily do the calculation of the probit values. The mean or median is set to 5 and the standard deviation to 1, i.e., the formula will look like this:

=NORMINV (relative response; 5 ;1)

By writing the relative response into the formula, Excel will return the probit value.

Note that the probit of 0.5 (50% response) is 5, and the probit of 0.9 (90% response) is 6.282. The reader should try other values if Excel is available. Note also that the probit of 0 is $-\infty$, whereas the probit of 1 is $+\infty$. Values of 0 or 100% response are therefore useless in this plot. Figure 2.2 and Figure 2.3 show the essence of some dose–response curves.

Figure 2.3a and b shows a case with sensitive and resistant flies mixed 50:50. The same data are used in both plots.

Figure 2.2a to c and Figure 2.3a and b show that the transformation of the doses to log dose, and the use of probit units for responses, makes it much easier to interpret the graphs. However, there are several difficulties with dose–response graphs.

Mathematically, the probit of a value P is Y in the integral

$$P = \frac{1}{\sqrt{2\pi}} \int_{-\infty}^{Y-5} e^{-\frac{1}{2}u^2} du$$

It cannot be expressed as a simple function, and some mathematical skill is necessary to interpret its meaning. Therefore, the much simpler logit transformation $L = \ln\{P/(1 - P)\}$ is often used. The logit values (L) can be calculated from the relative response values (P) with a pocket calculator. The logit transformation also gives almost straight lines if the sensitivity is normally distributed. The most serious problem with dose–response graphs, however, is not this mathematical inconvenience. The low reproducibility is a more serious problem. As an example, if you know exactly the LD50 (lethal dose in 50% of the population) and give this to 10 animals, the probability that 5 die is only 0.246. The confidence intervals of the responses for the same dose, or for the doses calculated to give a specified response (e.g.,

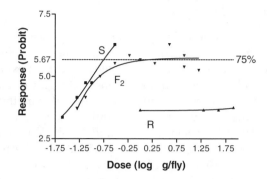

Figure 2.4 Dose–response relationships of *Stomoxys calcitrans* treated with the DDT analogue DDD. S, susceptible strain; R, resistant strain; F$_2$, second generation from crosses of S and R.

resulting F2 generation was tested. As seen from Figure 2.4, these flies had a very heterogeneous sensitivity against DDD. About 75% (probit value of 5.674490) were quite sensitive, whereas 25% were almost impossible to kill with DDD. This result is expected if just one (recessive) gene is involved in the resistance mechanism. The other DDT group insecticides gave similar results.

2.2.3.2   Scatter in dose–response data

The figure of the *Stomoxys* strains also illustrates the wide scatter expected for the response data. Each point is based on 20 individuals, i.e., more than 400 flies plus controls (60) were used in this small experiment. The scatter is formidable in spite of the great number of flies used. The reason is the stochastic nature of the outcome. For instance, the probability (P) of getting exactly 15 dead flies by using a dose that should kill 75% is only P = 0.203. It is much more probable that we get another "wrong" value. These results may be calculated from the binomial formula

$$P = \frac{n!}{(n-r)! \times r!} \times p^r \times (1-p)^{(n-r)}$$

where n = 20, number of insects tested in a group; r = 15, number of insects dying; p = 75/100, the expected value of relative response when a huge number of insects was used; and ! is the faculty sign (e.g., n! = n × (n − 1) × (n − 2) ... 3 × 2 × 1). It may be calculated that P = 0.203, which is the probability of obtaining a response of r = 15 in an experiment where p = 0.75 and n = 20. An outlier (see Figure 2.4), as that obtained at 4 μg/fly (log dose = 0.602), with a response of r = 18 (90% mortality instead of the expected 75%), has a probability of 0.0669, i.e., is expected in as many as 7 experiments out of 100. Such uncertainties are inherent in dose–response relationships and have nothing to do with experimental errors, which may also be a source of scattering.

2.2.4 LD50 and related parameters

The statistical problems in making good dose–response curves can only be overcome by using many organisms in the experiment. A better method may be to determine one dose, for instance, the dose that is expected to kill or harm 50% of the individuals, and not to construct a graph. This can be done satisfactorily with much fewer individuals. The latter method is definitely better when studying vertebrates. Most countries have strict legislation concerning the use of vertebrates in research, and it is difficult to get permission to do experiments involving hundreds of animals. Furthermore, most vertebrates suitable for research are expensive. Therefore, we seldom find graphs of dose–response relationships on mammals in the more recent scientific literature. More often, we find a value called LD50 that can be determined with reasonable accuracy by using few individuals. LD50 is the dose expected to kill half of the exposed individuals. Sometimes we are interested in determining the doses that kill 90 or 10%, etc., and these doses are called LD90 and LD10, respectively. They can easily be determined from a dose–response curve, but these values are less accurate than LD50. If we study endpoints other than death, we use the term ED50 (effective dose in 50% of the population), and if we study *concentrations* and not doses, we use the terms LC50 (lethal concentration in 50% of the population) and EC50 (effective concentration in 50% of the population). Protocols for determination of LD50 for rodents are available in order to minimize the number of animals necessary for a satisfactory determination. According to Commission of the European Communities' Council Directive 83/467/EEC, 20 rats may be sufficient for an appropriate LD50 determination. LD50 values are often given as milligram of toxicant per kilogram of body weight of the test animals, assuming that twice as big a dose is necessary to kill an animal of double weight. It is therefore easier to compare toxicity data from different species, life stage, or sex. The LD50 values or the related values should not be taken as accurate figures owing to the intrinsic nature of these parameters, as well as the difficulties of determination. Even if you know the exact LD50 value, for example, of parathion to mice (LD50 =12 mg/kg according to *The Pesticide Manual*), and give these doses to a group of animals, for instance, n = 10, the probability that r = 5 will die is only P = 0.246. This can easily be calculated from the binomial formula. However, you can be confident that between 1 and 9 will die (P = 0.998). LD50 values are therefore very useful if you do not need to know the exact number of fatalities, but merely want to describe the toxicity of a compound by one figure. Complicated statistical methods are needed to determine the true confidence limits of LD50. Many statistical methods are described in the books of Finney (1971) and Hewlett and Plackett (1979). Data programs may be used, e.g., Sigmaplot® or Graphpad Prism®. A simple program in BASIC is available (Trevors, 1986), whereas Caux and More (1997) describe the use of Microsoft Excel®.

Table 2.1 shows how toxicants are classified according to their LD50.

Table 2.1 Common Classification of Substances

Toxicity Class	LD50 (mg/kg)	Examples, LD50 (mg/kg)
Extremely toxic	Less than 1.0	Botulinum toxin: 0.00001
		Aldicarb: 1.0
Very toxic	1–50	Parathion: 10
Moderately toxic	50–500	DDT: 113–118
Weakly toxic	500–5000	NaCl: 4000
Practically nontoxic	5000–15,000	Glyphosate: 5600
		Ethanol: 10,000
Nontoxic	More than 15,000	Water

2.2.5 Acute and chronic toxicity

An important distinction has to be made between acute and chronic toxicity.

Substances that are eliminated very slowly and therefore accumulate if administered in several small doses over a long time may, when the total dose is large enough, cause symptoms. A good example is cadmium that accumulates in the kidneys. Another example is organophosphates that in repeated small doses eventually inhibit acetylcholinesterase more than 80%, which will produce neurotoxic symptoms. Because the inhibition is partly irreversible, many small doses may cause poisoning even though the poison itself does not accumulate. Other poisons (e.g., ethanol) may be given in large, but sublethal doses for years before any sign of chronic toxicity is observed (liver cirrhosis), whereas the acute toxicity results in well-known mental disturbances. In many cases, acute or subacute doses may give chronic symptoms or effects many years after poisoning (cigarette smoking and cancer) or effects in the following generation (stilbestrol may give vaginal cancer in female offspring at puberty).

We use the following terms:

Acute dose — The dose is given during a period shorter than 24 h.
Subacute dose — The doses are given between 24 h and 1 month.
Subchronical dose — The doses are given between 1 and 3 months.
Chronical dose — The doses are given for more than 3 months.

These terms apply to mammals, whereas the times are shorter for short-lived animals or plants used in tests. The dose of a pesticide toward a pest will usually be acute, whereas the dose that consumers of sprayed food will be exposed to is chronic.

2.3 Interactions

One toxicant may be less harmful when taken together with another chemical. If we use *blindness* as an endpoint for methanol poisoning, then whisky or other drinks that contain ethanol would reduce the toxicity of methanol considerably. When ethanol is present, methanol is metabolized more slowly to formaldehyde and formic acid, which are the real harmful substances. Ethanol is therefore an important antidote to methanol poisoning. Malathion is an organophosphorus insecticide with low mammalian toxicity, but if administered together with a small dose of parathion, its toxicity increases many times. This is because paraoxon, the toxic metabolite of parathion, inhibits carboxylesterases that would have transformed malathion into the harmless substance malathion acid. In another example, a smoker should not live in a house contaminated with radon. Although smoking and radon may both cause lung cancer on their own, smoke and radon gas interact and the incidence will increase 10 times or more when smokers are exposed to radon. (Radon is a noble gas that may be formed naturally in many minerals. It may penetrate into the ground floor of houses and represents a health hazard.)

Two or more compounds may interact to influence the symptoms in an individual and change the number of individuals that get the symptoms in question. Interaction may be caused by simultaneous or successive administration.

2.3.1 Definitions

It is important but difficult to give stringent definitions of various types of interactions or joint action. Because the dose–response curve seldom is linear, and because the relative response to one or more substances given either alone or together cannot exceed 1, we cannot define additive interaction as cases where $p_{(a + b)} = p_a + p_b$.

This is often erroneously done. $p_{(a + b)}$ here is the relative response of two substances A and B, given together in doses a and b, while p_a and p_b are the expected relative responses when a and b are given separately. In cases where there are no interactions, but a joint action, i.e., the animals are exposed to two toxicants at the same time but they act independently, the organisms are killed by one or the other and the relative response may be

$$P_{(a + b)} = p_a + p_b - p_a \times p_b$$

Additive interaction is better defined as cases when half of the LD50 doses of A and B (i.e., LD50(A)/2 + LD50(B)/2) kills 50% when given together. As an example, we can use Parathion oil® and Bladan® and suggest that they have LD50 values of 12 and 10 mg/kg, respectively. A dose consisting of 6 mg/kg of Parathion oil and 5 mg/kg of Bladan will then kill 50%. (The two products have the same active ingredient — parathion.) If more than 50% are killed by such mixtures, we have a case of potentiation, or superadditive joint action, and if fewer are killed, we have antagonism, or subadditive joint action. If one substance is nontoxic alone, but enhances the toxicity of another, we have synergism, and if it reduces the toxicity of

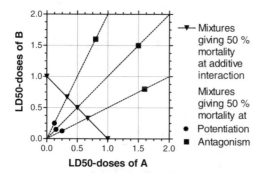

Figure 2.5 Isobolograms showing mixed doses giving 50% mortality in cases of additive interaction, potentiation, and antagonism. When given alone, LD50 = 1 unit for both substances.

the other, we have antagonism or an antidote effect. Endpoints other than 50% deaths may be used in similar considerations. The easiest way to test for interactions and define the various types of interactions is by making an *isobole diagram* (Figure 2.5).

2.3.2   Isoboles

Bolos (βολοσ) is a Greek word and may be translated as "a hit." *Isobole* may be translated as "similar hits." When making an isobole, we determine various mixtures of doses of A + B that together give the decided response, for instance, 50% kill. Many different mixtures should be tested in a systematic manner. The compositions of the mixtures given the wanted response are plotted in a diagram where the amount of (A) is given by the y-axis and the amount of (B) by the x-axis.

A typical experiment, where we want to see how A and B interact, using LD50 as the endpoint, may be carried out as follows. The LD50 values of each of the two substances are first determined. A mixture with the same relative proportion as LD50 values is made, e.g., 10 × LD50 units of each. A dilution series is made and the LD50 of the mixture is determined. Dilution series of mixtures with, for instance, 14 × LD50(A) + 7 × LD50(B) and 7 × LD50(A) + 14 × LD50(B) may also be tested. The compositions of the dilution series are marked with three dotted lines, and the compositions of the mixtures giving 50% kill are plotted as points in the diagram.

The location of the points is then compared to the location expected for mixtures with additive interaction, which is the straight diagonal line between points for A alone or B alone (e.g., $LD50_A$ and $LD50_B$). If the points fall outside the triangle, we have antagonism, whereas when inside, we have potentiation.

If one substance is nontoxic but modifies the toxicity of another substance, we get isoboles, as shown underneath. In this case, (B) is nontoxic but functions as a synergist or antagonist to (A).

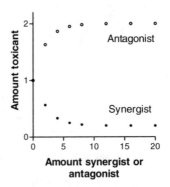

Figure 2.6 The composition of mixtures giving 50% kill in the case of synergism and antagonism when one substance is nontoxic.

The points in Figure 2.6 show isobolograms of mixtures giving 50% kill in the case of synergism and antagonism when one substance is nontoxic. The most important kind of interaction in pesticide toxicology is synergism, and piperonyl butoxide is the most widely used synergist. It inhibits the CYP enzymes in insects that are important for the detoxication of the pyrethrins, many carbamates, and other pesticides. By itself it has a low toxicity to insects or mammals, but its presence increases the toxicity of many pesticides toward insects. In some cases it also reduces the toxicity.

2.3.3   *Mechanisms of interactions*

When two substances react together chemically and the product has a different toxicity to the reactants, we have *chemical interaction*. A good example is poisoning with the insecticide lead arsenate ($PbHAsO_4$), which can be treated with the calcium salt of ethylenediaminetetraacetate and 2,3-dimercapto-1-propanol. These two antidotes react with lead arsenate and make less toxic complexes of lead and arsenate. The antidote atropine works through *functional interaction*. It blocks the muscarinic acetylcholine receptors and thus makes poisoning with organophosphates less severe. Another type of interaction is that one compound modifies the bioactivation or detoxication of the other. CYP enzymes may be induced or inhibited, the depots for glutathione may be depleted, or the carboxylesterases may be inhibited or kept busy with substrates other than the toxicant.

2.3.4   *Examples*

2.3.4.1   *Piperonyl butoxide*

Parathion and other phosphorothionates must be bioactivated to the oxon derivatives in order to be toxic. This is mainly done by the CYP enzymes

Table 2.2 Effect of Piperonyl Butoxide and SKF 525A Pretreatment on Organophosphate Insecticides' Toxicity in Mice

	24-h LD50 (mg/kg)		
Insecticide	Control (corn oil, 1 h)	Piperonyl Butoxide (400 mg/kg, 1 h)	SKF 525A (50 mg/kg, 1 h)
Parathion-methyl	7.6	330	220
Ethyl parathion	10.0	5.5	6.1
Azinphos-methyl	6.2	19.5	11.8
Azinphos-ethyl	22.0	3.4	9.1

Source: Based on data from Levine, B. and Murphy, S.D. 1977. *Toxicol. Appl. Pharmacol.*, 40, 393–406.

described later. Inhibition of the CYP enzymes with piperonyl butoxide or SKF 525A should therefore reduce the toxicity of parathion and other phosphorothionates. However, experiments with mice show that this is not the case. The symptoms and the time of deaths are delayed, but probably due to other oxidases (e.g., lipoxygenases); the same amount of paraoxon as in the control is gradually formed, only more slowly. Pretreatment with either of the two synergists increases the toxicity of parathion and azinphos-ethyl, but the two CYP inhibitors dramatically reduce the toxicity of the parathion-methyl. A similar pattern was shown for the two azinphos analogues (Table 2.2). The reason for this is that the methyl analogues have a fast route for detoxication through demethylation and therefore need quick bioactivation. If bioactivation is delayed, the detoxication route will dominate.

The involved reactions for parathion-methyl are

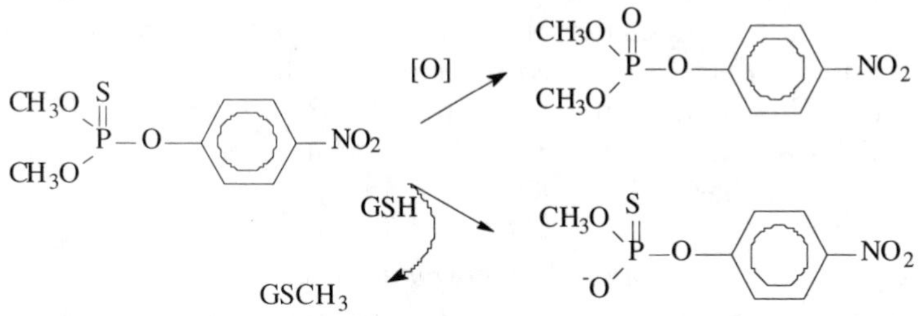

The oxidation, which is the bioactivation reaction, is inhibited by piperonyl butoxide, whereas the demethylation reaction catalyzed by glutathione transferase is not inhibited. Piperonyl butoxide is therefore an antagonist to methyl-parathion, but a synergist to most other pesticides, including carbamates and pyrethroids. Pyrethrins are very quickly detoxified by oxidation of one of the methyl groups, catalyzed by the CYP enzymes.

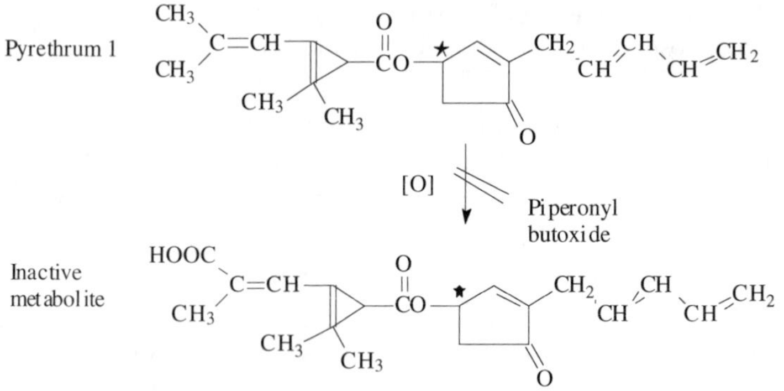

Pyrethrum 1

Inactive metabolite

[O]

Piperonyl butoxide

2.3.4.2 Deltamethrin and fenitrothion

Sometimes interactions may be detected even when an exact mechanism is unknown. As an example from real life, we can look at locust control in Africa.

Locust (*Locusta migratoria*) is an important pest in Africa. In order to find a suitable pesticide or pesticide mixture, fenitrothion or deltamethrin was tried alone or in combinations by B. Johannesen, a Food and Agriculture Organizaton (FAO) junior expert working in Mauritius. Dilution series of mixtures with different compositions were made and the LD50 values of these mixtures were determined. These values were plotted as shown in Figure 2.7. We see that the two pesticides potentiate each other.

The LD50 of deltamethrin alone was 1.2 µg/g of insects, whereas fenitrothion had an LD50 of 3.5 µg/g of insects. It is shown that the LD50 of mixtures of various compositions is lower than expected in cases of additivity. Hundreds of insects were used to determine the plotted LD50 doses of the mixtures. The great scatter illustrates the inborn uncertainty of such determinations. All the points are well inside the line for additivity, and some kind of potentiation is evident.

2.3.4.3 Atrazine and organophosphate insecticides

Sometimes more surprising examples of interaction may be observed.

The herbicide atrazine is not toxic to midge (*Chironomus tentans*) larvae but has a strong synergistic effect on several organophosphorus insecticides such as chlorpyrifos and parathion-methyl, but not to malathion. The increased rate of oxidation to the active toxicants, the oxons, is suggested as one of the mechanisms, and the level of CYP enzymes is elevated. Figure 2.8 shows the effect of the herbicide atrazine on the toxicity of chlorpyrifos. The data from Belden and Lydy (2000) show typical synergism. Altenburger et al. (1990) and Pöch et al. (1990) have described other examples of the use of isobolograms and how to interpret them.

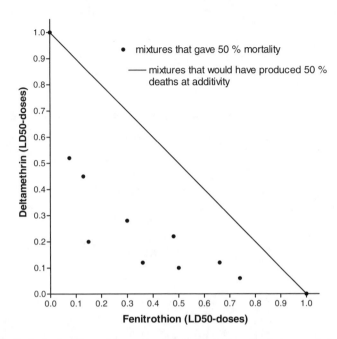

Figure 2.7 An isobologram of *Locusta migratoria* given mixed doses of deltamethrin and fenitrothion. Given separately, an LD50 dose of deltamethrin is 1.2 µg/g and of fenitrothion is 3.5 µg/g. The figure is based on data provided by Baard Johannessen and will be later published in full text.

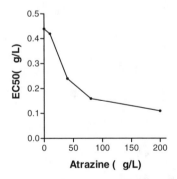

Figure 2.8 The effect of atrazine on the toxicity of chlorpyriphos. (Data from Belden, J. and Lydy, M. 2000. *Environ. Toxicol. Chem.*, 19, 2266–2274.)

chapter three

Pesticides interfering with processes important to all organisms

Energy production in mitochondria and the mechanisms behind cell division are very similar in all eukaryotic organisms. Furthermore, some inhibitors of enzymes have so little specificity that many different enzymes in a great variety of organisms may be targets. Many of the pesticides with such general modes of action have considerable historic interest. Not all of them are simple in structure, and many are used for purposes other than combating pests.

3.1 Pesticides that disturb energy production

3.1.1 Anabolic and catabolic processes

Green plants are anabolic engines that produce organic materials from carbon dioxide, other inorganic substances, water, and light energy. New organic molecules are made by anabolic processes, whereas organic molecules are degraded by catabolic processes. Plants are also able to degrade complicated organic molecules, but the anabolic processes dominate. Animals, bacteria, and fungi may be called catabolic engines. Their task is to convert organic materials back to carbon dioxide and water. Most of the energy from the catabolism is released as heat, but much is used to build up new molecules for growth and reproduction. Almost all of the energy required for these many thousand chemical reactions is mediated through adenosine triphosphate (ATP), which is broken down to adenosine diphosphate (ADP) and inorganic phosphate in the energy-requiring biosynthesis. ADP is again rebuilt to ATP with energy from respiration and glycolysis. The basic catabolic processes that deliver ATP are very similar in all organisms and are carried out in small intracellular organelles, the mitochondria. We should suppose, therefore, that pesticides disturbing the processes are not very selective, which is indeed the case. We find very toxic and nonselective substances such as arsenic, fluoroacetate, cyanide, phenols, and organic tin compounds, but also substances with some selectivity due to different uptake and metabolism in various organisms. Examples are rotenone, carboxin, diafenthiuron, and dinocap.

3.1.2   Synthesis of acetyl coenzyme A and the toxic mechanism of arsenic

Acetyl coenzyme A (Ac-CoA) plays a central role in the production of useful chemical energy, and about two thirds of all compounds in an organism are synthesized via Ac-CoA. Degradation of sugars leads to pyruvate, which reacts with thiamine pyrophosphate, and the product reacts further with lipoic acid. The acetyl–lipoic acid reacts with coenzyme A to give Ac-CoA and reduced lipoic acid. Lipoic acid, in its reduced form, has two closely arranged SH groups that easily react with arsenite to form a cyclic structure that is quite stable and leads to the removal of lipoic acid (Figure 3.1). Arsenic is toxic to most organisms because of this reaction. It is not used much as a pesticide anymore, but in earlier days, arsenicals, such as lead arsenate, were important insecticides. Natural arsenic sometimes contaminates groundwater, which led to a tragedy in Bangladesh. Wells were made with financial support from the World Health Organization (WHO), but their apparent pure and freshwater was strongly contaminated with the tasteless and invisible arsenic and many were poisoned. In Europe, arsenic is perhaps best known as the preferred poison of Agatha Christie's murderers, but it is also valuable for permanent wood preservation, together with copper and other salts. This use, however, also seems to have been terminated because of arsenic's bad reputation as a poison and carcinogen.

3.1.3   The citric acid cycle and its inhibitors

3.1.3.1   Fluoroacetate

Fluoroacetate is produced by many plants in Australia and South Africa and has an important function as a natural pesticide for the plants. It is highly toxic to rodents and other mammals. In certain parts of Australia, where such plants are abundant, opossums have become resistant to fluoroacetic acid. Good descriptions are presented by several authors in Seawright and Eason (1993).

The mode of action of fluoroacetate is well understood: it is converted to fluoroacetyl-CoA, which is thereafter converted to fluorocitric acid. This structure analogue to citric acid inhibits the enzyme that converts citric acid to cis-aconitic acid, and the energy production in the citric acid stops. Citric acid, which accumulates, sequesters calcium. α-Ketoglutaric acid and therefore glutamic acid are depleted. These changes are, of course, detrimental for the organism. The nervous system is sensitive to these changes because glutamic acid is an important transmitter substance in the so-called

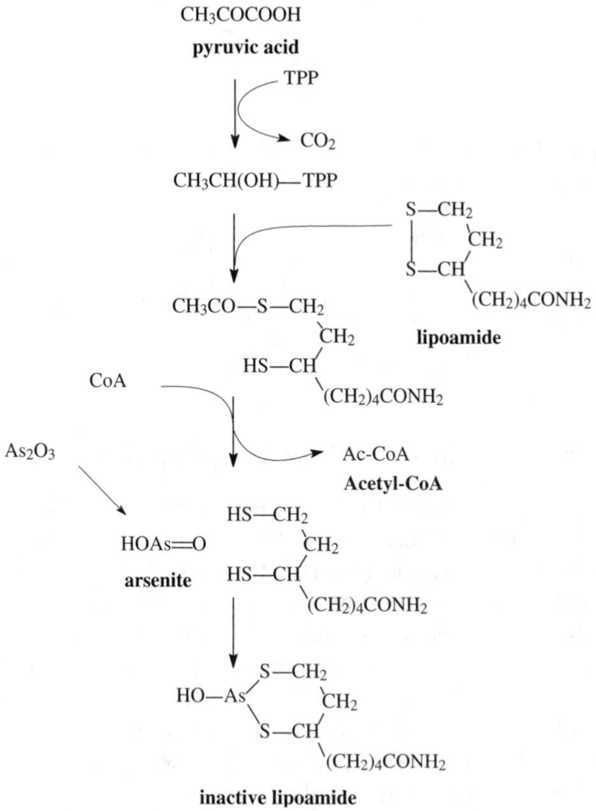

CH$_3$COCOOH

pyruvic acid

TPP

CO$_2$

CH$_3$CH(OH)—TPP

S—CH$_2$

CH$_2$

S—CH

(CH$_2$)$_4$CONH$_2$

CH$_3$CO—S—CH$_2$

CH$_2$ **lipoamide**

HS—CH

CoA   (CH$_2$)$_4$CONH$_2$

Ac-CoA

Acetyl-CoA

As$_2$O$_3$

HS—CH$_2$

HOAs=O   CH$_2$

arsenite   HS—CH

(CH$_2$)$_4$CONH$_2$

S—CH$_2$

HO—As   CH$_2$

S—CH

(CH$_2$)$_4$CONH$_2$

inactive lipoamide

Figure 3.1 The mode of action of arsenic.

glutaminergic synapses, and calcium is a very important mediator of the impulses. Furthermore, the halt of aerobic energy production is very harmful.

3.1.3.2   *Inhibitors of succinic dehydrogenase*

Inhibitors of succinic dehydrogenase constitute an important group of fungicides. In 1966, carboxin was the first systemic fungicide to be marketed. A systemic pesticide is taken up by the organism it shall protect and may kill sucking aphids or the growing fungal hyphae. The older fungicides are active only as a coating on the surface of the plants and do not fight back growing mycelia inside the plant tissue. Carboxin and the other anilides, or oxathiin-fungicides, as they are often called, inhibit the dehydrogenation of succinic acid to fumaric acid — an important step in the tricarboxylic acid cycle. The toxicity to animals and plants is low in spite of this very fundamental mode of action. The fungicides in this group are anilides of unsaturated or aromatic carboxylic acids. The first compound in this group to be synthesized was salicylanilide, which since 1930 had a use as a textile protectant.

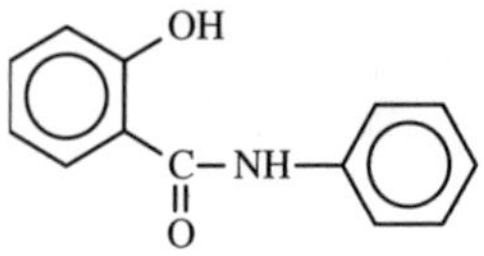

Other phenylamides with the same mode of action are fenfuram, flutalonil, furametpyr, mepronil, and oxycarboxin.

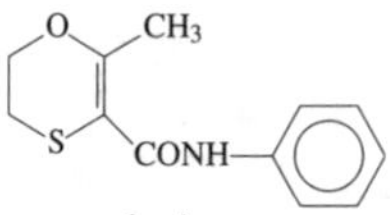

carboxin 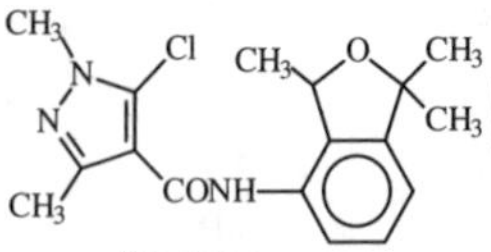   furametpyr 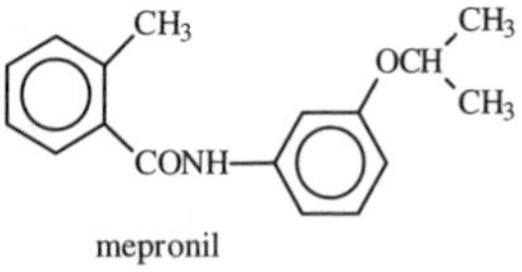   mepronil

3.1.4   The electron transport chain and production of ATP

When compounds are oxidized through the tricarboxylic acid cycle (Figure 3.2) to carbon dioxide and water, electrons are transferred from the compounds to oxygen through a well-organized pathway, which ensures that the energy is not wasted and, more importantly, that electrons are not taken up by compounds that make them into reactive free radicals. The electrons are first transferred to nicotineamide-adenine dinucleotide (NAD⁺) and flavine adenine dinucleotide (FAD), and from these co-substrates the electrons are passed on to ubiquinone and further on to the cytochromes in the electron transport chain. Their ultimate goal is oxygen, which is reduced to water. The energy from this carefully regulated oxidation is used to build up a hydrogen ion gradient across the inner mitochondrian membrane, with the lower pH at the inside. This ion gradient drives an ATP factory.

3.1.4.1   Rotenone

Rotenone is an important insecticide extracted from various leguminous plants. It inhibits the transfer of electrons from nicotineamide-adenine (NADH) to ubiquinone.

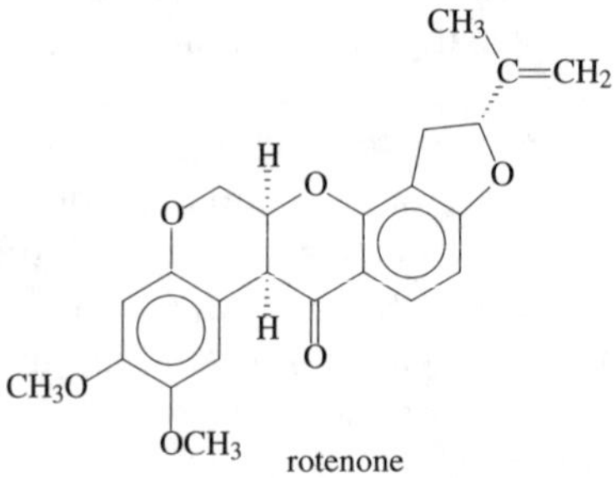

rotenone

It is also highly toxic to fish and is often used to eradicate unwanted fish populations, for instance, minnows in lakes before introducing trout, or

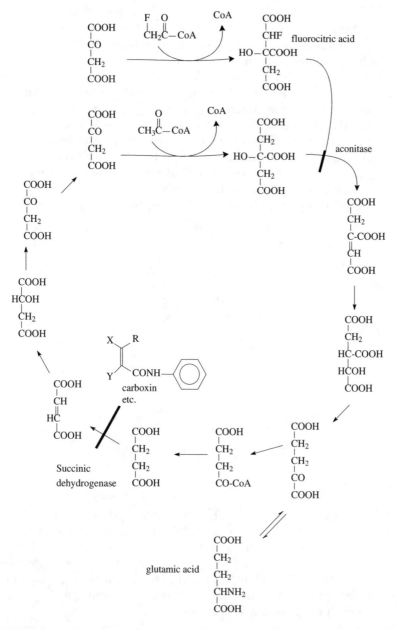

Figure 3.2 A simple outline of the citric acid cyclus and the sites of inhibition by the insecticide/rodenticide fluoroacetic acid, and the fungicide carboxin.

to eradicate salmon in rivers in order to get rid of *Gyrodactilus salaries*, an obligate fish parasite that is a big threat to the salmon population. The noninfected salmon coming up from the sea to spawn will not be infected if the infected fish present in the river have been killed before they arrive.

3.1.4.2   Inhibitors of electron transfer from cytochrome b to c_1

The strobilurins are a new class of fungicides based on active fungitoxic substances found in the mycelia of basidiomycete fungi. The natural products, such as strobilurin A and strobilurin B, are too volatile and sensitive to light to be useful in fields and glasshouses. However, by manipulating the molecule, notably changing the conjugated double bonds that make them light sensitive, with more stable aromatic ring systems, a new group of fungicides have been developed in the last decade. At least four are on the market (azoxystrobin, famoxadone, kresoxim-methyl, and trifloxystrobin). Their mode of action is the inhibition of electron transfer from cytochrome b to cytochrome c_1 in the mitochondrial membrane. They are supposed to bind to the ubiquinone site on cytochrome b.

The reactions inhibited by strobilurin fungicides:

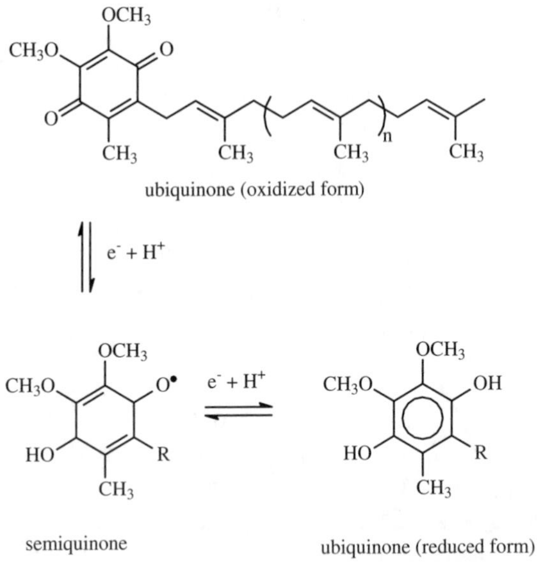

ubiquinone (oxidized form)

$e^- + H^+$

semiquinone   ubiquinone (reduced form)

The fungicides are very versatile in the contol of fungi that have become resistant to the demethylase inhibitor (DMI) fungicides described later. They have surprisingly low mammalian toxicity, but as with many other respiratory poisons, they show some toxicity to fish and other aquatic organisms. They may also be toxic to earthworms. In fungi they inhibit spore germination. The structures show the natural products strobilurin B and azoxystrobin, which has been marketed since 1996.

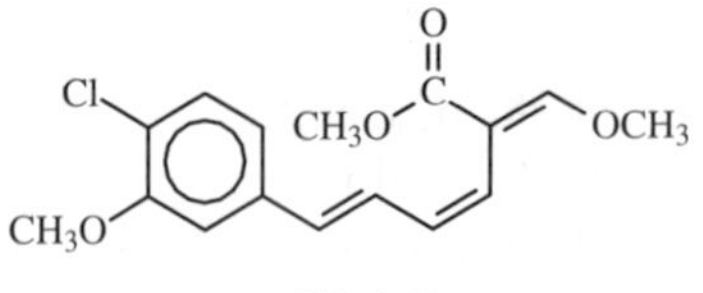

strobilurin B 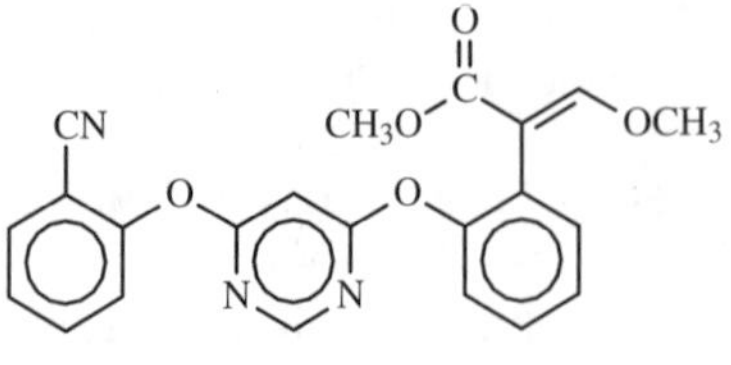   azoxystrobin

3.1.4.3   Inhibitors of cytochrome oxidase

Cyanide may still have some use against bedbugs and other indoor pests in spite of its high toxicity to man, but in the past it was used much more. In the 19th century, doctors prescribed it as a sedative and, of course, caused a lot of fatal poisoning (Otto, 1838). The recommended treatment was to let the patient breathe ammonia. Today we have very efficient antidotes, such as sodium nitrite and amyl nitrite. They cause some of the Fe^{++} of hemoglobin to be oxidized to Fe^{+++}, which then binds the CN^- ion. Cyanide inhibits the last step in the electron transport chain catalyzed by cytochrome oxidase by binding to essential iron and copper atoms in the enzyme. Cyanide is very fast acting and blocks respiration almost totally.

Phosphine is used extensively as a fumigant and is very efficient in the control of insects and rodents in grain, flour, agricultural products, and animal foods. It is used to give continual protection during shipment of grain. The gas is flammable and very unstable and is changed into phosphoric acid by oxidation. By using pellets of aluminum phosphide at the top of the stored product, phosphine is slowly released by reacting with moisture. Other phosphine salts are also used. Phosphine is reactive and is probably involved in many reactions, but the inhibition of cytochrome oxidase is the most serious. The gas is very toxic to man, but residues in food cause no problems because it is oxidized rapidly.

$$AIP + 3H_2O \rightarrow AI(OH)_3 + PH_3$$

3.1.4.4   Uncouplers

As discussed in Chapter 2, Section 1.4, uncoupling energy production and respiration is one of the fundamental toxic mechanisms. Weak organic acids or acid phenols can transport H^+ ions across the membrane so that energy is wasted as heat, and not used to produce ATP.

The name *uncouplers* arose from their ability to separate respiration from ATP production. Even when ATP production is inhibited, the oxidation of carbohydrates, etc., can continue if an uncoupler is present. Although the uncouplers are biocides, in principle toxic to all life-forms, many valuable pesticides belong to this group. However, few of them are selective, and they have many target organisms. The inner mitochondrial membranes are their most important sites of action, but chloroplasts and bacterial membranes will also be disturbed.

Figure 3.3 shows how weak acids can transport H^+ ions across the membrane.

Pesticides with this mode of action include such old products as the dinitrophenols (dinitroorthocreosol [DNOC], dinoterb, and dinoseb) and other phenols such as pentachlorophenol and ioxynil. DNOC is a biocide useful against mites, insects, weeds, and fungi. The mammalian toxicity is rather high, with a rat oral LD50 (lethal dose in 50% of the population) of 25 to 40 mg/kg of the sodium salt. The typical symptom is fever, which is

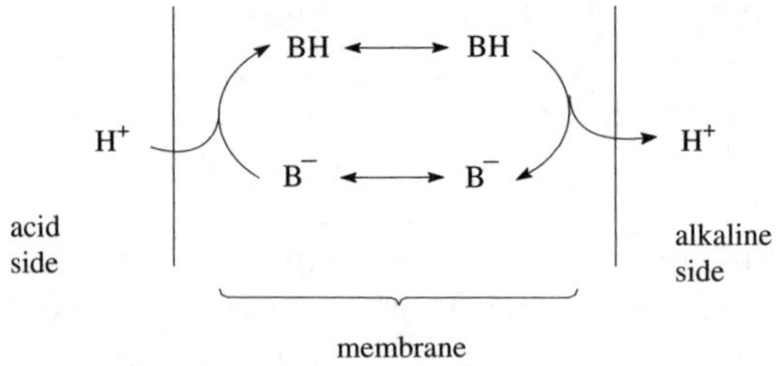

Figure 3.3 Transportation of H^+ ions across a biological membrane by a weak acid.

in accordance with its biochemical mode of action. The uncouplers have been tried in slimming treatments with fatal consequences.

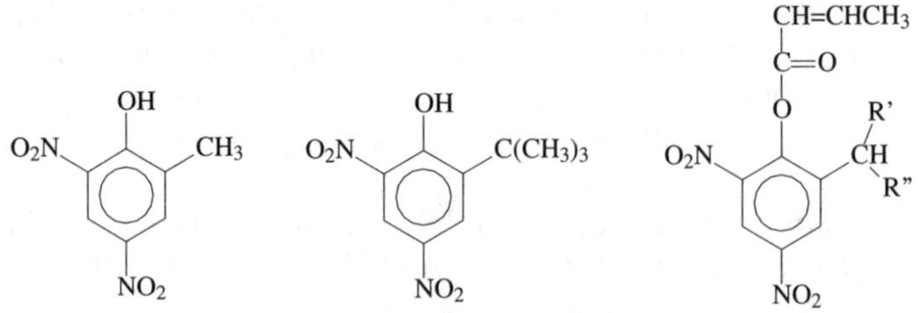

DNOC dinoterb dinocap

Dinocap is an ester that is taken up by fungal spores or mites. It is hydrolyzed to the active phenol. It has low toxicity to plants and mammals. Dinocap is a mixture of several dinitrophenol esters, and the structure of one is shown.

Ioxynil is a more important uncoupler that is widely used as an herbicide. It acts in both mitochondria and chloroplasts. Bromoxynil is similar to the ioxynil, but has bromine instead of iodine substitutions.

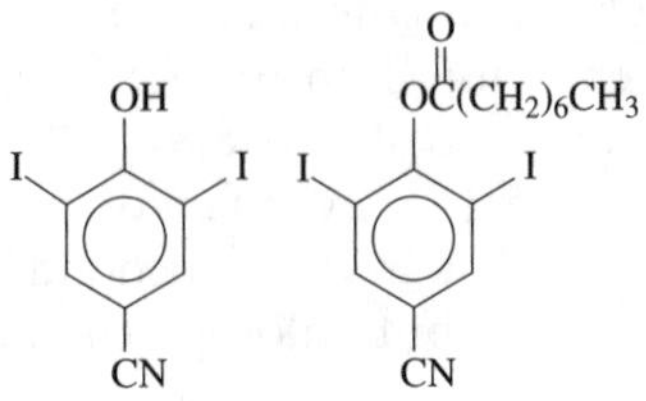

ioxynil ioxyniloctanoate

3.1.5 *Inhibition of ATP production*

ATP is produced from ADP and phosphate by an enzyme, ATP synthase, located in the inner mitochondrian or chloroplast membrane. The energy is delivered from a current of H^+ ions into the mitochondrian matrix. Some important pesticides inhibit this enzyme, leading to a halt in ATP production.

Table 3.1 Diafenthiuron and Organotin Compounds Used as Pesticides

Pesticide	Fish (Various Species) LC50 (24–96 h) (mg/l)	Daphnia EC50 (48 h) (mg/l)	Rodents (Various Species or Sex) Oral LD50 (mg/kg)
Cyhexatin	0.06–0.55 (24 h)	—	540–1000
Azocyclotin	0.004 (96 h)	0.04	209–980
Fentin (acetate)	0.32 (48 h)	0.0003–0.03	20–298
Tributyltin	0.0021 (96 h)	0.002	—
Diafenthiuron	0.0013–0.004 (96 h)	<0.5	>2000

Note: LC50 = lethal concentration in 50% of the population; EC50 = effective concentration in 50% of the population.

Source: Data from Tomlin, C., Ed. 2000. *The Pesticide Manual: A World Compendium.* British Crop Protection Council, Farnham, Surrey. 1250 pp.

3.1.5.1 Organotin compounds

Organotin compounds have been used extensively as pesticides for special purposes. At least some of them owe their mode of action to the inhibition of ATP synthase in the target organism. Tricyclohexyltin (cyhexatin) and azocyclotin are used as selective acaricides. Cyhexatin is toxic to a wide range of phytophageous mites, but at recommended rates it is nontoxic to predacious mites and insects. Triphenyltin acetate or hydroxide may be used as fungicide, algicide, or molluscicide. The toxicity of these compounds to fish is very high, but they have moderate toxicity to rodents. The data in Table 3.1 are taken from *The Pesticide Manual* (Tomlin, 2000).

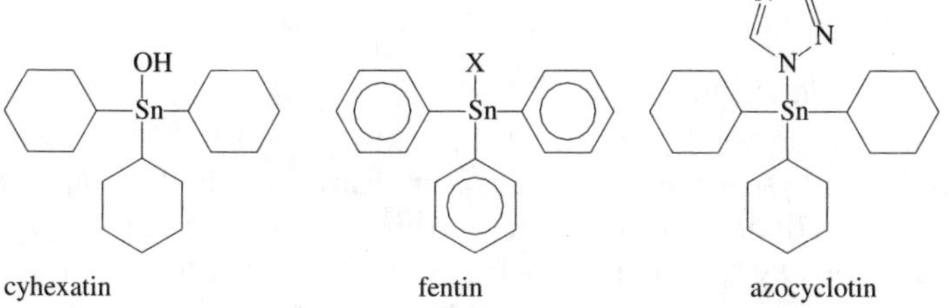

cyhexatin   fentin   azocyclotin

Tributyltin and tributyltin oxide are still used on boats and ships to prevent growth of barnacles. They are extremely toxic for many invertebrates in the sea, notably some snails whose sexual organs develop abnormally. In these snails the female develops a penis. In oysters and other bivalves, their shells become too thick. Tributyltin must be regarded as one of the most serious environmental pollutants, but contrary to the lower analogues, tri-methyltin and triethyltin, they are not very toxic to man and other mammals. Trimethyltin is of considerable interest for neurotoxicologists because it leads specifically to atrophy of the center for short-term memory, the hippocam-pus. The ethyl analogue has other serious detrimental effects on the brain.

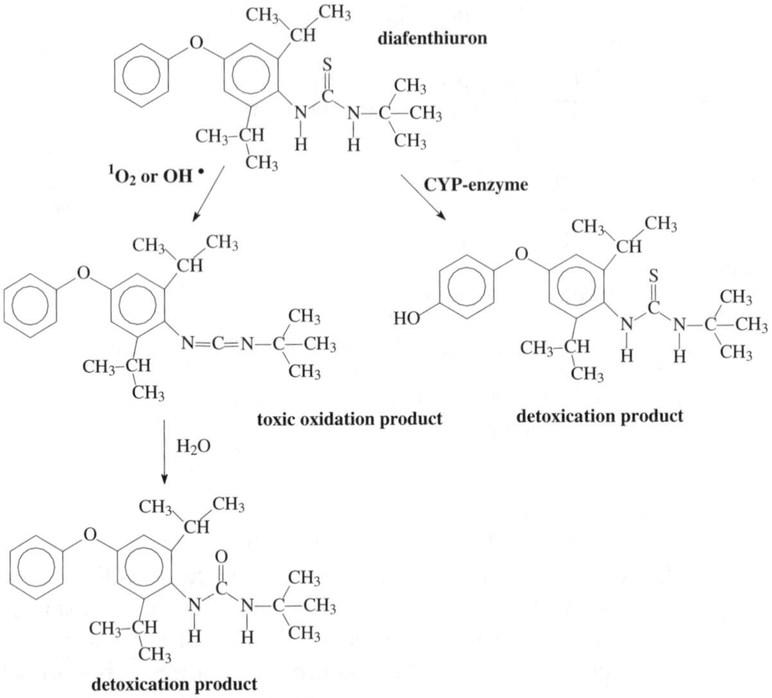

Figure 3.4 Activation and detoxication of diafenthiuron.

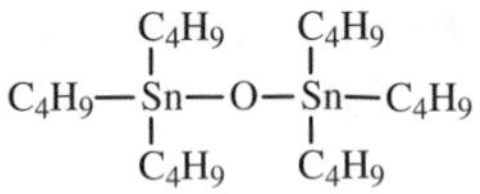

3.1.5.2 Diafenthiuron

Diafenthiuron inhibits ATP synthesis in the mitochondria (Ruder et al., 1991). This pesticide is interesting because, as is the case for the phosphorothioates, it needs to be activated by oxidation, which can occur abiotically by, for instance, singlet oxygen generated by sunlight or inside the organism by hydroxyl radicals generated by the Fenton reaction:

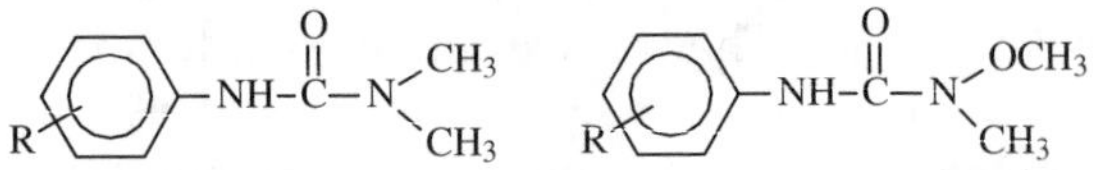

H_2O_2 may be produced as a by-product in the catalytic cycle of the CYP enzymes described later. Diafenthiuron therefore becomes more active in sunshine, and piperonyl butoxide that inhibits CYP enzymes makes diafenthiuron less toxic. However, some CYP enzymes are also important in the detoxication of diafenthiuron, as shown in Figure 3.4. Diafenthiuron may

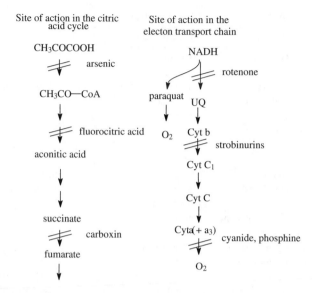

Figure 3.5 The site of inhibition of various pesticides in the citric acid cycle and the electron transport chain.

be used against mites, aphids, and other insects on several crops such as cotton, vegetables, and fruit. Table 3.1 shows that it has very high fish toxicity.

3.1.5.3   Summary

The mitochondrial poisons include such a variety of compounds with so many different activities on the organismic level that a summary may help. Figure 3.5 and Table 3.2 may help to identify the site of reaction.

In Figure 3.5, the arrows show the electron flow. When reaching oxygen the normal way, water is formed, while in the sideline via paraquat, super-oxide radicals are formed.

3.2   Herbicides that inhibit photosynthesis

About half of all herbicides inhibit photosynthesis. Most of them disturb one particular process, i.e., the transfer of electrons to a low molecular quinone called plastoquinone. The inhibition occurs by the binding of the inhibitor to a specific protein called D_1 that regulates electron transfer. This protein has 353 amino acid residues and spans the thylakoid membrane in the chloroplasts. In atrazine-resistant mutants of certain plants, a serine residue at position 264 in the D_1 protein of the wild type has been found to be substituted by glycine. It is now possible to replace the serine 264 with glycine by site-directed mutagenesis in its gene and to reintroduce the altered gene to engineer atrazine resistance in plants.

The inhibitors of photosynthesis are all nitrogen-containing substances with various structures. They may be derivatives of urea, s-triazines, anilides,

Table 3.2 Site of Action of Some Mitochondrial Poisons

Site of Action	Compounds	Toxic for
Inhibition of acetyl-CoA synthesis	Arsenic	Most animals
Inhibition of akonitase	Fluoroacetic acid (fluorocitrate)	Most animals
Inhibition of succinic dehydrogenase	Salicylanilide and oxathiin fungicides	Fungi
Inhibition of NADH dehydrogenase	Rotenon	Insects, fish
Inhibiting cytochrome b	Strobinurins	Fungi
Inhibiting cytochrome oxidase	Cyanide Phosphine	All aerobic organisms
pH gradient in mitochondrial membranes (uncouplers)	Phenols	Most organisms
Inhibitors of ATP synthase in the mitochondrial membrane	Organonotin compounds Diafenthiuron metabolite	Fungi, mites, aquatic organisms; some have high mammalian neurotoxicity Insects, fish
Superoxide generators	Copper ions	Most organisms
Takes electrons from the transport chain and delivers them to O_2	Paraquat	Most aerobic organisms

as-triazinones, uraciles, biscarbamates, pyridazinones, hydroxybenzoeni-triles, nitrophenols, or benzimidazols. We shall describe just a few of them and give a very brief outline of the photosynthetic process. Textbooks of cell biology, biochemistry, and plant physiology (e.g., Alberts et al., 2002; Nelson and Cox, 2000; Taitz and Zaiger, 1998) describe the process in detail. The action of the herbicides may be read in more detail in Fedke (1982) or Devine et al. (1993).

In photosynthesis, light energy is trapped and converted to chemical energy as reduced coenzymes (e.g., nicotineamide-adenine dinucleotide phosphate [NADPH]), triphosphates (e.g., ATP), and O_2. Oxygen is a poisonous waste product in plants, although they also need some oxygen in mitochondrial respiration.

The chlorophyll takes up light energy (photons) directly or through so-called antennae molecules. (All colored substances take up light energy, but convert it to heat and not to chemical energy.) Electrons that jump to another orbit requiring more energy take up the energy and are said to have become excited. Such excited electrons may be lost by being taken up by an acceptor molecule, leaving chlorophyll as a positively charged ion. According to this scheme, chlorophyll will have three different states: the normal form that can pick up light energy, the excited molecule that is a very strong reducing substance, and the positively charged ion that is a very strong oxidant. The reducing power in the excited chlorophyll molecule is used to produce ATP and NADPH, while the oxidation power of the chlorophyll ion is used to

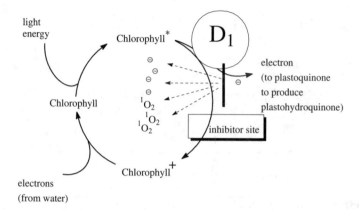

Figure 3.6 Blocking the passage of electrons from light-excited chlorophyll to plastoquinone by herbicides leads to production of singlet oxygen and electrons that may produce free radicals.

produce ATP and oxygen. ATP production is an indirect process coupled to the pH gradient between the inside and outside of the thylakoid membrane.

The photosynthetic apparatus is situated on and in the thylakoid membrane.

Four different and complicated protein complexes carry out the necessary chemical reactions: photosystem II, the cytochrome b_6f complex, photosystem I, and ATP synthase. These complexes are precisely oriented and fixed in the membrane. In addition, there are the plastoquinones, which easily undergo a redox cycle and can swim in the membrane's lipid phase. A manganese-containing complex in photosystem II is involved in the splitting of water and the generation of electrons and molecular oxygen. A small, copper-containing protein, plastocyanine (PC), is involved in transfer of electrons from $cytb_6f$ to photosystem I.

The thylakoid matrix is an extensive internal membrane system inside the chloroplasts, which are small organelles in the plant cells. The inner lumen of the membrane system maintains a pH of 5, while the outer compartment, called stroma, has a pH of 8. The energy picked up from the photons is used to establish and maintain this difference.

The chlorophyll pigments in photosystem II, organized in a structure called P680, catch energy from a photon and become excited. The electron is then transferred to a molecule called pheophytin and then to a tyrosine residue in protein D_1 called the reaction center. The oxidized form of plastoquinone (PQ) has a specific binding site on protein D_1, where it is reduced and then diffuses to the lumen side of the membrane (now as plastohydroquinone (PQH_2)). Here it binds to an iron–sulfur protein in the cytochrome b_6f complex and reduces it. The hydrogen ions released in this process are delivered to the inside of the membrane. The plastoquinone/plastohydroquinone is thus functioning as a proton pump driven by light-excited electrons. A summary of the process is shown in Figure 3.7.

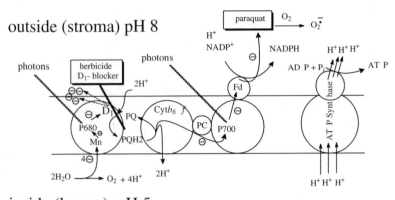

inside (lumen) pH 5

Figure 3.7 Schematic representation of photosynthesis and the site of action of D_1 blockers and paraquat.

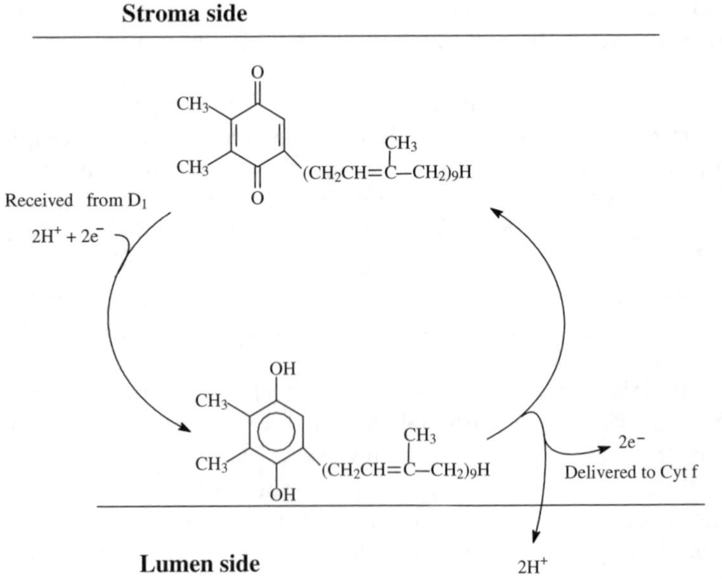

Stroma side

Received from D_1

$2H^+ + 2e^-$

Lumen side

$2H^+$

$2e^-$
Delivered to Cyt f

Figure 3.8 Schematic representation of the redox cycle of plastoquinone.

A simplified scheme of the redox cycle of plastoquinone is shown in Figure 3.8. The quinoid structure and the isoprene side chain make it possible for plastoquinone to take up one electron at a time, producing rather stable semiquinone radicals (which is not shown in the figure), and there are probably at least two plastoquinones involved.

The reduced cytochrome f delivers the electron to plastocyanine, a copper-containing, low-molecular-weight soluble protein, and then to special chlorophyll pigments in photosystem I (P700). The P700 can be excited by a new photon and deliver the electron to an iron-containing protein called ferredoxin. The reduced ferredoxin delivers the electrons to $NADP^+$ to produce NADPH, or to a minor pathway that reduces nitrate to ammonia at the outer membrane surface. Some important herbicides (paraquat and diquat) can snatch the electrons before the delivery to ferredoxin and generate free radicals.

The chlorophyll ion in $P680^+$ takes electrons from water, via a manganese-containing enzyme complex, and is reduced to the neutral unexcited state, ready to pick up new photons. Water is then split to oxygen and hydrogen ions. Oxygen is a toxic waste product, while the hydrogen ions contribute to the buildup of the pH difference across the membrane.

ATP is produced from ADP and phosphate by ATP synthase, an enzyme located in the membrane. The difference in hydrogen ion concentration between the inside and outside is used as the energy source. Because there are approximately 1000 times more hydrogen ions on the inside than on the outside of the membrane, the hydrogen ions will tend to diffuse out. This would waste energy, so instead, hydrogen ions are forced to flow through a special proton channel in the ATP synthase that uses the energy from the hydrogen ion flow to produce ATP. ATP synthase is very similar in chloroplasts and mitochondria.

In summary, there are four main types of herbicides that disturb the photosynthetic apparatus:

1. Weak organic acids that destroy the hydrogen ion concentration gradient between the two sides of the membrane
2. Free radical generators
3. Compounds that bind to the D_1 protein at (or near) the plastoquinone-binding site
4. Substances that destroy or inhibit synthesis of protecting pigments such as carotenoids

3.2.1   Weak organic acids

Weak organic acids with a pK value between pH 5 and 8, or close to these values, will cause leakage of hydrogen ions if the acid dissolves in the thylakoid membrane. Ammonia also has this effect as a result of the reaction $NH_4^+ \leftrightarrows NH_3 + H^+$. Instead of producing ATP, heat will be generated. These so-called uncouplers will act similarly in mitochondria, chloroplasts, and bacterial cell membrane. They may therefore also be toxic for animals and microorganisms, and some of them are described under mitochondrial poisons.

3.2.2   Free radical generators

These are a type of herbicide that is able to steal the electron on its long route from water to $NADP^+$. The most important herbicides in this category

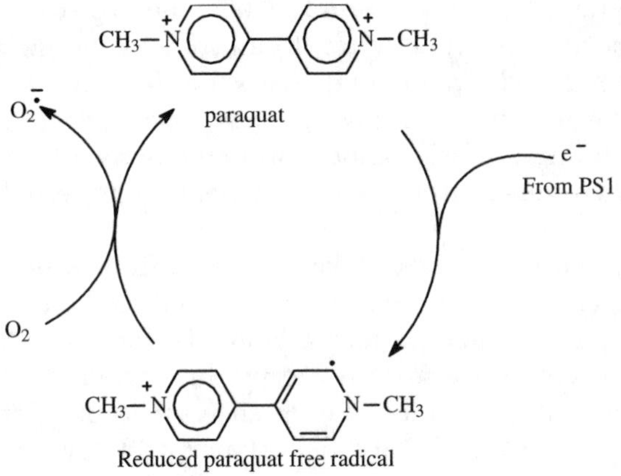

Figure 3.9 The toxic cycle of paraquat.

are paraquat (Figure 3.9) and diquat. They take up the electron at some stage before ferredoxin and deliver it to oxygen to produce the superoxide radical. Many other natural processes form superoxide radicals, and the cells have very efficient enzymes called superoxide dismutases that detoxify the superoxide radicals. However, the detoxication is not complete because another very reactive substance, namely, hydrogen peroxide (H_2O_2), is produced. Hydrogen peroxide must be detoxicated by catalases or glutathione peroxidases. If this does not happen fast enough, H_2O_2 may react through the Fenton reaction, producing the extremely reactive hydroxyl radical. Interestingly, paraquat is more toxic for plants when they are placed in light and is less toxic for bacteria when they are grown anaerobically, as shown by Fisher and Williams (1976). The superoxide generators are toxic to animals as well as to plants. Characteristically, the lung is the critical organ for paraquat in mammals, even when administered by mouth.

The partial detoxication of superoxide anions by superoxide dismutase is as follows:

$$2O_2 + 2H^+ \rightarrow O_2 + H_2O_2$$

The Fenton reaction produces the extremely reactive hydroxyl radical:

$$H_2O_2 + Fe^{++} \rightarrow OH\cdot + OH^- + Fe^{+++}$$

The detoxication of hydrogen peroxide with glutathione peroxidase is as follows:

$$H_2O_2 + 2GSH \rightarrow 2H_2O + GSSG$$

Paraquat chloride has been marketed since 1962. It is nonselective as an herbicide and rather toxic to animals (rat oral LD50 is between 129 and 157 mg/kg). Its valuable properties are that it is fast acting, is quickly inactivated through binding to soil and sediment (in spite of its high water solubility), and can be used for destruction of potato haulm before harvest. Its high human toxicity, with the lungs being the most severely affected, has led to many fatalities. Diquat dibromide started to be marketed at approximately the same time as paraquat and has approximately the same uses. It has a slightly lower acute toxicity to rats (LD50 ≈ 234 mg/kg). None of the bipyridylium herbicides have high dermal toxicity. They are placed in WHO toxicity class II.

3.2.3 D_1 blockers

The herbicides that act at D_1 have a low toxicity to animals. As explained, they block a specific site only present in photosynthesizing plants. The binding site is, however, the same or nearly the same for all plants, and little degree of selectivity is expected inside the plant kingdom. The inhibitors are more active in strong sunshine and in warm and dry weather with good moisture in the soil — the reason being that the strong transpiration stream in the plant takes up the herbicides, and the conditions are good for active photosynthesis. Plants adapted to low illumination are very sensitive to a combination of herbicide and strong light. The cause of death is definitely not lack of energy due to inhibition of photosynthesis, but is rather due to production of reactive oxygen species. When excited chlorophyll cannot transfer the energy to plastoquinone, it is forced to react with oxygen. Singlet oxygen ($^1O_2$) is formed. This can destroy beta-carotene and lipids in the thylakoid membrane. Excited chlorophyll can also react directly with unsaturated lipids.

3.2.3.1 Urea derivatives
This group is easy to recognize by their formulae:

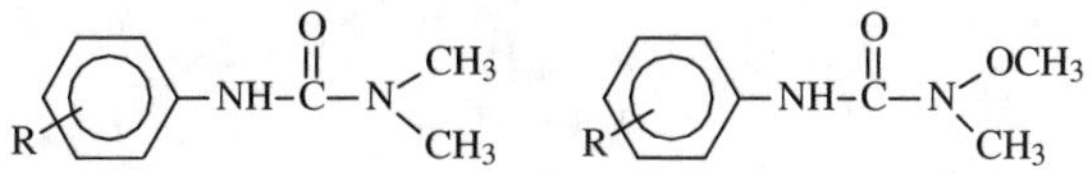

Urea is substituted with one aryl group in one nitrogen and two methyl groups or a methoxy and a methyl group on the other nitrogen.

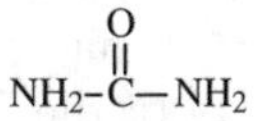

urea

The aryl group may be an unsubstituted simple phenyl ring, as in fenuron, or may be substituted with halogens, alkyl groups, and ring structures. The possibility of varying the aryl group and still retaining its activity makes it possible to modulate the properties, such as water solubility, stability, and uptake in plants. Most ureas have very low toxicity to birds and mammals, but fish and crustaceans may be sensitive (linuron has an LC50 for the fathead minnow of ≈1 to 3 mg/l, and for *Daphnia*, ≈0.1 to 0.75 mg/l). The herbicides are, of course, very toxic to photosynthesizing algae, and leakage into lakes and rivers and contamination of groundwater must therefore be avoided.

Linuron (left structure) was first marketed in the 1960s and has been one of the more popular herbicides in the culture of potatoes and vegetables:

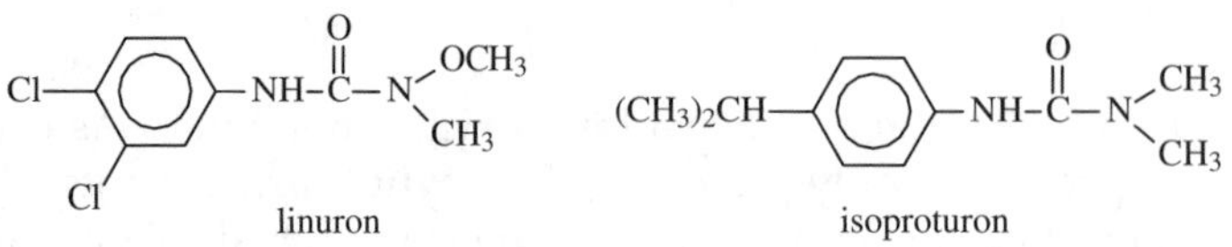

linuron   isoproturon

The plants take it up by roots and leaves, and it has a high persistence in humus-rich soil in cool climates, with a half-life of 2 to 5 months. Micro-organisms in soil degrade isoproturon, which can be used selectively in various cereal crops, and half-lives of 6 to 28 days have been reported under field conditions, depending on the microbial activity. The aliphatic substitu-tion in the aryl ring is sensitive to microbial oxidative attack.

3.2.3.2   Triazines

Most of the triazines are derivatives of the symmetrical 1,3,5-triaz-ine-2,4-diamine, but other possibilities also exist. In position 6 there is a methylthio (the -tryns), a methoxy (the -tons), or a chloro group (the -azines).

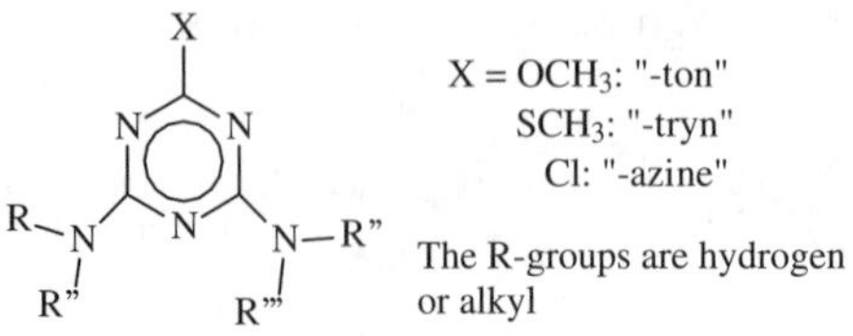

$X = OCH_3$: "-ton"
SCH_3: "-tryn"
Cl: "-azine"

The R-groups are hydrogen or alkyl

The triazines are also interesting because there is no negative correlation between water solubility and soil adsorbance, because of their cationic char-acter. Atrazine has been used a lot in maize because maize is less sensitive due to a glutathione transferase that inactivates atrazine. The same mecha-nism, as well as the mutant variety of the D_1 described in the beginning of this chapter, that reduces the binding may also be the cause of resistance in weeds. The substituents in the 6 position and the 2- and 4-amino groups greatly influence important properties such as soil-binding capacity, water solubility, microbial degradation, and other factors of importance. Besides atrazine, simazine, simetryn, dimethametryn, terbumeton, terbuthylazin, terbutryn, and trietazin are all available on the market.

3.2.4   Inhibitors of carotene synthesis

Some herbicides act by inhibiting the synthesis of carotenoids that protect chlorophyll from being destroyed by photooxidation. Amitrole is not selective, whereas aclonifen has valuable selective properties.

3.2.4.1   Amitrole

This once very promising herbicide with very low acute toxicity is carcinogenic and has been reported to increase the incidence of soft-tissue cancers in people engaged in thicket clearing along railways tracks in Sweden. Amitrole and aclonifen may cause enlarged thyroid in high doses. In plants, amitrole inhibits lycopene cyclase, an enzyme necessary for the synthesis of carotenoids. It is not selective, as opposed to aclonifen.

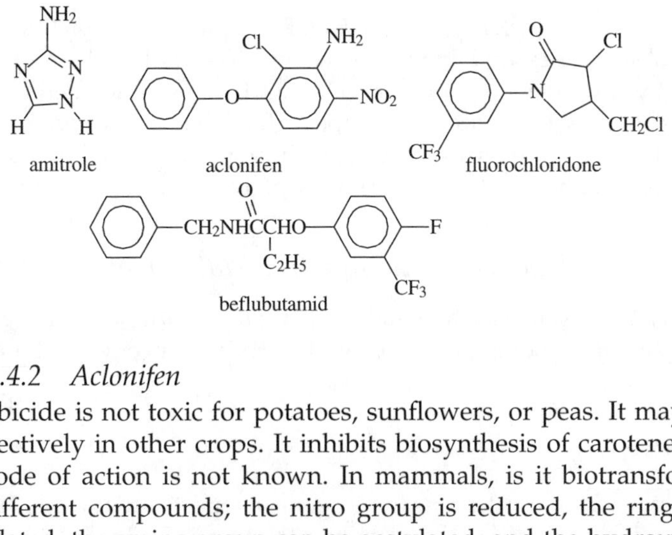

amitrole   aclonifen   fluorochloridone

beflubutamid

3.2.4.2   Aclonifen

This herbicide is not toxic for potatoes, sunflowers, or peas. It may also be used selectively in other crops. It inhibits biosynthesis of carotene, but the exact mode of action is not known. In mammals, is it biotransformed to many different compounds; the nitro group is reduced, the rings can be hydroxylated, the amino group can be acetylated, and the hydroxy groups formed by ring hydroxylation can be conjugated to sulfate or glucuronic acid. Its acute toxicity is very low. In mice and rats it may produce some kidney injury at high doses (25 mg/kg), but the no-observed-effect level (NOEL) for 90 days in rat is 28 mg/kg of body weight, and France has determined an acceptable daily intake (ADI) of 0.02 mg/kg.

3.2.4.3   Beflubutamid

Beflubutamid is a newly described carotenoid synthesis inhibitor, inhibiting phytoene desaturase. It has a very low toxicity to animals but is very toxic to algae and plants. It is not mutagenic or teratogenic in standard tests.

3.2.5 Protoporphyrinogen oxidase inhibitors

Acifluorfen is used in soybeans, peanuts, and rice, which are more or less tolerant to this herbicide.

acifluorfen

fluoroglycofen-ethyl

bifenox

oxyfluorfen

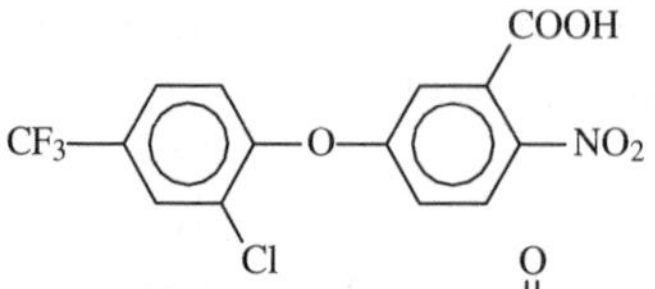

Bifenox, fluoroglycofen-ethyl, HC-252, lactofen, and oxyfluorfen have analogous structures and modes of action. The carboxyl group may be replaced by an ether group or an ester, and many other herbicides with related structures have been developed. Tetrapyrrol and protoporphyrin accumulate, act as photosensitizers, and cause photooxidation and necrosis. They are contact herbicides and are more active in strong sunlight.

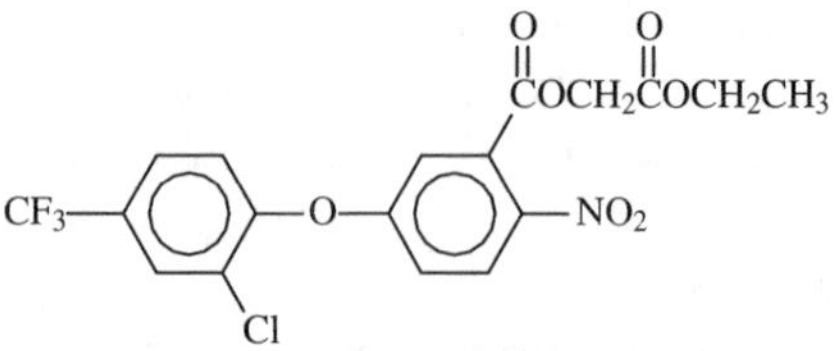

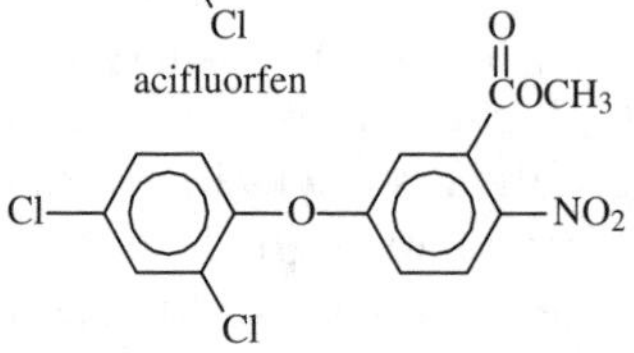

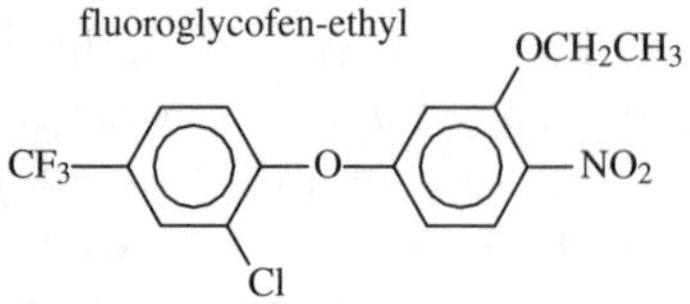

3.3 General SH reagents and free radical generators

Sulfhydryl groups are reactive and very often important in the active sites of many enzymes. Some pesticides with rather unspecific action are often SH reagents.

3.3.1 Mercury

We remember from our lessons in inorganic chemistry that HgS is insoluble (the solubility product for the reaction $Hg^{++} + S^{2-} \leftrightarrows HgS$ is 1.6×10^{-52} at 25°C). The very high affinity of Hg^{++} to SH groups is also the reason for the high toxicity of mercury compounds. Almost all organisms may be killed by mercurials. Resistance in fungi is therefore very rare, but may occur and result from an increased level of glutathione in the fungal cells that trap the Hg compounds.

The Pesticide Manual still has a few entries with mercurials, although the significance of mercury as a poison and as a general environmental pollutant is widely recognized. Today crematoriums have to install air-cleaning devices, dentists cannot use amalgam anymore, mercury-free thermometers are to be used, etc. The public concern about chronic mercury poisoning is very high. Organized groups of patients are convinced that their pains and problems are due to mercury released from their teeth even though most metal toxicologists believe that teeth amalgam gives too low a level of exposure to mercury to cause the claimed problems and think that patients who suffer from "amalgamism" should seek other reasons for their sufferings.

Pesticides with mercury also have a mixed reputation. Organic mercury compounds were used quite extensively as seed dressings for various cereals and other seeds to protect against fungal diseases. Very small amounts were effective in controlling fungus diseases. According to Mellanby (1970), as little as 0.5 kg of an organomercury preparation, containing 1% of mercury (5 g), was sufficient for 1 bushel of wheat grain. Only about 1 mg of mercury will be added to each square meter by this treatment — far below the natural level. Poultry or cattle fed on a moderate amount of dressed grain did not seem to suffer. No reason to wonder that organomercurials were popular. However, several epidemics of methylmercury poisoning have been reported, the most notable in Japan (1950s) and in Iraq. The Japanese case was due to mercury effluents from a factory. Microorganisms converted inorganic mercury to methylmercury, which poisoned fish, cats, and humans. The largest recorded epidemic of methylmercury poisoning took place in the winter of 1971–72 in Iraq, resulting in over 6000 patients and over 500 deaths (Goyer and Clarkson, 2001; WHO, 1974). The exposure was from bread containing wheat imported as seed grain and dressed with methylmercury fungicide. Several other serious accidents have been recorded. In Sweden, pheasant ate dressed seed and pike were poisoned or strongly contaminated by mercury from the chlorine–alkali process and by biocides used in the pulp industry. Mercury contamination was the big issue and changed public opinion and policy on effluents and agrochemicals in the late 1960s (Berlin, 1986; Borg et al., 1969; Fimreite, 1970, 1974; Mellanby, 1970).

Mercuric chloride ($HgCl_2$), mercuric oxide (HgO), and mercurous chloride (Hg_2Cl_2) still have some limited use as fungicides. They are very toxic (WHO toxicity classes Ia, Ib, and II). Mercurous chloride has a lower toxicity (rat oral LD50 = 210 mg/kg) than $HgCl_2$ (LD50 = 1 to 5 mg/kg) because it has a very low solubility.

Salts of methylmercury were used as a fungicide but may also be formed by biomethylation.

Hg^{++} is methylated by several sulfate-reducing bacteria (Desulfobacter) by a reaction with the methyl–vitamin B_{12} complex, which the bacteria normally use for producing some special fatty acids. The detection of methylmercury in the environment led to a shift from methylmercury fungicides such as Agrostan, Memmi, and Panogen 15 to ethoxyethylmercury as Panogen new, the methoxyethylmercury silicate Ceresan, and phenylmercurials such as Ceresol, Phelam, and Murcocide. Because mercury compounds are all very toxic to fungi, it was easy to make new alternatives, and many compounds and trade names were on the market.

The target organ in mammals and birds is the central nervous system. Symptoms include tunnel vision, paresthesias, ataxia, dysarthria, and deafness. Phenyl- and alkoxyalkylmercurials are absorbed through the skin and are almost as dangerous as the methyl analogue. Neurons are lost in the cerebral and cerebellar cortices. In the fungi, phenyl and alkoxyalkyl mercurials react with essential SH groups important in cellular division. Methylmercury interacts with DNA and RNA and binds with SH groups, resulting in changes of secondary structure of DNA and RNA.

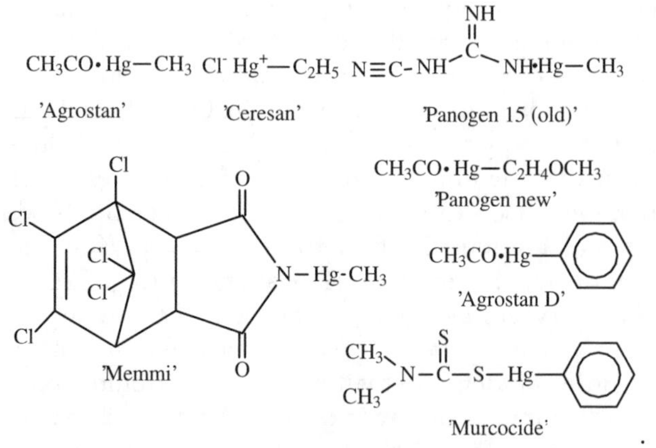

$$CH_3CO \cdot Hg—CH_3 \quad Cl^- Hg^+—C_2H_5 \quad N{\equiv}C\text{-}NH\underset{C}{\overset{NH}{\|}}NH\text{-}Hg—CH_3$$

'Agrostan'   'Ceresan'   'Panogen 15 (old)'

'Memmi'

$$CH_3CO \cdot Hg—C_2H_4OCH_3$$
'Panogen new'

$$CH_3CO \cdot Hg—\bigcirc$$
'Agrostan D'

$$\underset{CH_3}{\overset{CH_3}{\diagdown}}N{-}\underset{\overset{\|}{S}}{C}{-}S{-}Hg{-}\bigcirc$$
'Murcocide'

A selection of different organomerurials is shown above. They are taken from older issues of *The Pesticide Manual* (Worthing, 1979).

3.3.2   Other multisite fungicides

Dithiocarbamates, perhalogenmercaptans, sulfamides, copper salts, and ferric sulfate may be classified in this group. They all seem to be quite reactive against SH groups or are free radical generators. Detergents such as dodine and toxic salts such as sodium fluoride also have a multisite mode of action.

3.3.2.1   Perhalogenmercaptans
The fungicides in this group are good examples of pesticides that react with sulfydryl groups in many enzymes:

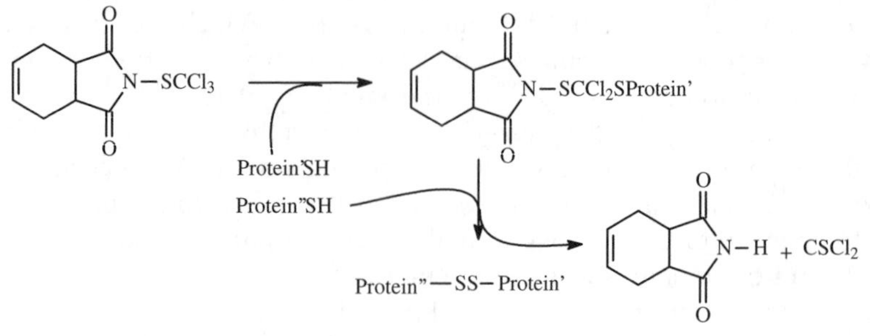

Other fungicides in this group are captafol, folpet dichlofluanid, and tolylfluanide. Captan has been widely used as a fungicide, but was blamed for being carcinogenic by the Environmental Protection Agency (EPA) in the U.S. in 1985. The perhalogenmercaptans' general toxicity toward animals is very low.

3.3.2.2 Alkylenebis(dithiocarbamate)s and dimethyldithiocarbamates

The pesticides in these groups are also regarded as unspecific SH reagents. Nabam is the disodium salt of alkylenebis(dithiocarbamate):

$$Na^+S\overset{\overset{\text{S}}{\|}}{C}NHCH_2CH_2NH\overset{\overset{\text{S}}{\|}}{C}SNa^+$$

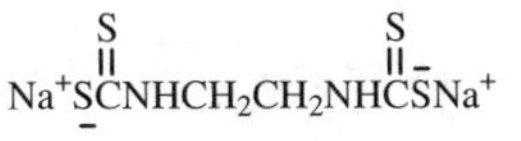

nabam

Nabam was originally described in 1943. When mixed with zinc or manganese sulfate, zineb or maneb is formed, respectively. The mixed salt of zinc and manganese, mancozeb is used quite extensively. The alkylenebis(dithiocarbamate)s have a low mammalian toxicity (e.g., LD50 = 8000 mg/kg for rats) but are considered to be carcinogenic, notably through the metabolite ethylenethiourea, which is formed by cooking.

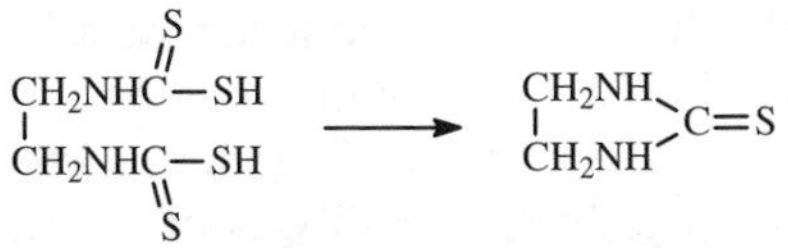

$$\begin{array}{l} CH_2NH\overset{\overset{\text{S}}{\|}}{C}-SH \\ CH_2NH\overset{}{C}-SH \\ \quad\quad\quad\underset{\text{S}}{\|} \end{array} \longrightarrow \begin{array}{l} CH_2NH \\ \quad\quad\quad\diagdown C=S \\ CH_2NH \diagup \end{array}$$

ethylenebisdithiocarbamic acid   ethylenethiourea

The dimethyldithiocarbamates, thiram ferbam and ziram, are disulfides, which were also described and made commercially available during the Second World War. It may be of interest to know that a structural analogue to thiram, disulfiram, is used as an alcohol deterrent because it inhibits aldehyde dehydrogenase, and when taken, ethanol is converted to acetaldehyde, which is then only slowly further metabolized. The high level of acetaldehyde in the body gives an extremely unpleasant feeling. A curious use of thiram is as a protectant against deer and stags in orchards. They dislike its scent and keep away, probably because the smell reminds them of some dangerous carnivores.

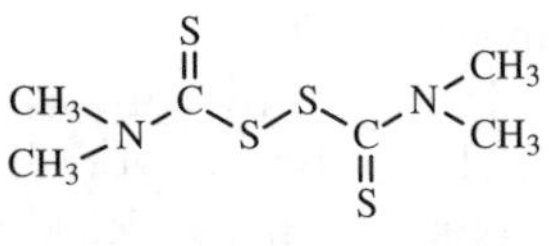

thiram 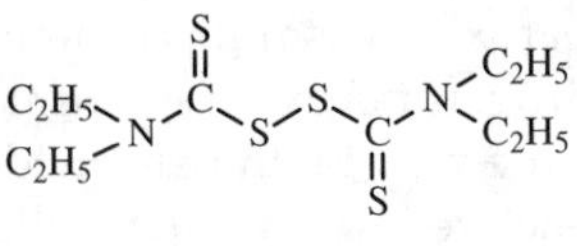   disulfiram

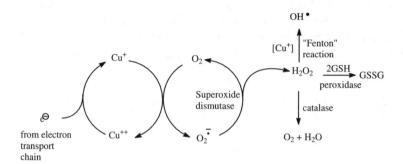

Figure 3.10 The toxic cycle of copper and the partial detoxication of superoxide by superoxide dismutase.

3.3.2.3 Fungicides with copper

Copper is an essential metal and all life-forms need it. It is a vital part of many enzymes such as cytochrome oxidase and the cytosolic form of super-oxide dismutase. Some organisms are very sensitive to it. Among them, we find such different organisms as sheep and goats, marine invertebrates, many algae, and fungal spores. Humans and pigs belong to the less sensitive species. The mechanism for this tolerance is at least partially due to a strict regulation mechanism of active copper concentration. A small cystein-rich protein called metallothionein can sequester metals like zinc, copper, cadmium, and mercury and plays a very important role in reducing the toxicity of these metals. The amount of metallothionein increases by exposure to the metals, and the difference in this ability is one of the mechanisms behind the great variation in copper sensitivity. Bordeaux mixture is a slurry of calcium hydroxide and copper(II) sulfate and has been used as an efficient spray to control *Phytophtora infestans* on potatoes, *Ventura inaequalis* on apples, *Plasmopara viticola* on vine, and *Pseudoperonospora humuli* on hops. It has a strong blue color and you cannot fail to recognize the blue-colored leaves on the grape vines during hikes through the vineyards on the Greek islands. It was introduced in France already in 1885 and soon played an important role in vine growing. It has surprisingly low toxicity to mammals, but when used for many decades in vineyards, it may give rise to an unwanted buildup of copper in the soil.

Bordeaux mixture and other copper salts owe their toxicity to the ions' ability to make one-electron exchanges (Cu^+ D $Cu^{++} + e^-$) (Figure 3.10). An electron is taken up from the electron transport chain and delivered to O_2 to form the superoxide anion. The anion radical is further transformed to H_2O_2 by superoxide dismutase catalysing.

The hydrogen peroxide is normally destroyed by catalase and peroxidases, but some may be transformed to OH·, the hydroxyl radical, which is extremely reactive and modifies all biomolecules in its vicinity. Membrane lipids are destroyed. The mode of action of copper salts and paraquat is in many ways the same, although their target organisms are quite different.

It may be of interest to know that the most efficient peroxidase contains selenium and is the reason why selenium is a necessary trace element. The most efficient superoxide dismutase in eukariotic cells contains copper. Copper is thus an important element for protection against free radicals but is also responsible for their formation. Copper ions form very stable SH compounds and would have been toxic without being superoxide generators.

3.4 Pesticides interfering with cell division

Bhupinder, P.S. et al. (2000) give a short and inspiring account of these pesticides, whereas a textbook in cell biology (e.g., Alberts et al., 2002) describes the function of tubulin in greater detail. Paclitaxel (Taxol®) is a very strong poison extracted from pacific yew (*Taxus brevifolia*), where it is present together with many other compounds with similar structure. The substance is also present in fungi (*Taxamyces adreanea* and *Pestalotiopsis microsporia*) associated with the Pacific and Himalayan yews, respectively. It has recently been detected in other species as well. Despite its toxicity, paclitaxel has become a very promising anticancer drug. The compound has a very complicated structure and is difficult, but not impossible, to synthesize. It has a strong fungicidal activity against oomycetes, i.e., fungi causing blights.

All garden enthusiasts know about the nice autumn crocus (*Colchicum autumnale*), which flowers in late autumn. It is not difficult to understand that this very conspicuous plant profits by containing a strong poison that protects it from pathogens and herbivores. It contains colchicine, which is very toxic and has a complicated structure. The substance is well known to plant breeders because it is used to double the number of chromosomes artificially in plants. A synthetic benzimidazole derivative, 1-methyl-3-dode-cylbenzimidazolium chloride, was developed in 1960 as a curative fungicide against apple scab. Thiabendazole, another synthetic benzimidazole derivative, has been used as an anthelmintic since 1962.

These synthetic and natural compounds are mentioned because they have a related mode of action. They react with *tubulin*, a protein that is the building block of the intracellular skeleton in eukaryotic cells. The shape and structure of a cell are dependent on microfilaments that keep the cell's constituents in the right place. It is different from a real skeleton because it is dynamic and changes structure. One important function of tubulin or, more precisely, the polymer of tubulin, called microtubules, is to make the spindle, a structure that pulls the chromosomes apart during mitosis. Two different tubulin subunits (α-tubulin and β-tubulin) make dimers that are stacked together and form the wall of hollow cylindrical microtubules. The α-/β-tubulin dimers are present in unpolymerized form in the cell, together with the polymeric microtubules, and a balance of assembly and disassembly maintains the microtubules, but the poisons mentioned above disturb this balance by binding to various sites of the β-tubulin, impairing normal cell division. Maintenance of cell shape, cell movement, intracellular transport, and secretion is dependent of the microtubuli. Therefore, it is not difficult to understand that strong poisons and anticarcinogenic compounds can be found among the substances interfering with the function of tubulin. Taxol acts by stabilizing the microtubule too much, whereas colchicin and most of the benzimidazoles act by inhibiting the formation of the microtubules. Although the mode of action of the two types of compounds is different, the net result will be the same: impairment of cell division. Binding of guanosine triphosphate (GTP) to β-tubulin stabilizes the polymeric form, whereas hydrolysis of GTP to guanosine diphosphate (GDP) destabilizes the micro-tubules. Benzimidazoles have a structural resemblance to the guanosine phosphates and may compete for the same binding site. This has been shown experimentally in preparations of rat testicles (Winder et al., 2001). Cell biologists studying tubulin and its function often use nocodazole.

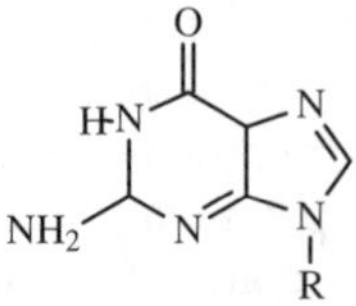

The biological activity of tubulin-interfering substances can be very high. The LC50 of taxol for cultured liver cells (HL-60 cells) is <0.001 μM, whereas thiabendazole at 80 μM completely inhibits mitosis in hyphae of *Aspergillus nidulans*, when growing in liquid culture.

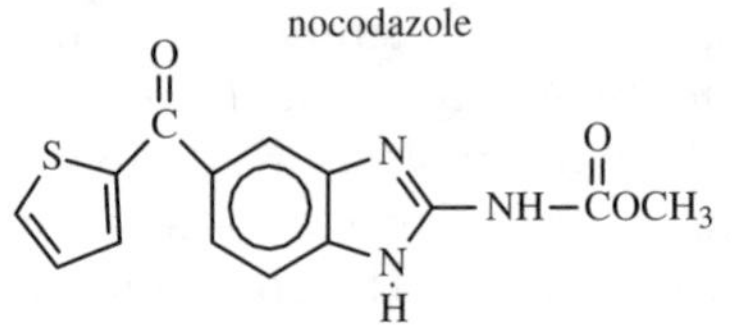

Benzimidazole fungicides with this mode of action include benomyl, carbendazim, dedacarb, fuberidazole, thiabendazole, and thiophan-ate-methyl. Other groups of fungicides that bind to tubulin include the fungicidal phenylcarbamate diethofencarb, swep and methyl 3,5-dichloro-phenyl-carbamate (MDPC), and the herbicidal carbamate carbetamide. A group of preemergence herbicides, the dinitroanilides, also bind to tubulin: this group includes benzfluralin, butralin, dinitramine, ethalfluralin, fluazinam, fluchlo-ralin, flumetralin, oryzalin, pendimethaline, prodiamine, and trifluralin.

Taxol and the other extremely biologically active derivatives of yews and fungi associated with yews have no value as fungicides, but may have evolved as natural pesticides:

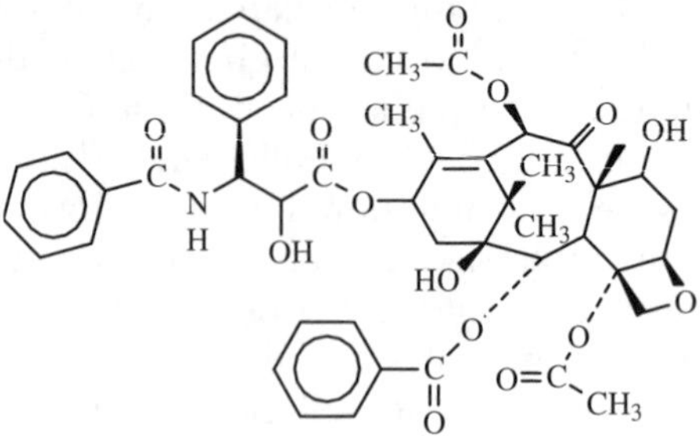

paclitaxel

3.4.1 Fungicides

3.4.1.1 Benomyl

As opposed to the older protective fungicides, the benzimidazole fungicides will kill growing mycelia and can therefore stop an infection already in progress. Benomyl has low water solubility (4 mg/l) but is degraded in soil and water to carbendazim and butylisocyanate with a half-life of less than 1 day. Carbendazim works, as described, as a tubulin poison, whereas butylisocyanate is very toxic and reacts with many cell constituents. Benomyl, or its degradation product butylisothiocyanate, may cause the plants to produce so-called phytoalexins. Attacted by patogens, plants have an inducible defense system in producing so-called fytoalexins, which are substances toxic for the fungi and protecting the plants against further attack. It has been shown that butylisocyanates induce production of such chemicals in plants and may be one of the mechanisms of the antifungal activity of benomyl.

Benomyl and the other benzimidazole fungicides are toxic to earthworms (Stringer and Wright, 1973) and may seriously disturb the earthworm population, for instance, in orchards, so that the leaf litter is not removed. Some benzimidazole fungicides (thiabendazole and mebendazole) may be used as anthelmintica.

3.4.1.2 Thiofanate-methyl

Thiophanate must also be transformed to carbendazim in order to be fungicidal. It is active against a wide variety of fungal pathogens and has low toxicity. In soil and plants it is slowly transformed into carbendazim.

1-methyl-3-dodecylbenzimidazolium chloride

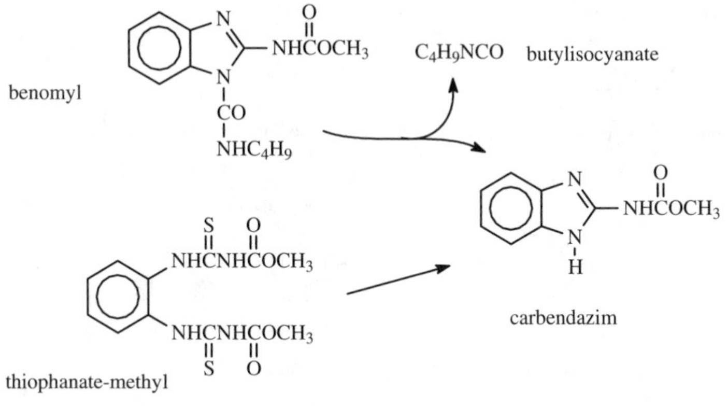

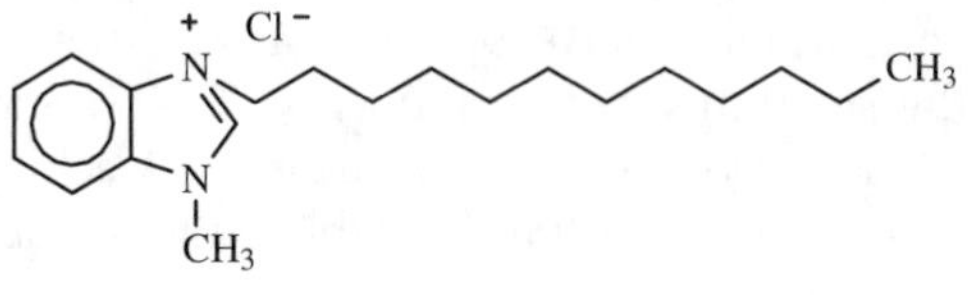

3.4.1.3 Carbendazim

Carbendazim was first described as a fungicide in 1973 and is active against a wide variety of fungi.

3.4.1.4 Thiabendazole

Thiabendazole's fungicidal properties had already been reported in 1964, but prior to that it had been used as an anthelmintic in human and veterinary medicine. In aqueous solution it is stable, but in mammals it is hydroxylated in the benzene ring by CYP enzymes.

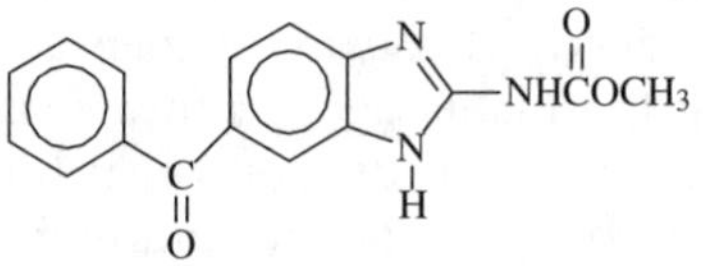

mebendazole
an anthelmintic used against nematodes

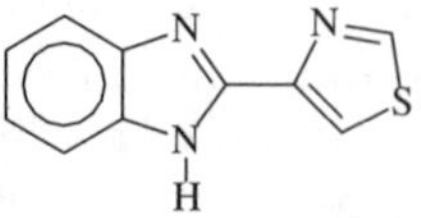

thiabendazole
anthelmintic and fungicide

3.4.1.5 Diethofencarb

Diethofencarb can be used against benzimidazole-resistant strains of *Botrytis* spp. It also inhibits mitosis. It is quickly degraded in soil and in animals through oxidation of the 4-ethoxy group.

diethofencarb

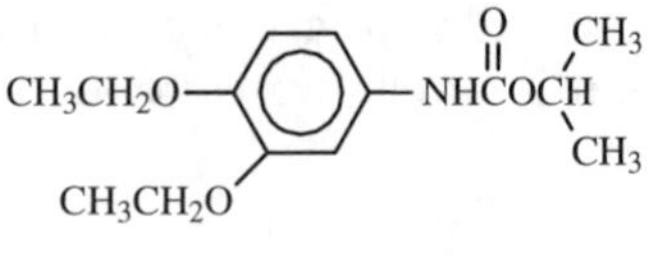

3.4.2 Herbicides

Many herbicides with a common structure inhibit cell division:

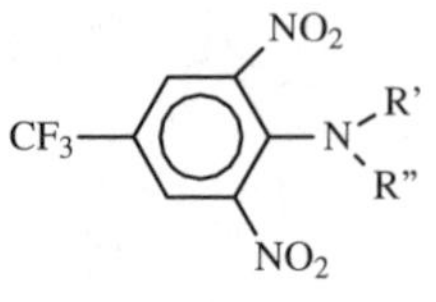

general structure

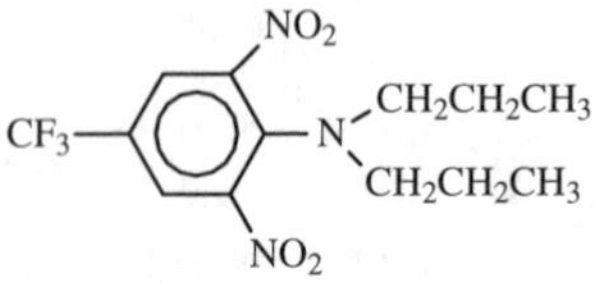

trifluralin

3.4.2.1 Trifluralin

Trifluralin has been marketed since 1961. It is a selective soil herbicide that acts by entering the seedlings in the hypocotyl region and disrupting cell division. It adsorbs in the soil and so is resistant to leaching, having a long residual activity. Trifluoralin is much more stable under aerobic than anaerobic conditions because microorganisms can reduce its nitro groups to amino groups anaerobically.

3.4.2.2   Carbetamide

This herbicide was first reported in 1963. It is highly selective against grasses and some broad-leaved weeds. Note that one enantiomer is regarded as the active ingredient:

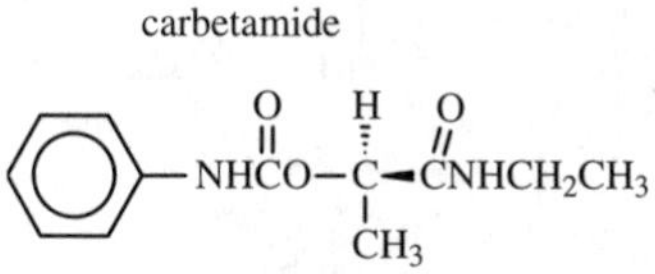

carbetamide

3.5   Pesticides inhibiting enzymes in nucleic acid synthesis

Important herbicides and fungicides may inhibit various enzymes in the synthesis of nucleic acids. One group of fungicides, the pyrimidinols, inhibits the synthesis of substances important for sporulation, whereas the acetanilides inhibit incorporation of nucleic acids in RNA.

3.5.1   Sporulation-inhibiting fungicides

Analogues of the insecticide diazinon were found to be active against powdery mildews, and studies of correlations between structure and activity led to the development of the pyrimidinol fungicides (ethirimol, bupirimate, and dimethirimol). The pyrimidinols appear to act by interfering with the metabolism of purines by inhibiting adenosine deaminase. The enzyme, which catalyzes the hydrolytic deamination of adenosine to inosine, is important in fungi but not in plants. Inhibition leads to a stop in the sporulation process. The fungicides are highly soluble in water and act systemically by being absorbed through the leaves and translocated. They are stable in soil, but their toxicity to animals is very low.

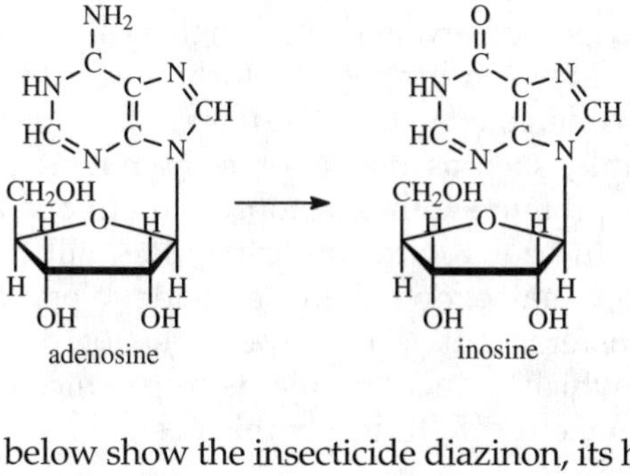

adenosine   inosine

The structures below show the insecticide diazinon, its hydrolytic product, and ethirimol. Bupirimate and dimethirimol have very similar structures.

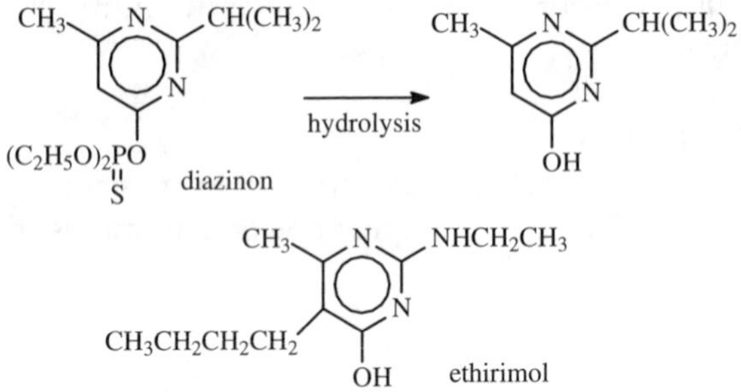

3.5.2 *Inhibition of incorporation of uridine into RNA*

Inhibition of the incorporation of uridine into RNA is caused by the herbicides referred to as the chloroacetanilides (e.g., acetochlor, alachlor, butachlor, and several others) and a group of fungicides referred to as phenylamides (metalaxyl, ofurace, and oxadixyl). They have similar structure and mode of action:

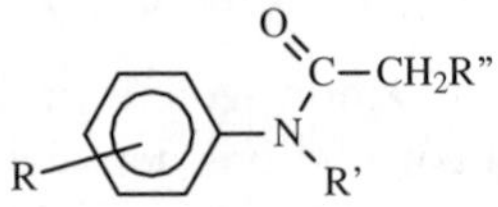

R is often one or two alkyl groups, R' may have various structures, and in the herbicides, R" is chlorine in accordance with the name chloroacetamides.

The discovery of the fungicides arose from the observation of the antifungal activity of chloracetanilide herbicides, and it proved possible to retain and improve this activity while minimizing the herbicidal effects. These compounds were commercialized during the 1970s. The herbicides are valuable because of their selectivity, and the fungicides are particularly useful against powdery mildew on grapes. Acetochlor is selective because it is detoxified by being conjugated to the SH-group of the tripeptide, glutathione (GSH) in some plants, such as maize, or to homoglutathione in soya. In sensitive plants the protein synthesis stops. Glutathione and the enzymes responsible for conjugations with this tripeptide will be described later. Many plants, notably maize, owe their herbicide tolerance to a high level of glutathione transferase. It is even possible to treat the seeds with so-called safeners, which induce the plants to produce even more of this enzyme and thus make it safe to use herbicides.

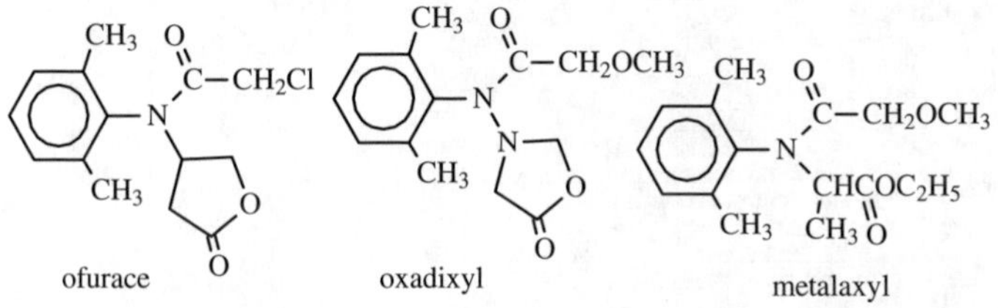
ofurace
oxadixyl
metalaxyl

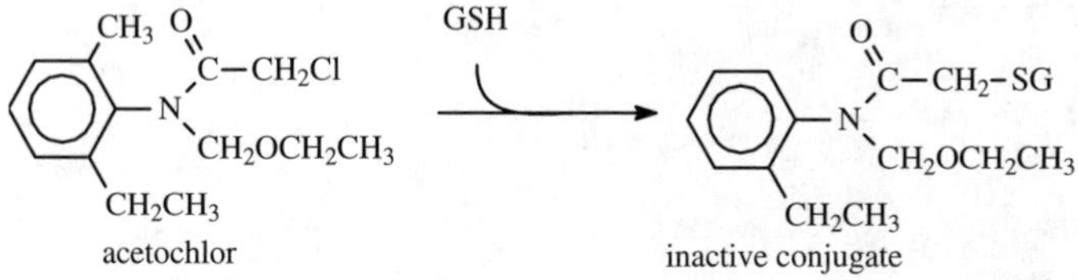
GSH
acetochlor
inactive conjugate

chapter four

Bacillus thuringiensis and its toxins

This chapter is based mainly on Glare and O'Callaghan (2000), Schnepf et al. (1998), and Crickmore et al. (1998). *Bacillus thuringiensis* (*Bt*) is now the most widely used biologically produced pest control agent. In 1995 world-wide sales of *B. thuringiensis* were $90 million, representing about 2% of the total global insecticide market. In 1998 nearly 200 insecticidal products were registered in the U.S. Studies have so far not shown any pathogenic action on mammals, birds, amphibians, or reptiles (Board of Agriculture and Natural Resources, 2000).

Bt was first described in 1911 when a bacillus was isolated from the Mediterranean flour moth, *Anagasta kuehniella*. It is a Gram-positive, rod-shaped, spore-forming bacterium that is closely related to *Bacillus cereus*, a bacterium causing gastroenteritis in humans, and to *Bacillus anthracis*, a very dangerous bacterium that can be used as a biological warfare agent. The first isolation of *Bt* was done more than 100 years ago by the Japanese biologist S. Ishiwata, who isolated a bacterium as the causal agent of a disease of silkworms, but not before 1928, when a project that tried to utilize *Bt* against a pest, the corn borer (*Ostrinia nubilalis*), was initiated. The presence of a parasporal body in sporulating *Bt* cells was noted as early as 1915, whereas its proteinaceous nature and its toxicity to silkworms were not recognized before 1954. The potential of *Bt* as a pesticide was recognized in the 1920s, and by 1992 more than 40,000 isolates were held in collections throughout the world. Today it is more than 60,000. Strains active against a wide variety of insects have now been isolated. These include Lepidoptera, Diptera, Coleoptera, Hymenoptera, Hemiptera, and Mallophaga. Nematoda and Protozoa are also susceptible to some strains. More than 170 genes encoding for *Bt* toxins have been identified.

The toxins fall into one of these groups:

- δ-Endotoxins or Cry toxins
- Smaller cytosolic endotoxins (Cyt toxins)
- β-Exotoxins

- Hemolysins
- Enterotoxins
- Vegetative insecticidal proteins (VIPs)
- Exoenzymes

The genes responsible for the crystal proteins are called *cry* genes, and the crystal toxin is called δ-endotoxin or Cry toxin, whereas the genes responsible for toxic cytosolic proteins (Cyt toxins) are called *cyt* genes. Many *Bt* strains are able to produce a number of such smaller cytosolic endotoxins in addition to the δ-endotoxins, but these are often deposited in inclusion bodies inside the crystals where they can comprise a considerable part of the crystal. Unlike the Cry toxins, the other toxins display a broader unspecific activity and may have some mammalian toxicity. They include the β-exotoxins, hemolysins, and enterotoxins.

The β-exotoxin is sometimes called the fly toxin, but it has a broad spectrum of activity, not confined to Diptera. It is thermostable and is not destroyed by heating at 70°C for 15 minutes. Because of the vertebrate toxicity, most commercial preparations use subspecies or isolates that do not produce β-exotoxin. However, the β-exotoxin can synergize the activity of δ-endotoxin against natural tolerant insects. This synergy may result from the inhibitory effect of β-exotoxin on the regeneration of midgut cells damaged by δ-endotoxin. A new variant of β-exotoxin that is toxic to aquatic mollusks has been described. It may be useful to control vectors of schistosomiasis and other snail-borne diseases, as well as snails in agriculture.

Hemolysins, which lyse vertebrate erythrocytes, are important virulence factors in vertebrate bacterial pathogens. Such toxins are also found in some *Bt* strains and seem to be identical to the hemolysin found in *B. cereus*. Some *Bt* isolates have been found to produce the same type of diarrhea-producing enterotoxins as *B. cereus*. *Bt* may therefore have implications in causing gastroenteritis.

Bt produces and secretes a number of enzymes, e.g., chitinases, proteases, and phospholipases, of importance to the pathogenicity. They disrupt the peritrophic membrane, providing access for the true toxins to the gut epithelium.

A new class of insecticidal toxins called vegetative insecticidal proteins has recently been isolated from *Bt*. They are produced during the vegetative growth stage. The proteins are different from other known proteins, and their function and versatility for insect control have yet to be elucidated.

4.1   The mechanism of action of δ-endotoxins

As mentioned, the δ-endotoxin (Cry protein) that makes up most of the conspicuous crystals is the main insecticidal component of *B. thuringiensis*. At sporulation, the majority of *Bt* strains produce crystalline inclusions that contain this insecticidal δ-endotoxin. The crystals account for 20% or more of the total bacterial protein at sporulation and may contain one or several endotoxins, which differ in activity. Many *Bt* toxin genes and genes for some endotoxins are encoded by extrachromosomal DNA, often located on large plasmids. (A plasmid is a small circular DNA molecule often present in bacterial cells. It replicates independently and can be transferred naturally or artificially to other bacteria.) One plasmid may have genes for more than one type of endotoxin. The *cry* genes can also be located on the bacterial chromosome. Even genes encoding the same protein may be located both on plasmids and on the chromosome. The amino acid sequences of the toxins are now easier to obtain experimentally than they were earlier; hence, it follows that the classification and nomenclature could be based on sequence similarity that reflects phylogeny, and not the biological activity.

The amino acid sequence is used for classification, first in superfamilies Cry1, Cry2, etc. If the sequence is more similar, a letter is added (Cry1A, Cry1B, etc.) and the individual proteins are called Cry1A1, Cry3B3, etc. Other older classification and designation systems, not based on amino acid sequence, are still in use.

The δ-endotoxins and the other *Bt* toxins function only as stomach poisons. A susceptible insect must eat them, and the crystals dissolve in the midgut. This solubilization is sometimes dependent on chymotrypsin-like enzymes. The protoxin is further attacked by proteolytic enzymes present in the gut that convert it to the active toxin. An interesting and unexpected finding is that DNA is intimately associated with the toxin crystal and appears to play a role in the proteolytic processing.

The active toxin has two main functional entities, responsible for receptor binding and ion channel activity, respectively. The activated toxin binds to receptors, which seem to be of different types, on the midgut microvilli of the susceptible insects. Different toxins seem to bind to different receptor proteins that may be an enzyme such as aminopeptidase or alkaline phosphatase, or a cadherin-like membrane protein. (The cadherins are proteins that are important in keeping the cells together by mediating Ca^+-dependent cell–cell adhesion in animal tissue.) The toxins are anchored to the outer epithelial cell membrane in such a way that the membrane is perforated by pores or channels where ions can freely pass. This model proposes that an influx of water, along with ions, results in swelling and lysis. The epithelium is destroyed and the insect rots.

Consumption of food treated with the endotoxins, or engineered plants that produce them, results in cessation of feeding of Lepidoptera larvae and paralysis of the gut that retards the passage of food and allows the spores to germinate. The larvae suffer a general paralysis and die. *Bt israelensis*-treated mosquito larvae cease feeding within 1 hour of treatment, show reduced activity after 2 hours, and general paralysis after 6 hours. In beetles death may take longer.

Engineering of Cry proteins to create better pesticides is possible. A mutant of Cry4B resulted in a threefold increase of toxicity against the mosquito, perhaps by removal of a site sensitive to proteolytic instability. More often, increased receptor binding causes the increased efficiency of mutant types.

4.2   Biotechnology

Crystal genes have been introduced into other bacteria such as *Escherichia coli*, *Bacillus subtilis*, *Bacillus megatorium*, and *Pseudomonas fluorescens*. Fermentations of recombinant *Pseudomonas* have been used to produce concentrated aqueous biopesticide formulations consisting of Cry inclusions encapsulated in dead cells. Engineered forms of Cry proteins may show improved potency or yield and may make them a more attractive and practical alternative or supplement to other traditional pesticides. The gene has also been introduced into a bacterium (*Clavibacterium xyli*) that lives inside plants. When corn is infested with this bacterium, the crop is protected against corn borer. Other endophytic microorganisms (*Azospirillum* spp., *Rhizobium*, etc.) have also been engineered, and *cry* genes specifically active to Diptera have been introduced into other bacteria, including Cyanobacteria.

4.3   Engineered plants

In order to get an optimal production of toxins in plants, the *cry* gene from *Bt* must be modified extensively before being introduced into the plant's genome, and only the part of the gene that codes for the active part of the *Bt* toxin is used. Full-length unmodified *cry* genes give quite inefficient toxin production. Varieties of potato, cotton, corn, and many other plants containing modified *cry* genes are now available on the market in some countries. The marketing started in 1996 and the proponents for such plants are very enthusiastic. Because the toxins are produced continuously and apparently persist for some time in the plant tissue, fewer, if any at all, chemical pesticides must be used. Beneficial insects are not harmed. However, the public opinion, notably that of environmentalists, is skeptical. There are fundamental beliefs that such gene manipulations are to play the role of God, or are to intervene with natural selection in an unethical and unacceptable way. Other more trivial arguments may be more relevant to the rational mind, i.e., that there may be some unknown health or ecological effects. It is outside my field of knowledge to offer a qualified opinion about these important questions.

4.4   Biology

A biologist will want to know something about the relationship of *Bt* to other bacteria, the natural function, if any, of the toxins, as well as the natural occurrence of the *Bt* strains. Numerous strains have been isolated from a wide range of habitats such as soil, mushroom compost, and stored products. *Bt* may be present in samples from beach deserts and tundra, and may multiply in cadavers of insects and other animals. Insecticidal activity does not correlate with the origin of the isolate, and many of them do not seem to be toxic for insects. *Bt* may be regarded as a ubiquitous soil microbe and common on vegetation. Their spores may be viable for years in the soil environment. Of viable bacterial spores present on foliage on various decid-uous and conifer trees, *Bt* often accounts for 30 to 100%. Viable spores are also quite common on grasses and herbs. The natural function of the toxins may be connected to this habitat. If insects visiting the plant surfaces are infected, they are killed by the toxins and make an excellent food source for the bacterium.

As mentioned, *Bt* makes a taxonomic group together with *Bacillus anthra-cis* and *B. cereus. Bt* and *B. cereus* are so closely related, both serologically and by various methods using DNA sequences, that they may be regarded as one species. The much feared toxin of *B. anthracis* is different from the δ-endotoxin of *Bt*. The *B. anthracis* toxin produces three factors — lethal factor, protective antigen, and edema factor — that are encoded by three different genes. The protective antigen recognizes and binds to certain recep-tors on the cell membrane, where they form a pore. The lethal factor and edema factor bind to the protective antigen and are taken into the cell by a process called endocytosis. After uptake, the lethal factor acts as a protease that specifically splits an enzyme important for internal signaling in the cell (MAP kinase, or MAPKK1 and MAPKK2). MAP kinase is an abbreviation of "mitogen-activated protein kinase," which is described in Alberts et al. (2002) and other cell biology textbooks. The edema factor acts also as an enzyme (adenylate cyclase) that inhibits immune response. The anthrax toxins are therefore different from the δ-endotoxin of *Bt*, although the making of pores is a related mechanism.

4.5 Commercial products

The first commercial product that appeared in 1938 for use against lepi-dopterous larvae was Sporeine, which was produced by several companies. Names of other old products are Dipel WP, a wettable powder, and Dipel LC, a liquid concentrate used for aerial low-volume application. The *Bt*-based microbial pesticides have so far been the most successful biopesti-cides. The rapid growth of *Bt*-based pesticides is occurring as replacements of chemicals that were banned or phased out in environmentally sensitive areas, in consumer and export markets in which concerns about food residue are high, and in organic food production. The current sales are at least $140 million. Today the commercial products are based on various strains of the more than 26 different subspecies.

The several strains of the subspecies *Bt kurstaki*, or strains of *Bt kurstaki* with various *cry* genes from *Bt aizowai, Bt morrisoni*, and *Bt kumatoensis*, are marketed for use against lepidopterous larvae, or in some cases against Colorado beetles. *Bt* is now sold under about 30 trade names. Preparations with *Bt tenebrionis* are used against Colorado beetle, and *Bt japonensis* may be used against soil-inhabiting beetles in turf and ornamentals. Even more inter-esting is *Bt israelensis*, which is used in aerial applications against mosquito and blackfly larvae. It is also sold under many different trade names.

Various preparations of δ-endotoxin (Cry1A(c), Cry1C, or Cry3A) are also marketed. The toxins are produced by engineered *Pseudomonas fluorescence* and are formulated as microcapsules or as granular formulations. They are used against Lepidoptera, armyworms, Colorado beetles, and corn borers.

National approval authorities in some countries hesitate to approve *Bt* preparations because of the similarities between *B. thuringiensis* and the pathogenic species *B. anthracis* and *B. cereus*. However, no ecotoxicological or human toxicity seems to be associated with *Bt* itself or its toxins. The half-life in soil is short, and it is destroyed by sunlight. The questions about possible ecological consequences of *Bt* plants are not yet settled.

chapter five

Specific enzyme inhibitors

Some pesticides, such as the herbicides inhibiting synthesis of amino acids in plants, are extremely selective between plants and animals and very potent. The chitin synthesis inhibitors used as insecticides are also extremely selective, because only insects and crustaceans (and fungi) make chitin. The fungicides first described are also efficient and have a high degree of selectivity, but are likely to produce effects in animals and plants because they inhibit enzymes of great importance to many types of organisms.

5.1 Inhibitors of ergosterol synthesis

Sterols are important building blocks in the cell's membrane system, and many sterols are important hormones. In animal tissues *cholesterol* is quantitatively most important, whereas in fungi we find *ergosterol* and in plants *stigmasterol* and *β-sitosterol*. Most eukaryotic organisms seem to be able to synthesize sterols with acetyl-coenzyme A (CoA) as the starting material: exceptions are insects and some fungi. The pathway is complex, with many steps and many enzymes involved. Some steps in the synthesis need oxygen and, for example, yeast cannot produce sterols when grown completely anaerobically. Therefore, yeast fermenting cannot go on forever without oxygen because the oxygen is needed as a co-substrate in sterol synthesis.

In spite of the similarity of sterol synthesis in plants, fungi, and animals, the pathway is an excellent target for fungicides. Inhibitors of ergosterol synthesis are the largest group of fungicides with the same target. Most of these fungicides, however, have various effects on plants and animals as well, but have low lethal toxicity.

The biosynthesis of sterols is extremely complicated and a good textbook in biochemistry should be consulted (e.g., Nelson and Cox, 2000). Let us recapitulate the process:

1. Three molecules of acetyl-CoA condense to form mevalonate.
2. Mevalonate is converted to isoprene units (isoprene pyrophosphate having five carbons).
3. Six isoprene pyrophosphate molecules are converted to squalene (having 30 carbon atoms).

4. Squalene is converted to squalene epoxide and then to lanosterol.
5. Lanosterol is converted to stigmasterol (in plants), cholesterol (in animals), and 24-methylenedihydrolanosterol (24-MDL) (in fungi), which is further converted to ergosterol.

All the steps involve many enzymes — oxidations, reductions, isomerizations, methylations, and demethylations.

The steps that are of greatest importance as targets for inhibitors are:

- The formation of mevalonate from β-hydroxy-β-methyl-glutaryl-coenzyme A (HMG-CoA)
- Epoxidation of squalene
- Removal and addition of methyl groups in lanosterol and other sterols that are precursors of cholesterol and ergosterol
- Isomerization reactions

5.1.1 Inhibition of HMG-CoA reductase

Acetyl-CoA is first transferred through many steps to HMG-CoA, which is then reduced to mevalonate by HMG-CoA reductase:

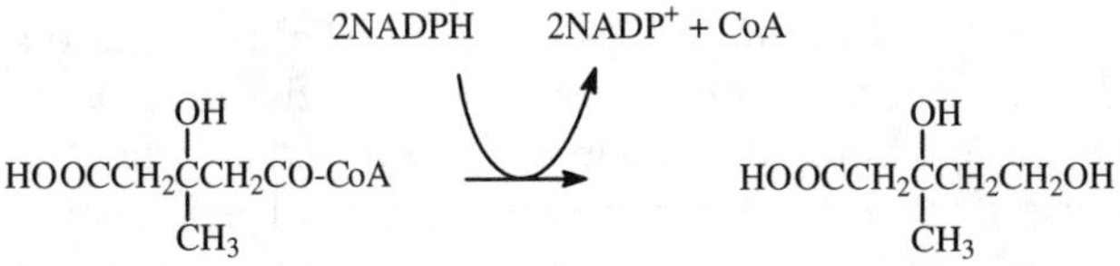

$$2NADPH \qquad 2NADP^+ + CoA$$

$$\underset{\underset{\displaystyle CH_3}{|}}{\overset{\overset{\displaystyle OH}{|}}{HOOCCH_2CCH_2CO\text{-}CoA}} \xrightarrow{\hspace{2cm}} \underset{\underset{\displaystyle CH_3}{|}}{\overset{\overset{\displaystyle OH}{|}}{HOOCCH_2CCH_2CH_2OH}}$$

HMG-CoA   mevalonic acid

HMG-CoA reductase is the rate-determining enzyme of sterol synthesis, and its activity is regulated by competitive inhibition by compounds that bind to the same site as HMG-CoA. It is also regulated by substances that bind to other (allosteric) sites on the enzyme molecule. Inhibitors of this enzyme (e.g., simvastatin) are used as medicines to reduce cholesterol in patients whose cholesterol levels are too high. Through feedback inhibition, cholesterol is a strong inhibitor of the enzyme itself. No fungicides with this mode of action have yet been developed, but the possibility that they will be exists.

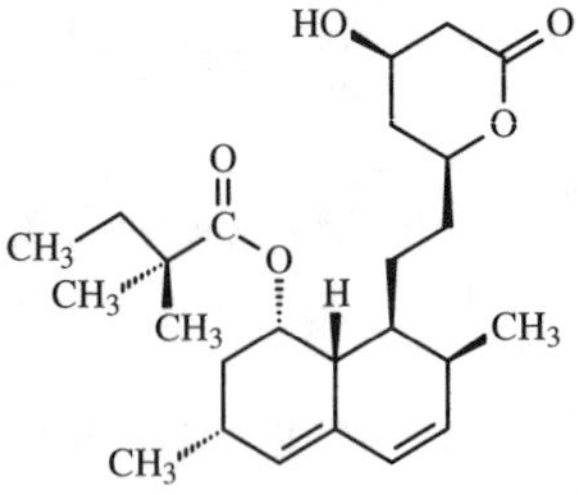

Simvastatin

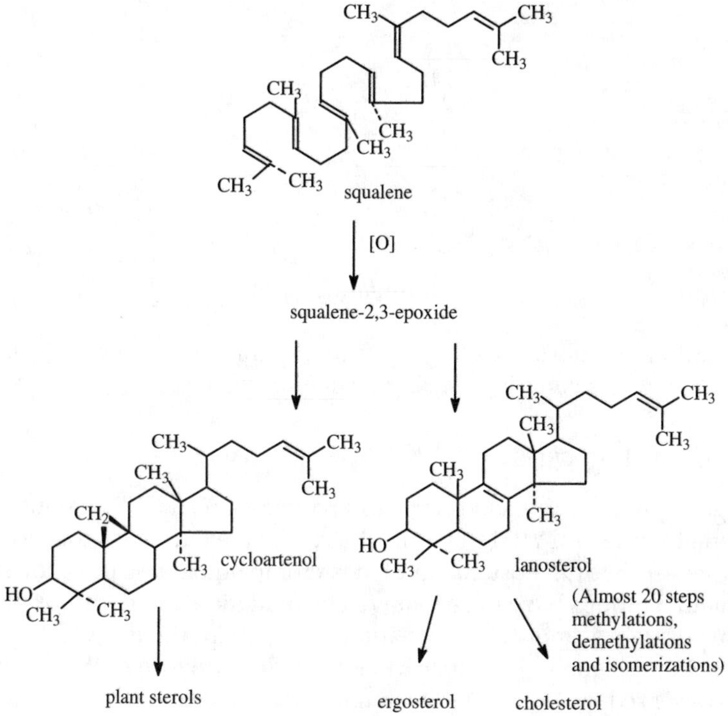

Figure 5.1 Formation of sterols in plants, fungi, and animals.

5.1.2   Inhibition of squalene epoxidase

Mevalonate is first phosphorylated and decarboxylated through four steps to give isopentenyl pyrophosphate and dimethylallyl pyrophosphate. Through three new steps these compounds react with each other to give squalene, an aliphatic hydrocarbon with 30 carbons and 6 double bonds. A hydroxyl group is introduced into squalene and formation of the typical ring system of the sterols takes place (Figure 5.1).

A group of fungicides that inhibit squalene epoxidation has been developed primarily for use against pathogenic fungi in medicine. Epoxidation of squalene is catalyzed by squalene epoxidase (a flavoprotein) that starts the complicated cyclization of squalene. The squalene-2,3-epoxide formed by this enzyme is further metabolized to a protosterol cation intermediate, which is transformed to either cycloartenol in plants (cycloartenol synthase) or lanosterol (lanosterol synthase). Cycloartenol is the precursor to plant sterols, whereas lanosterol is the precursor of cholesterol and the other sterols in animals, and to ergosterol in plants.

Terbinafine, which also has a complicated structure, is an example of a fungicide that inhibits this enzymatic step. It is used as a fungicide against systemic and dermal infections in humans.

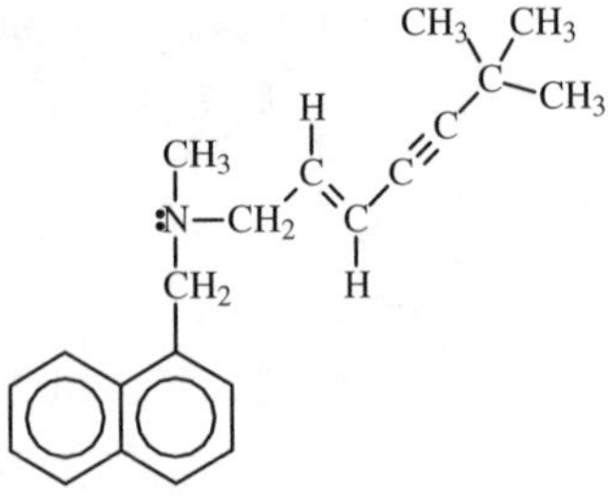

terbinafine

Several other substances toxic to fungi inhibit squalene epoxidase, the key enzyme in this complicated ring formation.

5.1.3 DMI fungicides

The largest group of fungicides inhibits an oxygenase, a CYP enzyme called 14-α-demethylase or CYP51. It has a vital role in the pathways transforming 24-methylenedihydrolanosterol and lanosterol to ergosterol and cholesterol. Three methyl groups have to be removed by oxidation and decarboxylations (two in position 4 and one in position 14). This particular CYP enzyme removes the 14-α-methyl group. The amino acid sequence of the enzyme is highly conserved and is similar in fungi, plants, and animals. It is the only family of CYP enzymes recognizable across all eukaryotic phyla.

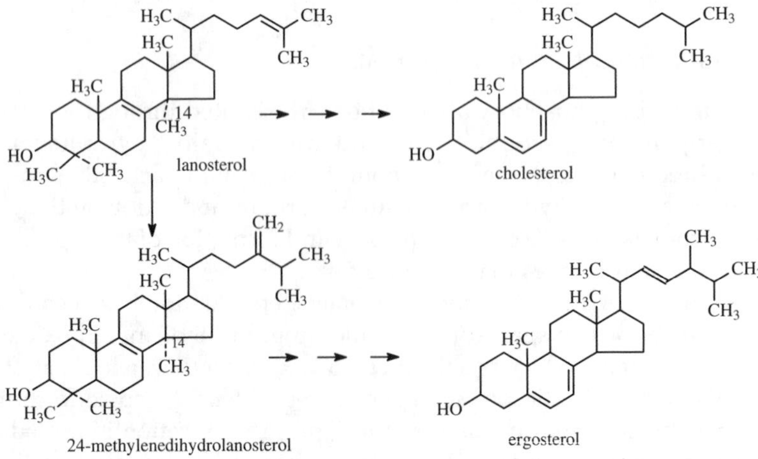

lanosterol

cholesterol

24-methylenedihydrolanosterol

ergosterol

There are approximately 20 enzymatic steps from lanosterol to cholesterol or ergosterol, and probably as many from 24-methylenedihydrolanosterol to ergosterol.

The fungicides that inhibit fungal CYP51 are often called demethylase inhibitor (DMI) fungicides, but the group is chemically very diverse. A DMI

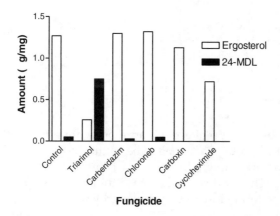

Fungicide

Figure 5.2 The effects of some fungicides on the sterol composition in sporidia. This figure is based on some data presented at the 7th British Insecticide and Fungicide Conference (1973) and shows the effect of the concentration of ergosterol and 24-methylenedihydrolanosterol in sporidia of a fungus. It is evident that triarimol was the only fungicide of those tested that reduced ergosterol and increased 24-MDL significantly.

fungicide always has a heterocyclic N-containing ring, as in pyrimidines, pyridines, piperazines, and azoles. As a consequence, they are not too difficult to recognize by formula. Characteristically, they also have at least one enantiomeric C atom. The CYP enzymes have an important iron atom that can bind to one of the N atoms with a free electron pair, thus competing with the binding of oxygen.

The DMI fungicides do not appear to affect CYP enzymes in general but may, of course, inhibit other CYP enzymes than CYP51, and may inhibit CYP51 in organisms other than fungi, thereby interfering with their normal development. The CYP51 enzyme involved in sterol synthesis in plants does not seem to be seriously inhibited, or it does not seem to matter if it is.

The DMI fungicides cause intermediates, e.g., sterols with methyl groups such as 24-methylenedihydrolanosterol, to accumulate (Figure 5.2). The amount of free fatty acids also increases because acetyl-CoA is no longer used to produce sterols and the phospholipids in the membrane are degraded. The symptoms in the fungi correspond to these biochemical changes, resulting in the disturbance of the cell membrane. The fungal spores may start growing as normal but change in their appearance as the hyphae swell and branch.

The DMI fungicides have interesting effects on plants that are not related to sterol synthesis but to gibberellin synthesis. Some of them are therefore more useful as plant growth regulators than as fungicides. Ancymidol is a typical example of a DMI used as a plant growth regulator. The superseded fungicides triarimol and triamedifon also inhibit plant growth. The leaves of triarimol-treated plants become dark green, and the growth becomes slower. The reason for these effects is not due to inhibition of ergosterol synthesis but is caused by an inhibition of gibberellin synthesis.

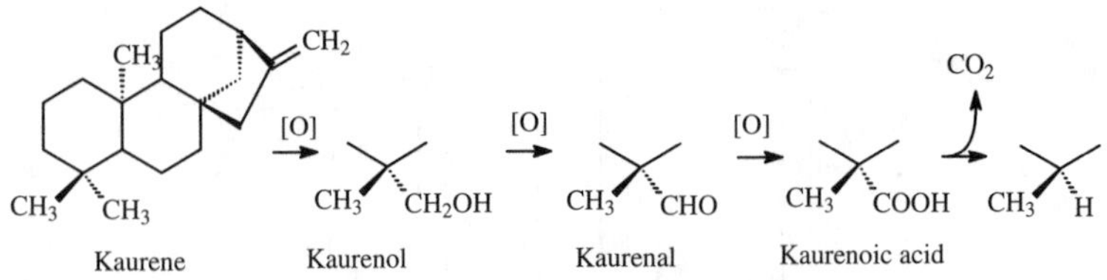

Kaurene   Kaurenol   Kaurenal   Kaurenoic acid

Gibberellins, a group of growth hormones, are produced via intermediates with methyl groups that need to be eliminated by oxidation. More than 60 gibberellins are known, but the most important is gibberellic acid or gibberellin A_3. The DMI fungicides also inhibit this step, and not enough gibberellins are formed to give maximal growth.

5.1.4 Examples of DMI fungicides from each group

5.1.4.1 Azoles and triazoles

This is the biggest group, and the 12th edition of *The Pesticide Manual* describes 5 fungicidal imidazoles and 22 triazole fungicides (Tomlin, 2000). We take two examples: *Imazalil* is regarded as especially valuable against benzimidazole-resistant plant-pathogenic fungi. *Flusilazol*, a stable fungicide, is interesting because the central atom is silicon and not carbon. It has some solubility in water and is systemic in plants. It is used against a wide variety of fungi.

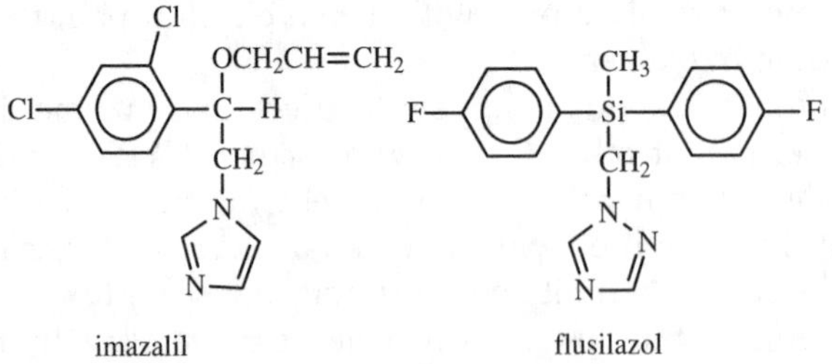

imazalil   flusilazol

5.1.4.2 Pyridines and pyrimidines

In this group we find ancymidol, which is mainly used as a plant growth regulator, and a few fungicides.

Pyrifenox is relatively rapidly degraded in soil and metabolized in animals and plants.

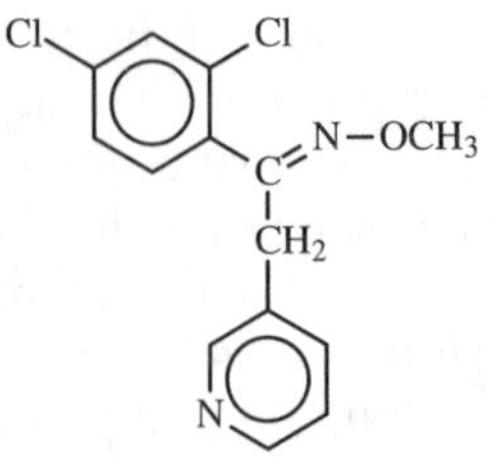

Triarimol, a superseded fungicide/plant growth regulator, was introduced in 1969 and is included here because of its importance in much of the fundamental research on DMIs.

Fenarimol may be used against powdery mildews and other plant pathogens. Leaves become abnormal and dark green if the dose is too high. It decomposes rapidly in sunshine but is very stable in soil.

Ancymidol is classified as a plant growth regulator and has a wide application. It is taken up and translocated in the phloem and inhibits internode elongation by inhibiting the CYP enzyme in the biosynthetic pathway of gibberellins. The structures of the three above-mentioned compounds are reasonably similar:

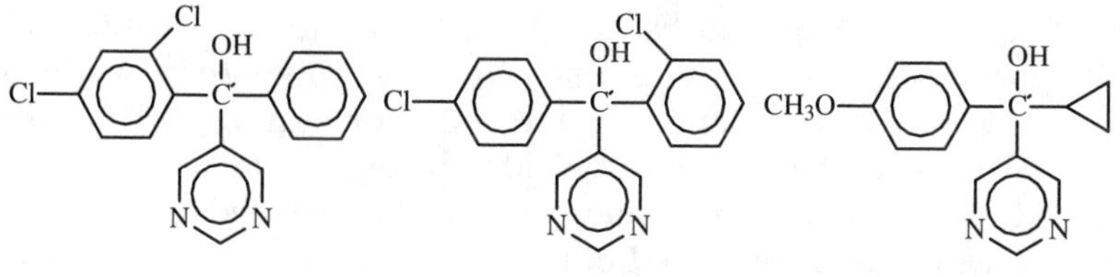

triarimol   fenarimol   ancymidol

5.1.4.3 Piperazines

Triforine is metabolized in plants to many products that are not toxic to fungi according to *The Pesticide Manual* (Tomlin, 2000). It is regarded as environmentally safe.

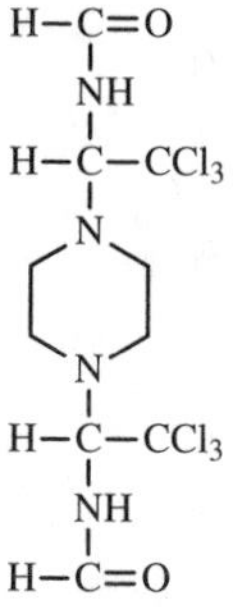

5.1.4.4 Amines

CYP-inhibiting amines (e.g., SKF 525A) have been used to control elevated levels of cholesterol in humans. They are also toxic to fungi by the same mechanism. SKF 252A has been extensively used as a specific inhibitor of CYP enzymes in research and is a particularly strong inhibitor of CYP51, but has not been used as a commercial fungicide.

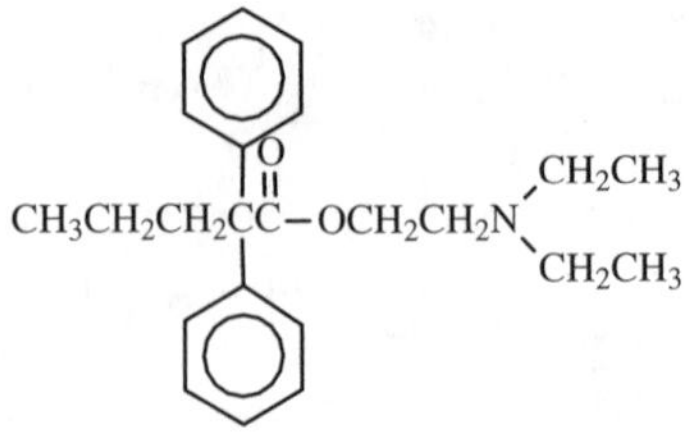

$$CH_3CH_2CH_2\overset{\overset{O}{\|}}{C}-OCH_2CH_2N\overset{CH_2CH_3}{\underset{CH_2CH_3}{\diagdown}}$$

SKF 525A

5.1.4.5 Morpholines

Enzymes later in the pathway, from desmethyl-24-methylenedihydrolanos-terol to ergosterol, may also be targets for fungicides. The morpholines inhibit enzymes called Δ^{14}-reductase, which saturate the double bond between carbon 14 and carbon 15, and $\Delta^8 \rightarrow \Delta^7$-isomerase, which change the localization of a double bond. Fungicides belonging to this group were described in 1967, and the group may therefore be regarded as old, although its mode of action was elucidated much later.

Dodemorph has a 12-membered alkyl ring connected to the morpholine ring, whereas tridemorph has a 12- to 14-membered aliphatic chain.

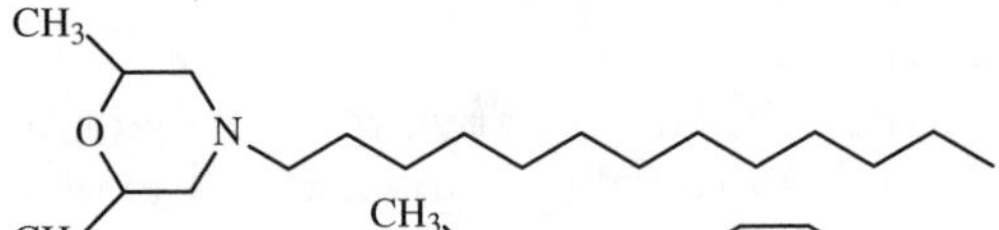

tridemorph

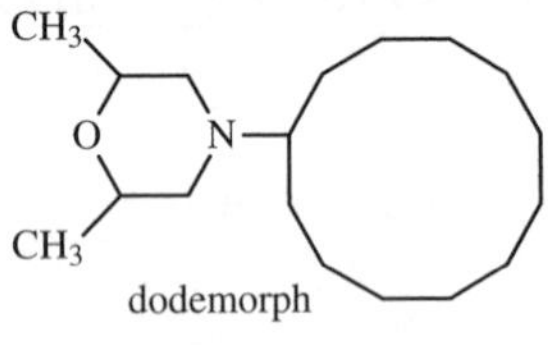

dodemorph

Fenpropimorph and spiroxamine have more complicated structures.

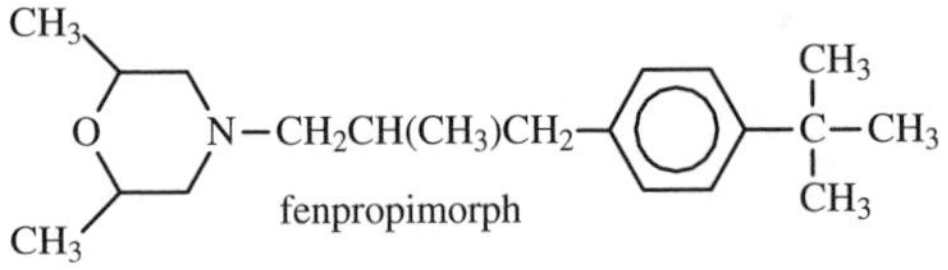

fenpropimorph

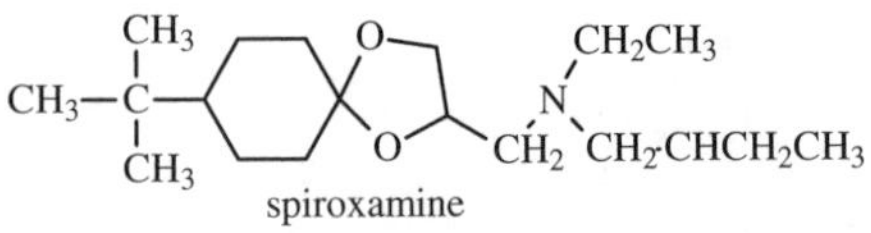

spiroxamine

Spiroxamine was first sold in 1997 and is reported to mainly inhibit Δ^{14}-reductase.

5.1.5   Conclusions

The ergosterol-inhibiting fungicides are systemic and are active against many different fungi, e.g., Ascomycetes, Deuteromycetes, and Basidiomycetes. Some of them are active in nanomolar concentrations. Although they disturb sterol synthesis in higher plants, as well as the synthesis of gibberellins, their phytotoxicity is low. The many steps catalyzed by a variety of enzymes are potential targets for many more biologically active substances waiting to be discovered. More about ergosterol-inhibiting fungicides is found in Khambay and Bromilow (2000) and Köller (1992).

5.2   Herbicides that inhibit synthesis of amino acids

Herbicides that inhibit enzymes important for amino acid synthesis account for 28% of the herbicide market. Just three enzymes are involved: the enzyme that adds phosphoenolpyruvate to shikimate-3-phoshate in the pathway leading to aromatic compounds, the enzyme that makes glutamine from glutamate and ammonia, and the first common enzyme in the biosynthesis of the branched-chain amino acids.

5.2.1   The mode of action of glyphosate

The amino acids tryptophan, phenylalanine, and tyrosine are products of the shikimic acid pathway. This pathway is present in plants and many microorganisms but is completely absent in animals, which acquire the aromatic amino acid in their diet. Conversely, plants must produce these essential amino acids to survive and propagate. The aromatic ring structure is also needed for synthesis of tetrahydrofolate, ubiquinone, and vitamin K, which are essential substances for plants and other life-forms. The cofactor tetrahydrofolate is required for biosynthesis of the amino acids glycine, methionine, and serine, and the nucleic acids. Aromatic ring structures are present in numerous secondary plant products such as anthocyanins and lignin. The important plant growth hormone indole–acetic acid is produced from tryptophan. As much as 35% of the ultimate plant mass in dry weight is produced from the shikimic acid pathway. It is not surprising that at least one chemical acting selectively on plants by inhibiting this pathway exists. It is more surprising that only one such compound, useful as an herbicide, has been found. This herbicide, glyphosate, was introduced in 1971 by Monsanto and has been extremely useful. Although many environmental scientists and human toxicologists have searched for side effects, this herbicide is still regarded as safe. It is interesting that the herbicidal effect of glyphosate was found prior to the full elucidation of the shikimic acid pathway. Its interference with the synthesis of aromatic acid synthesis was also found after its introduction as an herbicide. Jaworski (1972) described the inhibition of plant aromatic amino acid biosynthesis in 1972, whereas Amrhein et al. (1980) first demonstrated identification of the specific site of action in 1980.

The target enzyme is 5-enolpyruvoylshikimate-3-phosphate synthase (EPSPS). The enzyme catalyzes the reaction between shikimate-3-phosphate (S3P) and phosphoenolpyruvate (PEP). Jaworsky (1972) showed that when *Lemna gibba* (duckweed) was kept with glyphosate added to its medium, its growth ceased. If shikimate, shikimate-3-phosphate, or other compounds are added together with glyphosate, duckweed still will not grow. But if chorismate, prephenate, or the amino acids phenylalanine, tyrosine, and tryptophan are added, the inhibitory effect of glyphosate is removed.

All plant, fungal, and most bacterial EPSPSs that have been isolated and characterized to date are inhibited by glyphosate, but EPSPS from various sources may have very different sensitivity. Glyphosate binding is competitive with the substrate phosphoenolpyruvate but binds to the enzyme only after the enzyme has complexed with the other substrate, shikimate-3-phosphate. Plant enzymes are inhibited by concentrations of $<1 \mu M$ glyphosate. Some other enzymes in the shikimate pathway are also inhibited but at concentrations more than a thousand times higher. If genes coding for more glyphosate-tolerant EPSPSs are introduced into susceptible plants, they become more tolerant to this herbicide. The amino acid sequences of EPSPSs from different sources (e.g., *Escherichia coli*, tomato, and petunia) are very similar. Between the two plants the similarity is as much as 93%, and between petunia and *E. coli* it is 55%, whereas the similarity between the fungus *Aspergillus nidulance* and *E. coli* is much less (38%). The target enzyme and the other enzymes in the shikimate pathway are localized in the chloroplasts of the plant cells. EPSP is synthesized in the cytoplasm as a preenzyme, which has an extra tail of 72 amino acids that is important for its transport into the chloroplast, but this is cut off when inside. Interestingly, glyphosate at $10 \mu M$ inhibits the import of this pre-EPSPS into the chloroplasts.

Naturally the reactions involved in the synthesis of 5-enolpyruvoyl-shikimate-3-phosphate and its inhibition by glyphosate have been studied extensively, and many thousands of publications are available. In spite of this, only glyphosate is in use as a commercially relevant compound. Many other compounds that inhibit EPSPS or other important enzymes in the shikimate pathway have been found, but none of these seem to be suitable as herbicides. The situation is therefore very different for the EPSPS-inhibiting pesticides than for many other groups of enzyme inhibitors used as pesticides, such as the acetylcholinesterase-inhibiting insecticides, which constitute many hundreds of organophosphorus insecticides in current use. In contrast with many contact herbicides, the phytotoxic symptoms of glyphosate injury often develop slowly. Death can take several days or even weeks to occur. Glyphosate is translocated via the phloem throughout the plant but tends to accumulate in the meristematic regions. The most common symptom observed after application of glyphosate is foliar chlorosis, followed by necrosis. Signs of injury include leaf wrinkling or malformation and necrosis of the meristems, including the rhizomes and stolons of perennial plants.

The diagram shows the pathway from shikimate to chorismate and the step inhibited by glyphosate.

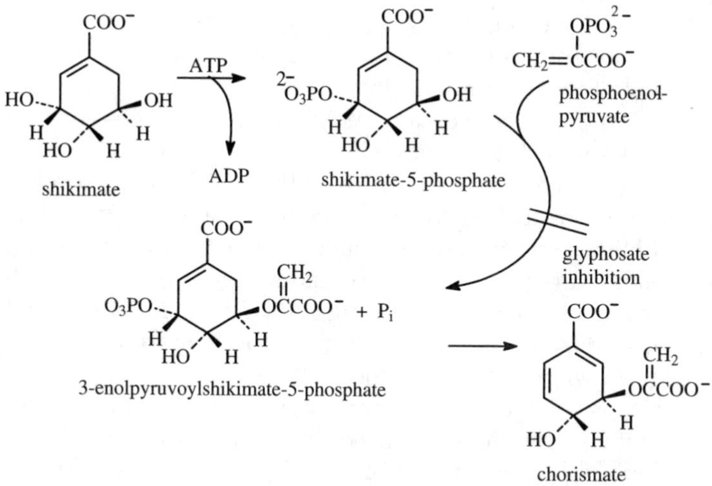

5.2.2   Degradation of glyphosate

The C–P bond in glyphosate is not very common in biomolecules, but in spite of this, some bacteria split it easily. In plants, glyphosate is quite stable, but microflora efficiently degrade it to simple nitrogen and carbon metabolites and several microorganisms are able to use it as a phosphorus source. The most important degradation pathway is probably through the formation of aminomethylphosphonic acid (AMPA), followed by the split of AMPA to inorganic phosphate and methylamine. Microorganisms such as *Arthrobacter atrocyaneus* and *Pseudomonas* spp. seem to be important degraders. Glyoxalate is metabolized further in the glyoxalate pathway. The C–P bond in glyphosate may also be split by a glyphosate lyase present in some microorganisms.

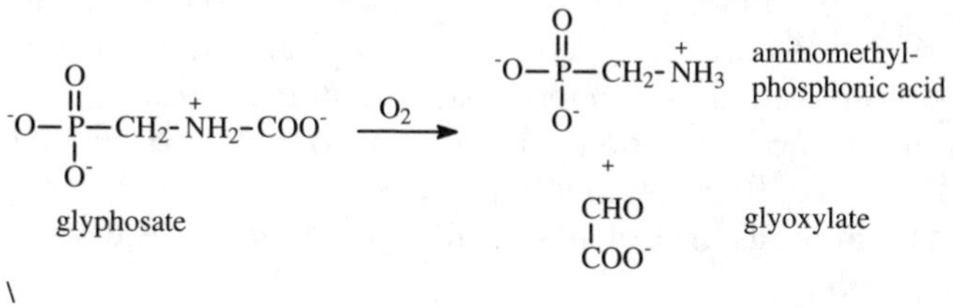

5.2.3   Selectivity

The selectivity between animals and plants is extremely high for glyphosate, although phosphoenolpyruvate is a substrate for many enzymes in plants and animals. However, glyphosate does not seem to inhibit enzymes other than EPSPS, which is completely absent in animals. Although glyphosate is a metal chelator, this property does not play a role in the inhibition process and it is not a general inhibitor of metal-requiring enzymes. Although glyphosate inhibits EPSPS from a wide variety of organisms, high selectivity between plants is possible to obtain. As mentioned, the sensitivity of EPSPS from various sources differs markedly. Some bacteria (e.g., *Agrobacterium tumefaciens*) have a glyphosate-insensitive EPSPS, and commercially successful glyphosate-insensitive soybean and cotton plants that have had the *A. tumefaciens* EPSPS gene introduced have been made. Glyphosate-tolerant sugar beet plants carry, in addition to this transgenic construct, a bacterial gene encoding a glyphosate-degrading enzyme. The transformed plants show no deleterious effects from the application glyphosate and the herbicide can be used without harming them.

Glyphosate is soluble in water but not in waxes and lipids. The uptake and therefore the sensitivity of plants with waxy cuticles are thus low. Furthermore, glyphosate is inactivated in soil by forming insoluble salts with soil minerals, and this property can be exploited in selective usage. When used during the summer months, white anomones and some other dormant spring flowers are not harmed and will flourish the next year.

In 1995 over \$1.7 billion worth of glyphosate was sold. (The total world market for herbicides is estimated to be \$14 billion.) This herbicide thus makes up more than 12% of the herbicide market. It has been more than 30 years since the phytotoxic properties of glyphosate were first described, and it is still an herbicide with great unexploited potential through the use of genetically engineered crop plants resistant to it. Whether such techniques are ethically acceptable and favorable for the chemical environment and biodiversity is another question. The debate about this will probably continue for another decade or so.

5.2.4   Mode of action of glufosinate

Glutamine synthase (GS) is an important enzyme in nitrogen assimilation and photorespiration in plants. In animals the enzyme is of special importance because glutamate is a neurotransmitter that is inactivated through conversion to glutamine by glutamine synthase. Consequently, inhibitors of glutamine synthase may be toxic for plants and animals. The enzyme from plant cytosol, chloroplasts, bacteria, and mammals differs in amino acid composition, but the 13 amino acids thought to make up the active site are identical. Therefore, there is no *a priori* reason to believe that a great selectivity between animals and plants should be found for glutamine synthase inhibitors, and it is not difficult to understand that such substances must be toxic. However, the exact mode of action and the critical effect that causes death are not so easy to point out. Ammonia is toxic to cells because it functions as an uncoupler and disturbs normal membrane function. The high ammonia level caused by inhibition of glutamine synthase may therefore contribute much to the toxicity. Furthermore, inhibition causes a strong decrease in the free pools of glutamine, glutamate, aspartate, alanine, serine, and glycine because all these amino acids are made from the corresponding keto-acid through transamination reactions with glutamate. These are necessary to build up proteins and many other processes. There is a higher level of glyoxalate, the precursor of glycine, which inhibits the enzyme responsible for CO_2 fixation (ribulose-1,5-bisphosphate carboxylase). This may be the most serious consequence of glutamate synthase inhibition and the reason for the fast-acting property of the herbicide. When CO_2 fixation is stopped while light energy is still being harvested, free radicals are formed. Furthermore, the assimilation of NO_3^- into glutamate requires a large input of electrons — two to reduce nitrate to nitrite from nicotineamide-adenine dinucleotide (NADH), six to reduce nitrite to ammonia (from reduced ferredoxin), and two (from reduced ferredoxin) to incorporate ammonia to make glutamate from glutamine and 2-oxoglutamate. The last reaction also requires one molecule of adenosine triphosphate (ATP). If light is still absorbed so that electrons flow from water via chlorophyll to ferredoxin but are not used to produce glutamine, they may be available to make free radicals.

The best-known inhibitors are glufosinate and methionine sulfoximine (MSO). Bilanafos, trialaphos, and phosalacine are substances produced by various *Streptomyces* and other bacteria. They are not inhibitory to glutamate synthase as such, but are hydrolyzed to phosphinotricin (PPT). Glufosinate is the synthetic variant of PPT and is a mixture of the D and L forms. Note that these substances have direct bonds between phosphorus and carbon, which is seldom found in natural compounds.

No compound has yet been synthesized that has the inhibitory capacity of PPT, or has a comparable herbicidal activity.

The first inhibitor demonstrated for glutamate synthase was L-MSO. The compound may be synthesized but has also been found in the bark of the *Cnestis glabra* tree and is therefore sometimes called glabrin. It is used in neurochemical research as an inhibitor of the glutamine synthase that terminates the effect of glutamate as a neurotransmitter. Many other glutamate synthase inhibitors that have been synthesized are found in various microorganisms. They are often phosphinotricin attached to a peptide chain. One such herbicide is bilanafos. It is produced by *Streptomyces hygroscopicus* during fermentation. It translocates in the phloem and xylem and is metabolized in the plants to glufosinate. It is almost nontoxic to aquatic animals, has a very low toxicity to mammals, and is regarded as nonmutagenic and nonteratogenic.

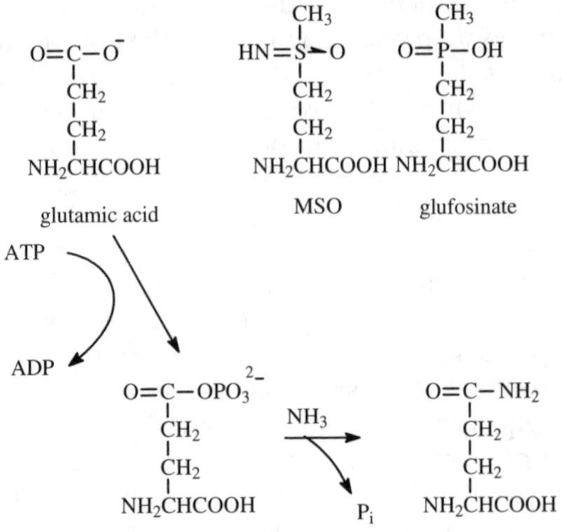

$$O=C-O^-$$
$$|$$
$$CH_2$$
$$|$$
$$CH_2$$
$$|$$
$$NH_2CHCOOH$$

glutamic acid

$$\begin{array}{c}CH_3\\ |\\ HN=S{\to}O\\ |\\ CH_2\\ |\\ CH_2\\ |\\ NH_2CHCOOH\end{array}$$

MSO

$$\begin{array}{c}CH_3\\ |\\ O=P-OH\\ |\\ CH_2\\ |\\ CH_2\\ |\\ NH_2CHCOOH\end{array}$$

glufosinate

ATP
ADP

$$O=C-OPO_3^{2-}$$
$$|$$
$$CH_2$$
$$|$$
$$CH_2$$
$$|$$
$$NH_2CHCOOH$$

NH_3

P_i

$$O=C-NH_2$$
$$|$$
$$CH_2$$
$$|$$
$$CH_2$$
$$|$$
$$NH_2CHCOOH$$

The figure shows the steps catalyzed by glutamine synthase and the structural similarity between glutamic acid, MSO, and glufosinate.

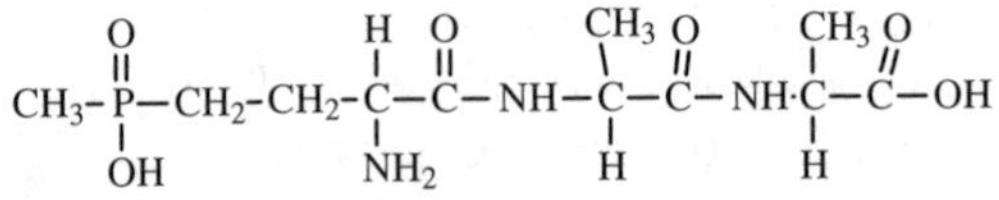

$$\begin{array}{ccccc}&O&&H\ O&CH_3\ O\quad CH_3\ O\\&||&&|\ \ ||&|\ \ \ ||\quad\ |\ \ \ //\\CH_3-P&-CH_2-CH_2-C&-C-NH-&C-C-NH{\cdot}C-C-OH\\|&&|&\ \ \ |\quad\quad\ |\\OH&&NH_2&\ \ \ H\quad\quad\ H\end{array}$$

bilanafos

5.2.5 Inhibitors of acetolactate synthase

A great number of herbicides that work through the inhibition of acetolactate synthase (ALS) have been commercialized. They belong to four chemical groups: sulfonylureas (23), triazolopyrimidines (2), imidazolinones (5), and pyrimidinyloxybenzoic analogues (3). (The number of active ingredients in parentheses is taken from *The Pesticide Manual*.) Also in this case, potent herbicides were developed (e.g., chlorsulfuron) before the site of action was found.

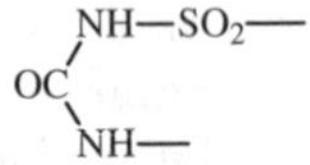

$$\begin{array}{c}NH-SO_2-\\ OC\\ NH-\end{array}$$

sulfonylureas

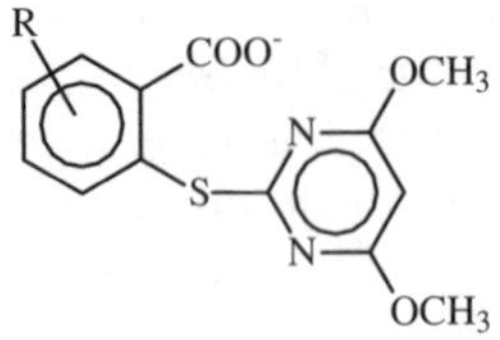

pyrimidinyloxybenzoic acid

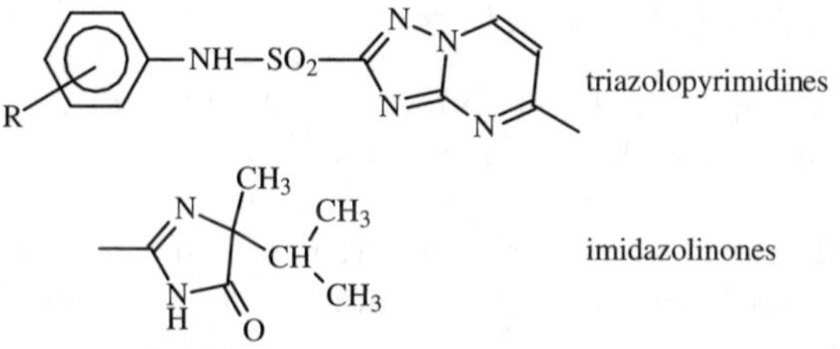

triazolopyrimidines

imidazolinones

Chemical structures found in acetolactate synthase inhibitors from commercial herbicides

Some of these inhibitors are extremely potent and as little as 2 g/ha may control weeds. They may be used both pre- or postemergence. The toxicity to other higher organisms is very low because of their high specificity as inhibitors of an enzyme not present in insects, mammals, or other animals, which have to get the branched-chain amino acids through the diet. Chlorsulfuron, for instance, has an apparent K_i of about 0.004 μM for acetohydroxyacid synthase and gives a 50% growth reduction of corn at 0.8 g/ha. The extreme toxicity of chlorsulfuron on pea seedlings or other plants can be abolished if valine, leucine, and isoleucin are added to the medium. Phenylalanine and threonine do not have any effect. The first symptom of acetolactate synthase inhibition is growth arrest. Cell division in pea root tips is inhibited by chlorsulfuron. Similar effects are seen by other acetolactate synthase inhibitors (e.g., imazapyr) on other systems (e.g., corn seedlings).

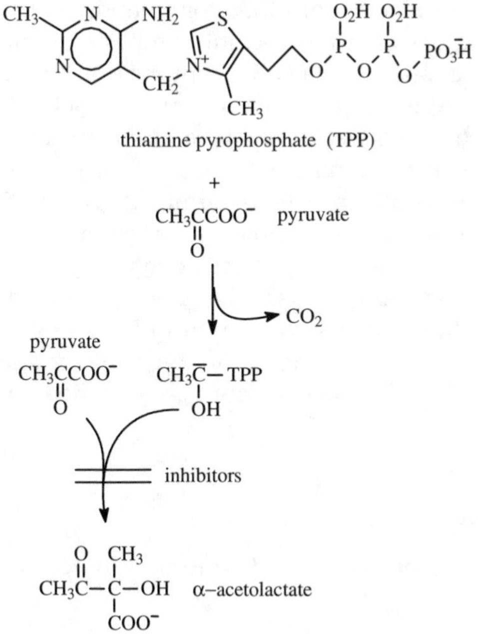

thiamine pyrophosphate (TPP)

+

CH_3COO^- pyruvate

CO_2

pyruvate

inhibitors

α–acetolactate

5.3   Inhibitors of chitin synthesis

Chitin, next to cellulose, is the most abundant polysaccharide in nature, but is only distributed among arthropods and fungi, and is absent in plants and mammals. Chemicals that interfere with chitin biosynthesis could therefore *a priori* be excellent selective pesticides. They would in insects act primarily at the stage of metamorphosis by preventing the normal molting process and would probably not harm adult insects. Their usefulness would therefore be restricted compared to nerve poisons. Such compounds would probably be toxic to crustaceans and other arthropods having a chitinous skeleton. The same or similar compounds could be toxic for both fungi and arthropods, but be harmless for other creatures. They could therefore be excellent in integrated pest control programs.

Three series of compounds with this type of mode of action have been found, exemplified by polyoxin B, diflubenzuron, and buprofezin. Diflubenzuron belongs to the group called benzoylureas.

Curiously, the insecticidal activity of benzoylureas was found by sheer coincidence in a search for herbicides. Derivatives of dichlobenil, with some similarity to the urea herbicides, were tested. Daalen and co-workers (1972) in the Netherlands observed that the compound DU19111, under certain circumstances, was very active against insect larvae. Further studies unveiled that the mosquito larvae were extremely sensitive. Adult houseflies, Colorado potato beetles, and aphids were not affected. In spite of the fact that DU19111 is chemically related to the herbicides dichlobenil and diuron, no phytotoxicity was observed, and the mammalian toxicity was very low. With several insect species the death was invariably connected with the molting process, and a series of other compounds with similar structures were found. Their mode of action as chitin synthesis inhibitors was elegantly established by Hajjar and Casida (1978), but the exact mechanism is still not known. Hajjar and Casida (1978) made small vessels of the abdomen of newly emerged adult milkweek bugs (*Oncopeltus fasciatus*) and filled them with a reaction cocktail containing ^{14}C-glucose. Incorporation of radioactivity into insoluble chitin could then be determined. The ability of substituted benzoylphenylureas to inhibit ^{14}C-glucose was compared to their toxicity to fifth instar *O. fasciatus* nymphs. The correlation was very good. Interestingly, diflubenzuron or other compounds in this group do not inhibit incorporation of uridine diphospate-N-acetylglucosamine or N-acetylglucosamine (or glucose) into chitin in cell-free systems of chitin synthetase, but are potent inhibitors in tissue or cell systems from newly molted cockroaches (Nakagawa et al., 1993).

5.3.1   Insecticides

The Pesticide Manual (Tomlin, 2000) describes 10 insecticidal benzoylureas.

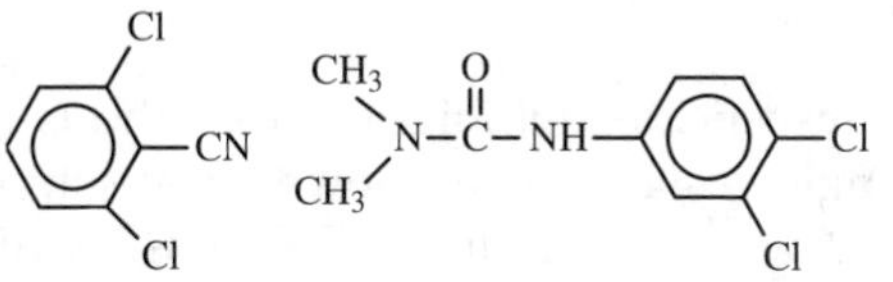

dichlobenil and diuron

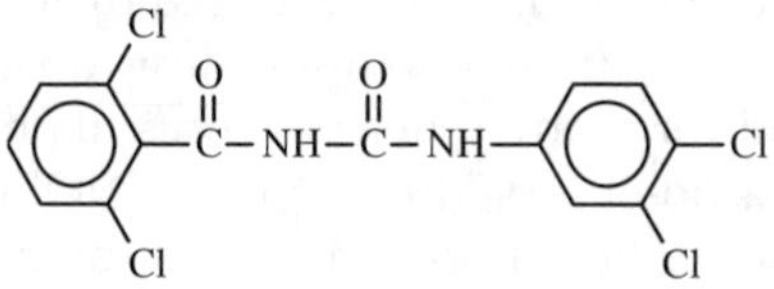

DU19111

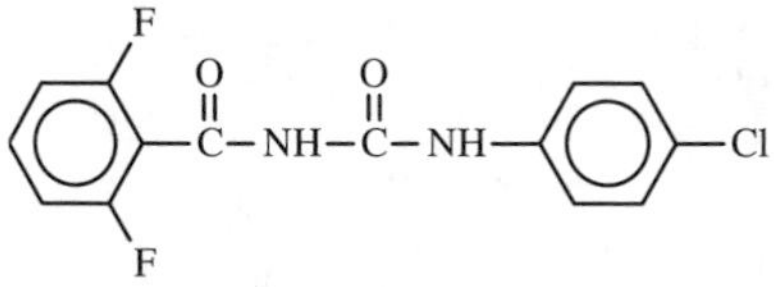

diflubenzuron

The two herbicides cichlobenil and diuron were built together in order to create a superherbicide but instead became the starting point for new insecticides

Buprofezin is a specific poison for Homoptera, but the mode of action is not known. It is included in this chapter because it probably interferes with molting or chitin synthesis in some way. It inhibits embryogenesis and progeny formation of some insects at very low concentrations (see Ishaaya, 1992). Cyromazil was first marketed in 1980 and is an insect growth regulator. Insect larvae, particularly fly larvae, develop cuticular lesions before they eventually die.

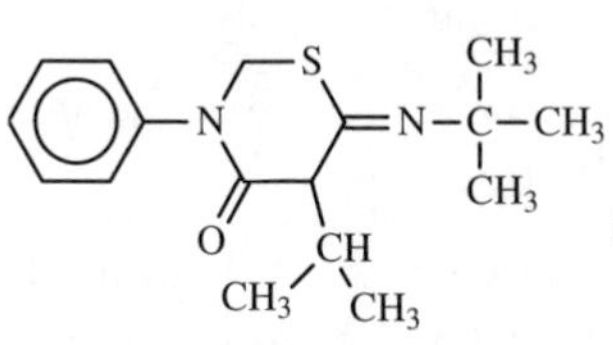

buprofezin 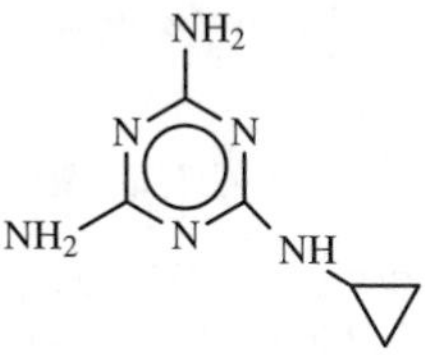   cyprodonil

Characteristically, their mammalian and fish toxicity are very low, and they have rather high acceptable daily intake (ADI) values. Tripathi et al. (2002) have made an extensive review on the chitin synthesis-inhibiting insecticides, including 156 references.

5.3.2 Fungicides

As mentioned, the insecticides inhibit chitin synthesis indirectly and they are not useful as fungicides. Polyoxins, however, are structural analogues to uridine diphospate-2-acetamido-2-deoxy-D-glucose, which is the substrate for chitin synthetase, and inhibit the incorporation of 2-aceta-mido-2-deoxy-D-glucose into chitin. It is produced by fermentation of *Strep-tomyces cacaoi* var. *asoensis*. It is used as a fungicide against powdery mildews in apples and pears, and for many other purposes. Its mammalian toxicity is very low, and it has a no-observed-effect level (NOEL) in rats of 44,000 mg/kg in diet in 2-year studies. The compounds inhibit chitin synthetase from insects, but are not toxic to insects *in vivo*.

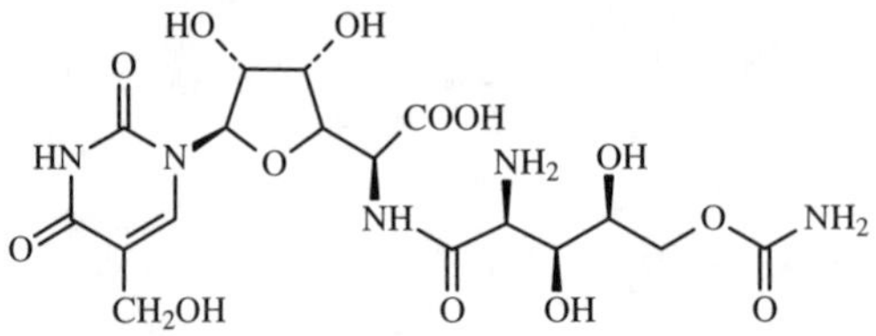

polyoxin B

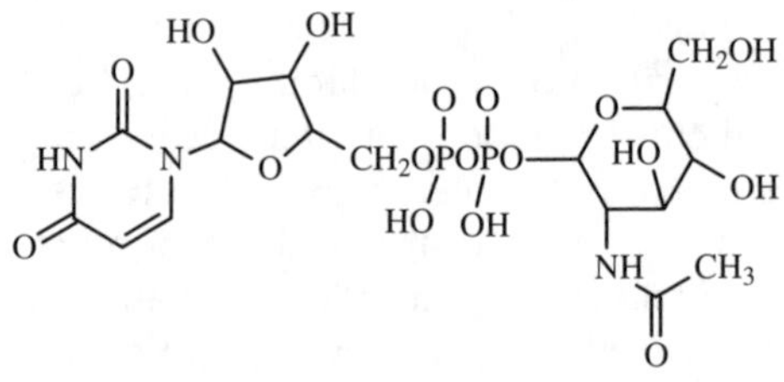

UDP-glucosamine — the substrate for chitin synthase,
drawn to show its similarity to polyoxin B

5.4 Inhibitors of cholinesterase

The great majority of insecticides are nerve poisons. The target for most of them is an enzyme called acetylcholinesterase (AChE). We will describe the enzyme and its inhibition in some detail because there are no other enzymes for which we know so much about the relationship between its structure and its activity. The cholinesterase-inhibiting insecticides, the warfare gases, and the target enzyme have been the objects of intense study by scientists for many years.

5.4.1 Acetylcholinesterase

Acetylcholinesterase does a simple job: it hydrolyzes acetylcholine, an ester, which is released when a nerve impulse is transmitted from one nerve cell

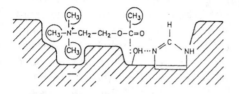

Figure 5.3 The classical model of the active site of acetylcholinesterase.

to another, from a nerve cell to a muscle, or to an endocrine cell. Acetylcholinesterase is found in significant concentrations throughout the nervous system in most animals but is also present in many nonnervous tissues. The function of the nonnervous enzyme is not known, but its presence in erythrocytes is often useful for the pesticide toxicologist because it may be readily accessible. Health servants can measure the activity of AChE in the erythrocytes and a related enzyme, butyrylcholinesterase (BuChE), in plasma taken from pesticide workers. If the level of the enzymes is below a certain threshold, the pesticide worker can be taken out of work until the normal value is restored and the environment in which he works has been changed in order to reduce the exposure. The properties of AChE have been studied in detail; its active site and catalytic properties are well understood and its physiological function in the nervous system is known. A good source of the enzyme is the electric organ of electric eel (*Electrophorus electricus*) and skate (*Torpedo marmorata*). The activity of the enzyme is easy to measure with acetylthiocholine as the substrate. Thiocholine is released and is measured continuously in a spectrophotometer by means of an added SH reagent. Acetylocholinesterase is primarily a membrane-bound enzyme but can easily be extracted from the membranes by detergent-containing buffers. Differential centrifugation of nervous tissues from various sources shows that most enzymes are connected to the synaptic membranes in the nervous system; however, the enzyme is also present in many body fluids. The hemolymph of mussels (*Mytilus*) has an AChE that is not membrane bound. Snake venom is also a rich source of AChE.

$$\underset{\text{acetylcholine}}{CH_3\overset{\overset{\displaystyle O}{\|}}{C}OCH_2CH_2\overset{\oplus}{\underset{\underset{\displaystyle CH_3}{|}}{\overset{\overset{\displaystyle CH_3}{|}}{N}}}-CH_3}$$

The most important part of the enzyme is its active site, where the acetylcholine and the many inhibitors bind. The classical model shown in Figure 5.3 (Nachmansohn and Wilson, 1951) is still very useful, although not exactly correct. The model says that acetylcholinesterase has two subsites in the active site called the esteratic and anionic sites. Because acetylcholine is an ester where the alcoholic part (choline) carries a positive charge, this part will seek the anionic site, whereas the ester bond will react with the esteratic site. The esteratic site is believed to resemble the catalytic subsites in other hydrolases with the amino acid serine in its active site.

A more complete model has recently been proposed (Axelsen et al., 1994; Koellner et al., 2000; Sussman et al., 1991), whereas Silver (1974), in a monography of almost 600 pages, describes the biological role of cholinesterases known at that time. The residues of the amino acids serine, histidine, and glutamate are still regarded as the most important in hydrolysis. They are located near the bottom of a narrow pocket named the active site gorge, which is about 20 Å deep. The wall of this gorge is lined by rings of 14 aromatic residues, which may contribute as much as 68% of its surface. It penetrates halfway into the structure and widens out close to its base. The active site gorge is filled with 20 water molecules, which have poor hydrogen-bonding coordination. Therefore, some of these molecules can easily move and be displaced by the incoming substrate. The acetylcholine molecule is actually too large to enter the gorge, but scientists think that the narrowest part of the gorge has large-amplitude size oscillations, thus making entrance possible during brief periods of time. The choline-recognizing site is near the opening and involves the side chain of the amino acids tryptophan and phenylalanine. Through studies using cationic and uncharged homologues of acetylcholine, the anionic subsite was in fact shown to be uncharged and lipophilic, not anionic. This anionic subsite binds the charged quaternary group of the choline moiety of acetylcholine, as well as other substances with quaternary ligands, such as edrephonium and N-methylacridinium, which act as competitive inhibitors. Quaternary oximes, which often serve as effective antidotes to organophosphate poisoning, are also bound here. In addition to the two subsites of the catalytic center, AChE has one or more additional binding sites for acetylcholine and other quaternary ligands. The binding of ligands leads to uncompetitive inhibition. Acetylcholine at high concentration therefore inhibits its own hydrolysis.

Considering its complicated structure and the many stages in the catalytic cycle, AChE possesses a remarkably high activity. The substrates, and most inhibitors, have to slip into the narrow gorge acylating a serine residue. The acyl group has to be displaced by a part of a water molecule, and the choline and the acetic acid have to escape the gorge. The outer architecture of the enzyme is also quite complex (Figure 5.4). Groups of four subunits are linked to a collagen-like tail. The most complex form has 12 subunits and is found in the electric organ of electric fish and in vertebrate muscles. The tail is tied to the outer surface of the postsynaptic membrane. In various other organs the catalytic subunits are linked together in less complex structures (Chatonnet et al., 1999; Chatonnet and Lockridge, 1989).

Let us look at the reaction kinetics between the enzyme and acetylcholine according to Aldridge and other pioneer workers (Aldridge and Reiner, 1972; O'Brien, 1976):

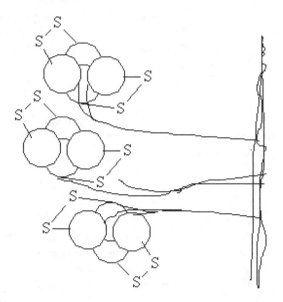

Figure 5.4 A simplified model of vertebrate acetylcholinesterase bound to the postsynaptic membrane. Each circle is a catalytic unit.

Acetylcholine reacts with the enzyme (EH) and makes a so-called Michaelis complex. The acetyl group reacts with the serine-hydroxyl in the enzyme, which forms the acetylated enzyme and releases choline. The acetylated enzyme is then hydrolyzed (reacts with water) and the enzyme with its serine-hydroxyl group is restored. These reactions occur extremely fast. The organophosphorus insecticides may be regarded as substrates, but because the reaction speed is so slow, they block the enzyme. If we call the ester AB, where A is the acyl part and B the alcoholic (or phenolic) part, often also called the leaving group, the enzyme EH, the acylated enzyme AE, and the Michaelis complex ABEH, we can describe the catalysis or inhibition by the following equation:

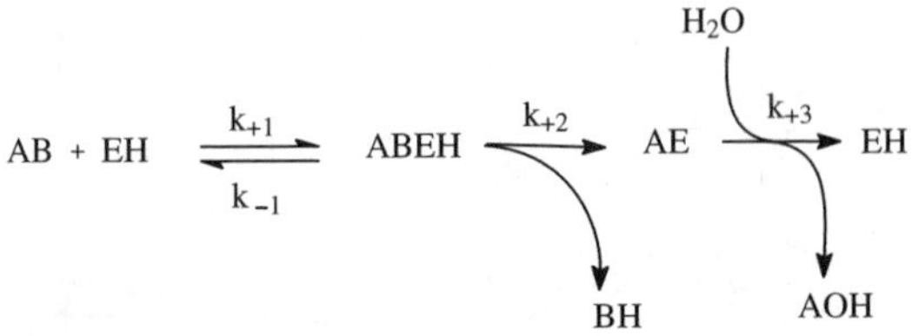

$$AB + EH \underset{k_{-1}}{\overset{k_{+1}}{\rightleftharpoons}} ABEH \overset{k_{+2}}{\longrightarrow} AE \overset{k_{+3}}{\longrightarrow} EH$$

with H_2O, BH, and AOH as byproducts.

The letter k with the indices (−1, +1, or +2) symbolizes the velocity constants of the reactions defined by the rates of changes in concentrations. Square brackets symbolize concentrations of the respective substances. The velocity of the first reaction step is so fast that it is impossible to determine the constants k_{-1} and k_{+1} separately, and the equilibrium constant K_d is more useful:

$$\frac{[AE][EH]}{[ABEH]} = \frac{k_{-1}}{k_{+1}} = K_d$$

k_{+2} and k_{+3} can often be measured and are defined by

$$\frac{d[AE]}{dt} = k_{+2}[ABEH] - k_{+3}[AE]$$

Organophosphorus insecticides give very stable acylated enzymes (AE does not hydrolyze because k_{+3} is very small). The potency of an inhibitor is much determined by how stable the Michaelis complex is (as expressed by the size of K_d) and how fast the acylated enzyme is formed (expressed as the size of k_{+2}). A constant including K_d and k_{+2}, called the bimolecular inhibition constant, is often used to describe the strength of an inhibitor because it is very easy to determine experimentally and it tells us how potent the inhibitor is:

$$k_i = \frac{k_{+2}}{K_d}$$

An empirical constant called I_{50} is also often presented. It is the concentration of the inhibitor that, under specified conditions, will reduce the enzyme activity to half.

5.4.2 Organophosphates

When AB is an organophosphate, e.g., paraoxon,

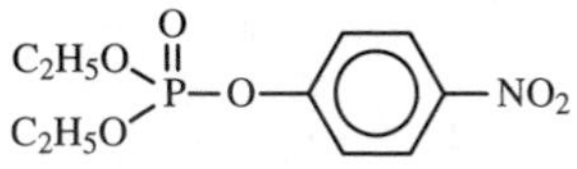

paraoxon

the first two reactions determined by k_{-1}, k_{+1}, and k_{+2} are quite fast, although not so fast as with acetylcholine, but the last reaction determined by k_{+3} is very slow. Many of the enzyme molecules end up with a diethylphosphoryl group attached to their serine residue, blocking the active site. The diethylphosphoryl group is very slowly removed by hydrolysis, a situation very different from what occurs after reaction with acetylcholine when the acetylated enzyme is hydrolyzed extremely fast. Carbamates are also bad substrates and react more as inhibitors because the carbamylated enzyme obtained is hydrolyzed very slowly. The formulae show acylated enzymes (AE) formed by organophosphorus insecticides (diethylphosphoryl enzyme, dimethylphosphoryl enzyme), the acetylated enzyme (formed by reacting with the true substrate), and a typical carbamylated enzyme (obtained by reaction with an insecticidal carbamate):

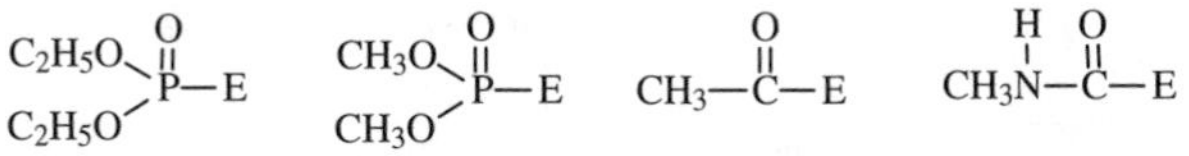

Different adducts symbolized as AE

A nucleophilic attack on the acylated enzymes by water is the important rate-limiting step in the catalytic cycle or the restoration of inhibited enzymes. The nucleophilic oxygen attacks the bond between phosphorus and the enzyme, leading to the restoration of the free enzyme (EH). As a side reaction, an alkyl group of the phosphorylated enzyme may be removed, as shown below. When this occurs, the enzyme cannot be restored.

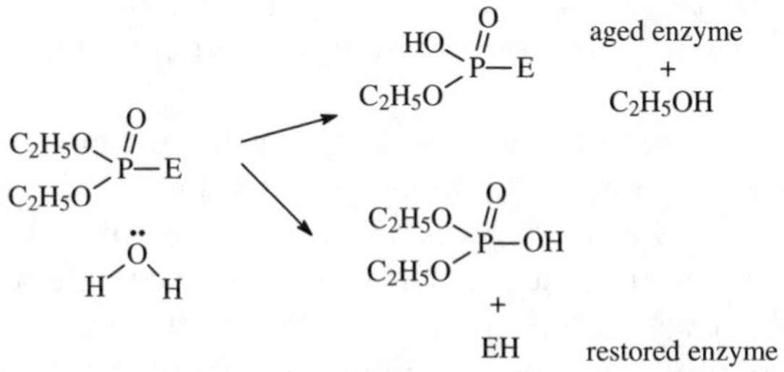

This reaction is called aging because the mechanism behind the gradual loss of ability to be reactivated was not known when first observed. It was said that the inhibited enzyme was "old." Stronger nucleophiles than water, such as the N-methylpyridinium-2-aldoxime ion, can increase the hydrolytic rate of phosphorylated enzymes and can be used as antidotes, provided they are used before aging has occurred.

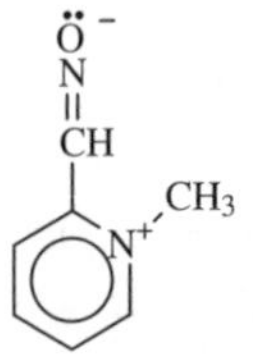

$$\ddot{O}^-$$
$$N$$
$$\parallel$$
$$CH$$

(structure of N-methylpyridinium-2-aldoxime ion with CH_3 group attached to N^+)

The antidotes *cannot* be used to cure poisoning by carbamates because they may react with the carbamate and make a more potent inhibitor. The easy methods available for determining the potency of inhibitors make it possible to test the relationship between structure and activity, and to try to find inhibitors that work better for insect acetylcholinesterase than for the mammalian enzyme.

The German chemist Gerhard Schrader (1951, 1963) made the first, and still very useful, model for an active organophosphorus insecticide during the Second World War when he worked at AG Farben, a big German company engaged in war industry:

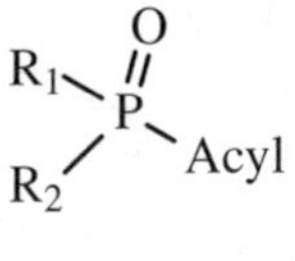

$$R_1 \diagdown \overset{O}{\underset{\parallel}{P}} \diagup$$
$$R_2 \diagup \diagdown Acyl$$

This scheme says that a phosphorus atom is bound to oxygen and to an acidic group as well as two other groups that can be almost anything. The structures of these groups determine, of course, the potency of the inhibitor toward acetylcholinesterase or other serine hydrolases, how they behave in soil and water, and how they are degraded by various enzymes in insects, nematodes, mammals, birds, etc. Gerhard Schrader himself, together with his colleagues, tested an enormous number of substances and even described their synthesis.

Good inhibitors are not always strong poisons for insects or mammals because there is a tendency for good inhibitors to be less stable. The leaving group (–B), called Acyl in Gerhard Schrader's model, must have the ability to withdraw electrons from the phosphorus atom in order to make it a good electrophile and active as an inhibitor. This property also makes the organophosphorus inhibitor unstable. Therefore, good inhibitors are often very unstable and not very suitable as pesticides. Furthermore, many very poisonous organophosphorus insecticides do not inhibit cholinesterase before they have been activated through oxidation. The insecticides very often have a sulfur atom attached to phosphorus and must be desulfurated before becoming active as inhibitors. Sulfur makes the compound more stable and somewhat less toxic to mammals.

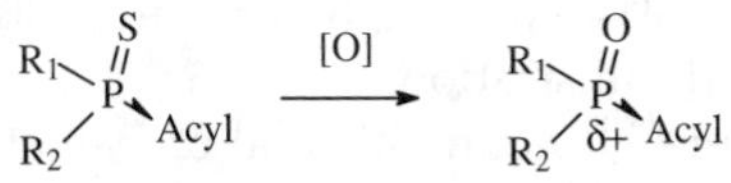

phosphorothioate
(must be activated)

phosphate
(active inhibitor)

The Pesticide Manual from 1977 describes 100 organophosphorus and 25 carbamates in common use, compared with a total of 543 pesticides, while the newer editions (1994) describe 72 organophosphorus insecticides out of a total of 515 pesticides (14% of all pesticides in common use are cholinesterase-inhibiting organophosphates).

5.4.2.1 Naturally occurring organophosphorus insecticides

Despite their simplicity in structure and simple mode of action, the triesters of phosphorus or thiophosphorus acids are typical xenobiotics. They are typical products of the chemical industry and not organisms. However, there are some cyclic esters that are present in a few organisms, and these are strong inhibitors of AChE. A soil bacterium (*Streptomyces antibioticus*) contains a substance that is highly active as an inhibitor of AChE and is very toxic to insects. The biological function for the bacteria is not known. However, the phosphorus atom is linked to four different groups and therefore gives rise to optical isomers. One of the carbon atoms is also linked to four different groups. The optical isomers have, of course, different biological activity, and *Streptomyces* seems to synthesize the most active. It is therefore reasonable to believe that its function is connected to its activity as an inhibitor (Neumann and Peter, 1987).

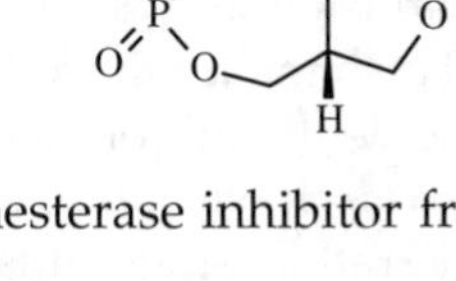

The acetylcholinesterase inhibitor from *S. antibioticus*

5.4.3 Carbamates

The carbamates are also typical xenobiotics. One important exception is eserine (physostigmine), a very toxic substance that is present in Calabar beans, the seeds of a Leguminosae from West Africa. Calabar beans were used in legal matters in West Africa. Suspected persons were given a drug made from the beans, and if they survived, they were not guilty. Eserine was the first known acetylcholinesterase inhibitor and soon became a valuable tool for the biochemist, because it inhibits all types of cholinesterases but no other serine hydrolases. Cholinesterases therefore are often defined as hydrolases that can be inhibited by eserine. It has an LD50 (lethal dose in 50% of the population) in mice of 4.5 mg/kg by oral administration and 0.64 mg/ kg by intraperitoneal administration.

The general structure of the insecticidal carbamates is

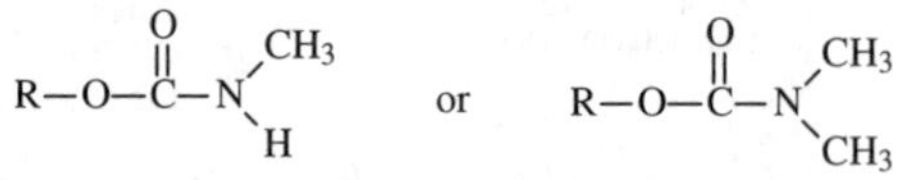

$$\begin{array}{cc}
\overset{\displaystyle O}{\underset{\displaystyle \parallel}{R-O-C-N}}\diagdown\!\!\!\!\!\begin{array}{c}CH_3\\H\end{array}
&
\text{or}
&
\overset{\displaystyle O}{\underset{\displaystyle \parallel}{R-O-C-N}}\diagdown\!\!\!\!\!\begin{array}{c}CH_3\\CH_3\end{array}
\end{array}$$

Most of the insecticides are of the type shown on the left. They are therefore quite easy to recognize by formula.

Eserine and other carbamates react at exactly the same active site as substrates and the organophosphorus insecticides. The only major difference is that the constant k_{+3} is much higher for carbamates than for organophosphates, but much lower than for the substrate. Whereas the carbon–oxygen–phosphorus bond in the phosphorylated enzyme is very stable, the carbamate group on the enzyme hydrolyzes off much faster:

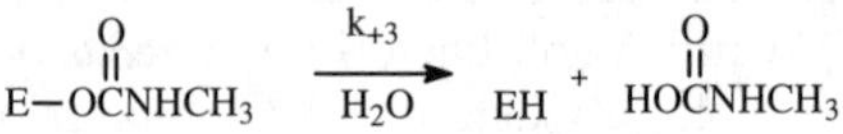

$$\underset{\displaystyle E-OCNHCH_3}{O\atop\parallel} \xrightarrow[H_2O]{k_{+3}} EH \;+\; \underset{\displaystyle HOCNHCH_3}{O\atop\parallel}$$

Some carbamates are very toxic for mammals, birds, and earthworms, but it is not easy to predict their toxicity from their formulae.

5.4.3.1   Molecular structure and potency of inhibition

Some examples of dissociation constants of Michaelis complexes, rate constants, and bimolecular inhibition constants are presented in Table 5.1 and Table 5.2. Table 5.1 shows name, structure, K_d values, k_{+2} values, and k_i values for some organophosphates and carbamates chosen to show the relationships between structure and potency as inhibitors. Most of the values are taken from the book by Aldridge and Reiner (1972). Paraoxon is a strong poison and has a high k_{+2} value that also makes the k_i value high. It is too toxic and unstable to be a good insecticide. Parathion does not inhibit acetylcholinesterase but is bioactivated by oxidation to paraoxon in insects and vertebrates, and is therefore very toxic. Parathion-methyl is not shown in the table but is very similar to parathion. Its toxicity is a little lower. Parathion-methyl's oxygen analogue is methyl-paraoxon, which has a lower affinity to AChE than paraoxon but the k_{+2} value is approximately the same.

Introducing a methyl group into the ring, as in fenitrothion, increases the hydrophobicity and increases the affinity to the active site of the enzyme, but reduces its reaction velocity. The net result on AChE is therefore much the same. However, the changes make it less toxic to mammals, while the toxicity to insects of fenitrothion is similar or even higher than that of parathion-methyl. Dimethylphenylphosphate does not inhibit AChE because the phenyl group has insufficient electron-withdrawing power to make the phosphorus atom electrophilic. Electronegative substituents such as nitro groups, halogens, the $-SO_2CH_3$ group, etc., must be introduced. Diisopropylphosphorofluoridate (DFP) is a nerve gas. It is not very active as an AChE inhibitor, but the inhibited enzyme cannot be reactivated ($k_{+3} = 0$) and is stable in the organism.

Amiton and amiton-methyl are organophosphates designed to resemble acetylcholine. In accordance to this similarity, amiton has a high affinity to AChE, but the k_{+2} value is low because the leaving group has low electronegativity. Replacement of the ethyl groups with methyl groups changes both reactivity and affinity to the same degree, so the outcome is the same. These organophosphates are not used as insecticides anymore.

The next three compounds are carbamates. Eserine is important because it is a very strong inhibitor of all cholinesterases and is used to verify that an esterase is a cholinesterase. Carbaryl also has a very high affinity to some cholinesterases, but a very low affinity to others. It is an important insecticide. Aldicarb was made to resemble acetylcholine. In spite of this, its affinity is very low (high K_d), but due to its high reaction rate (k_{+2} value), aldicarb is very toxic.

Table 5.2 shows some velocity constants for the hydrolysis of acylated acetylcholinesterases and the source of the enzyme. Most of the data are taken from Aldridge and Reiner (1972).

The half-lives and the corresponding k_{+3} values are dependent on the structure of, and therefore the source of, the enzyme and the structure of the group attached to the serine residue of the enzyme. Note the tremendous difference between the acetylated enzyme and the enzyme's phosphate and carbamate adducts. The half-lives are generally lower for the carbamylated enzymes but long enough to cause serious poisoning.

5.4.4 Development of organophosphorus and carbamate insecticides

A systematic study of the organic phosphorus compounds started in 1874 with the works of A. Michaelis (1847–1916) at the Technical High School at Karlsruhe, Germany, and later at the Technical High School in Aachen, Germany. He did not find any use for the compounds he and his co-workers made. The very strong toxicity of some of these compounds was not recognized until 1932, when W. Lange and G. v. Krueger published an article about the synthesis of dialkylfluorophosphoric acid in *Chemische Berichte* (Vol. 65, p. 1598). They wrote (translated from German):

> Interesting is the strong effect of monofluorophosphoric acid alkyl ester on the human organism. The vapours of the compounds smell pleasant and aromatic. However, just a few minutes after breathing there is a strong pressure against the head, connected

Table 5.1 Structures and K_d, k_{+2}, and k_i Values for Some Organophosphates and Carbamates, Showing the Relationship between Structure and Potency of Inhibitors of AChE

Name	Structure	K_d (μM)	k_{+2} (min^{-1})	k_i $(min^{-1}\mu M^{-1})$
Paraoxon	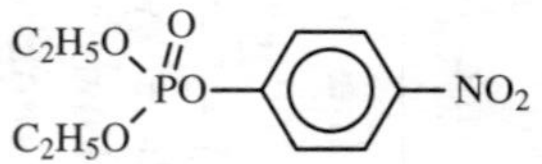 C_2H_5O–PO(=O)–O–C$_6$H$_4$–NO$_2$	360	43	$1.2 \cdot 10^{-1}$
Parathion	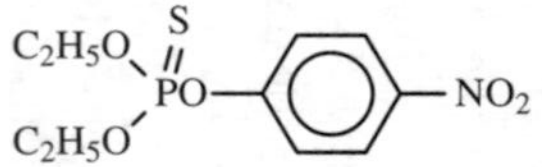 C_2H_5O–PO(=S)–O–C$_6$H$_4$–NO$_2$	—	0	—
Methyl-paraoxon	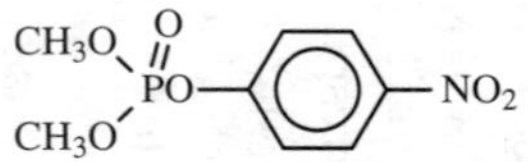 CH_3O–PO(=O)–O–C$_6$H$_4$–NO$_2$	880	50	$5.7 \cdot 10^{-2}$
Fenitrooxon	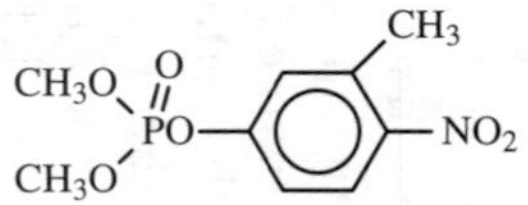 CH_3O–PO(=O)–O–C$_6$H$_3$(CH$_3$)–NO$_2$	67	5	$7.4 \cdot 10^{-2}$
Diethylfenyl-phosphate	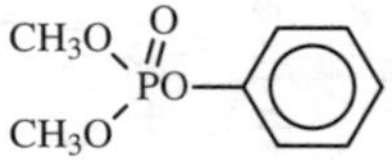 CH_3O–PO(=O)–O–C$_6$H$_5$	—	0	—
DFP	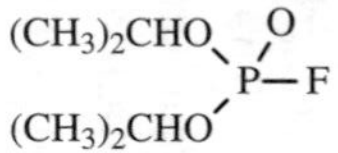 $(CH_3)_2CHO$–P(=O)(–F)–OCH(CH$_3$)$_2$	1600	12	$7.5 \cdot 10^{-3}$

Name	Structure			
Amiton	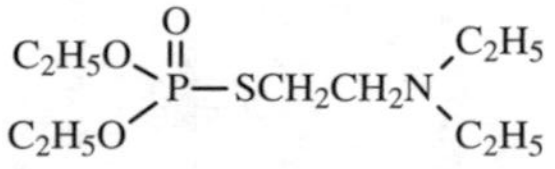	7.2	6.7	$9.3 \cdot 10^{-1}$
Amiton-methyl	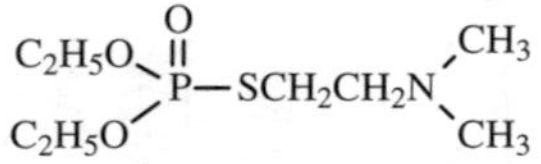	180	126	$7.0 \cdot 10^{-1}$
Eserine	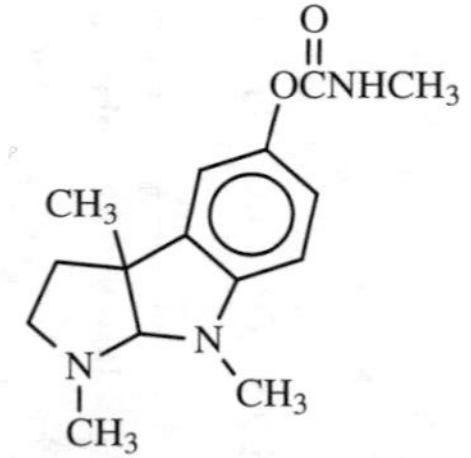	3.3	10.8	$3.3 \cdot$
Carbaryl	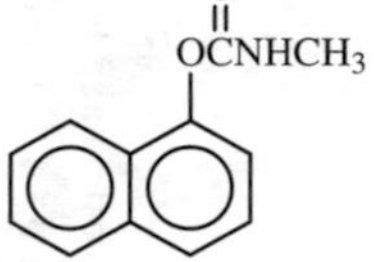	11	1.3	$1.1 \cdot 10^{-1}$
Aldicarb	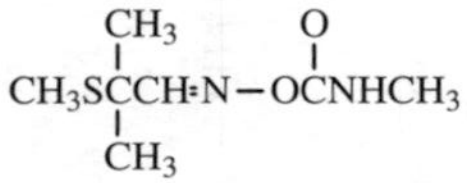	10,300	146	$1.2 \cdot 10^{-2}$

Table 5.2 Velocity Constants for the Hydrolysis of Acylated Acetylcholinesterases and the Source of Enzymes

Acylated Enzyme	Half-Life	k_{+3} (min^{-1})	Source of the Enzyme
CH_3O—$\overset{\overset{O}{\|}}{P}$—$E$ (with CH_3O)	80 min	$8.7 \cdot 10^{-3}$	Rabbit erythrocyte
	200 h	$5.5 \cdot 10^{-5}$	Rat serum
	∞	0	Housefly
C_2H_5O—$\overset{\overset{O}{\|}}{P}$—$E$ (with C_2H_5O)	500 min	$1.4 \cdot 10^{-3}$	Rabbit erythrocyte
$(CH_3)_2CHO$—$\overset{\overset{O}{\|}}{P}$—$E$ (with $(CH_3)_2CHO$)	∞	0	Rabbit erythrocyte
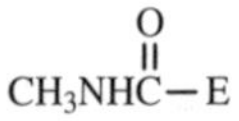	19 min	$3.6 \cdot 10^{-2}$	Bovine erythrocyte
	24 min	$2.9 \cdot 10^{-2}$	Housefly
$CH_3NH\overset{\overset{O}{\|}}{C}$—$E$	26 min	$2.6 \cdot 10^{-2}$	Bee
	38 min	$1.8 \cdot 10^{-2}$	Electric eel
	180 min	$3.8 \cdot 10^{-3}$	Horse serum
	156 min	$4.4 \cdot 10^{-3}$	Human serum
$CH_3\overset{\overset{O}{\|}}{C}$—$E$	2.3×10^{-6} min	$3.0 \cdot 10^{5}$	Not known

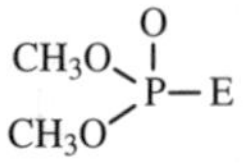

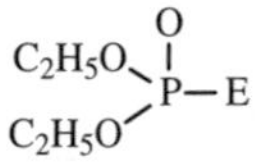

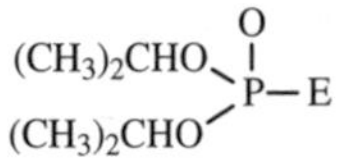

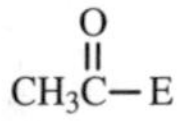

Note: The half-life may be calculated from the following formula: $t_{1/2} = \ln 2/k_{+3}$

Source: Most of the data are taken from Aldridge, W.N. and Reiner, E. 1972. *Enzyme Inhibitors as Substrates: Interactions of Esterases with Esters of Organophosphorus and Carbamic Acids*, Vol. XVI. North-Holland Pub. Co., Amsterdam. 328 pp.

with respiration trouble. Then a light unconsciousness and visual problems with painful hypersensitivity of the eyes against light appears. After several hours the symptoms disappear.

Their work was not much noticed in Germany, but in England the work with the phosphofluorin was continued and led to DFP and other nerve gases. In Germany the cyano-compounds were studied by G. Schrader, who synthesized the nerve gas and insecticide tabun.

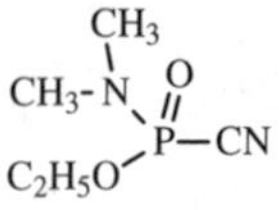

tabun 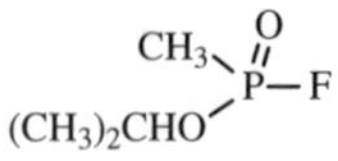   sarin 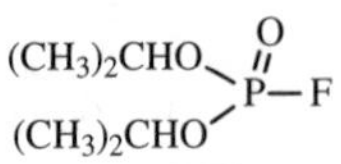   DFP

Other extremely toxic insecticides were also produced. Tetraethylpyro-phosphate (TEPP) is probably the most toxic substance ever used in agriculture.

$$C_2H_5O \underset{C_2H_5O}{\overset{\overset{O}{\underset{\underset{POP}{}}{\parallel}} \overset{O}{\parallel}}{}} \overset{OC_2H_5}{\underset{OC_2H_5}{}}$$

TEPP

The pesticides described are selected in order to illustrate some properties of the group.

5.4.4.1   Parathion and similar compounds

The dialkyl aryl phosphates and phosphorothionates are a good illustration of the connection or correlation between structure and activity — as cholinesterase inhibitors and poisons. Parathion was first described by Gerhard Schrader in 1944 and was soon marketed by American Cyanamid, USA; Monsanto Chem. Co., USA; and others. Farbenfabriken Bayer produced the compound under the name E 605, a name still well known by German farmers. Parathion is interesting because it does not show selective toxicity between animals. It is a strong poison for both sexes and all stages of insects and vertebrate life. Typical LD50 values for insects are 1.0 μg/g (*Apis mellifera*) and 0.5 μg/g (*Musca domestica*) at topical application. Rat oral male and female LD50 values are 13 and 3.6 mg/kg, respectively, and the dermal values are 21 and 6.8 mg/kg, respectively. For mouse, guinea pig, rabbit, cat, dog, and horse, the LD50 values are between 3 and 50 mg/kg. (All values are from G. Schrader, 1963.) Exposure through skin or through the mouth is equally effective. Parathion has a high vapor pressure and can poison insects and vertebrates through the respiratory system.

Many molecular modifications of parathion were tested at an early stage of the development of organophosphates. Aldridge and Davison (1952a, 1952b) found a close correlation between log k_i of erythrocyte cholinesterase inhibition by various diethyl aryl phosphates and the hydrolyzability of the esters. This fact illustrates that a certain degree of reactivity, and thus instability, is necessary for the cholinesterase inhibitors to be toxic. (The chlorinated hydrocarbons owe their toxicity to hydrophobic interactions with their receptor sites. No covalent bonds are formed and broken. They may therefore be very recalcitrant to degradation.)

Figure 5.5 shows the relationship between hydrolyzability and anticholinesterase activity for seven substituted diethyl phenyl phosphates. The hydrolysis constants were determined in a buffer of pH 7.6 at 37°C and have the unit min^{-1}.

G. Schrader (1951, 1963) also synthesized a lot of other derivatives of parathion. Parathion-methyl is of similar toxicity and mode of action but is

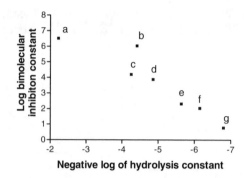

Figure 5.5 The relationship between hydrolyzability and anticholinesterase activity for seven substituted diethyl phenyl phosphates. a, tetraethyl pyrophosphate; b, paraoxon; c, ortho analogue to paraoxon; d, meta analogue to paraoxon; e, o-chlorophenyl diethylphosphate; f, p-chlorophenyl diethylphosphate; g, phenyl phosphate. (Data from Aldridge and Davidson, 1952a, 1952b.)

Table 5.3 Toxicity of Fenitrothion, Parathion, and Fenthion according to G. Schrader (1963)

Animal	Administration	LD50 (mg/kg)		
		Fenitrothion	Parathion	Fenthion
Mouse	Oral	870	9.5	—
Mouse	Subcutaneous	1000	11.5	—
Mouse	Intraperitoneal	280	6.0	—
Mouse	Dermal	3000	120	—
Rat ♂	Oral	242	8.5	215
Rat ♀	Oral	433	8.5	245
Cat	Oral	142	0.93	>100

Note: This table illustrates the high toxicity of parathion compared with two similar analogues. The insecticidal properties of fenitrothion are approximately similar to those of parathion.

somewhat unstable compared with parathion. A more dramatic change in biological activity is achieved when a methyl group is introduced into the phenyl ring, as in fenitrothion. Table 5.3 was made from some of Schrader's data.

Of the other older dialkyl-aryl phosphorothionates, fenthion is still widely used. Fenthion has a better residual efficiency than parathion and methyl parathion and is probably less toxic to fish — 0.01 mg/l at 24 hours' exposure results in 100% mortality in mosquitoes (*Aedes aegypti*), whereas 1 mg/l does not kill any guppy (*Lebistes reticulatus*) during 48 hours' exposure.

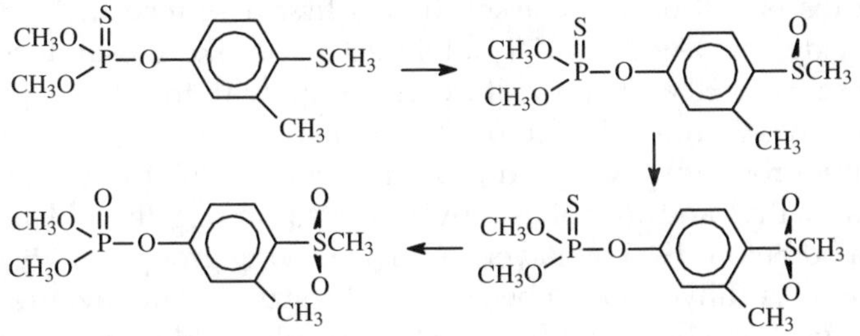

The above reactions show how fenthion is successively oxidized to the sulfon of the oxon, which is the most toxic metabolite.

Bromophos and trichloronate were once very popular. They have now been superseded because of the content of trihalogenephenol moiety, which makes it likely that preparations are contaminated with dioxins.

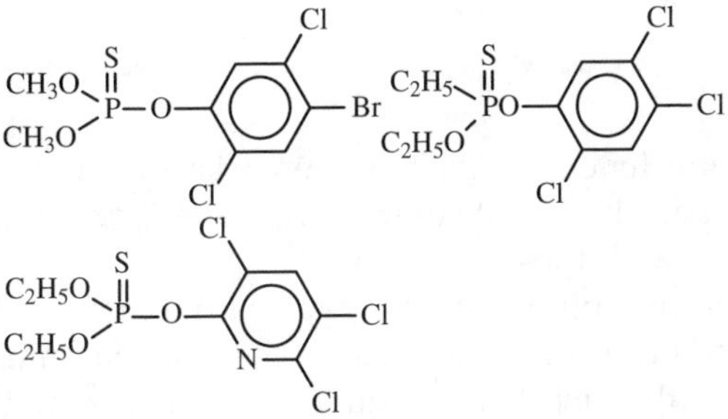

bromophos   trichloronate   chlorpyrifos

Bromophos is an interesting organophosphate because of its low toxicity toward mammals (rat oral = 3750 to 7700 mg/kg, topical > 4000 mg/kg). If the methyl groups are substituted with ethyl groups, the LD50 value is decreased to about 70 mg/kg (oral). The difference in toxicity is at least partly caused by differences in detoxication. Bromophos is demethylated by a reaction with glutathione to form desmethyl-bromophos and methylglutathione, whereas bromophos-ethyl cannot be de-ethylated that easily. Fenchlorphos is similar to bromophos but has three chlorine atoms instead of two chlorines and one bromine. The LD50 is high (oral LD50 for rats = 1750 mg/kg) and it has been used against ectoparasites on cattle and dogs, even by oral administration. Trichloronate has a high acute toxicity and causes delayed neurotoxicity. It was used in granules and as seed dressing against soil insects. Note that fenchlorphos and trichloronate have the same 2,4,5-trichlorphenol structure as the herbicide 2,4,5-T. Technical products may therefore be contaminated by the extremely toxic tetrachlorobenzoldioxines (TCDD). Trichloronate has four different groups connected to the phosphorus atom and is therefore a mixture of two stereoisomers that have different biological effects.

Chlorpyrifos is still on the market. It was first described in 1965 as a useful insecticide for a wide variety of insect pests like mosquitoes and household pests, as well as in agriculture. A variant with the ethoxy groups replaced by methoxy groups is also on the market.

The leaving group (explained on p. 94) is a chlorinated pyridine ring.

Azinphos-methyl and azinphos-ethyl are also among the old insecticides, developed before 1955 by Bayer AG. The leaving group is in this case the complicated 3,4-dihydro-4-oxobenzo-[1,2,3]-triazin-3-ylmethyl group. It got its name Gusathion because of its excellent results in Mexico against the cotton pest *Pectinophora gossypiella*, which is called *gusano* in Spanish.

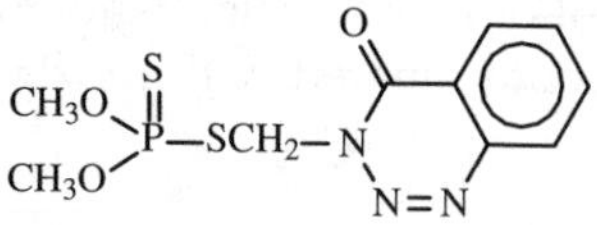

azinphos-methyl 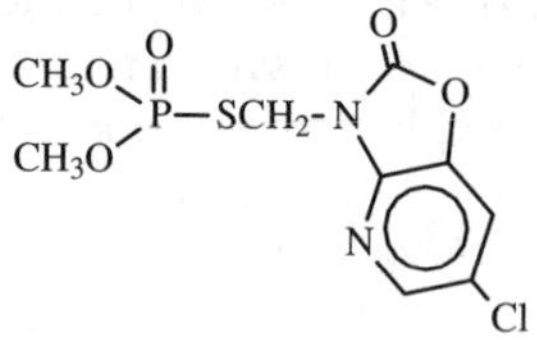   azamethiphos

Bromophos and fenchlorphos have remarkably low mammalian toxicities compared to parathion, but some newer organophosphorus insecticides may compare well with these.

Azamethiphos and phoxim have very low mammalian toxicities but are both more toxic to birds and crustaceans. The low mammalian toxicity is at least partly due to the fact that it is quickly metabolized. The figure shows some of the biotransformation products found in mice.

Azamethiphos has a sufficiently low fish toxicity to be used against parasitic crustaceans (salmon lice) in fish breeding, whereas phoxim is used a lot in public health.

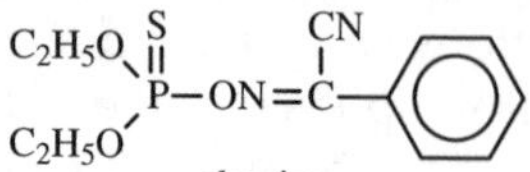

phoxim 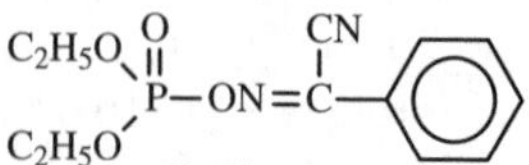   phoxim-oxon

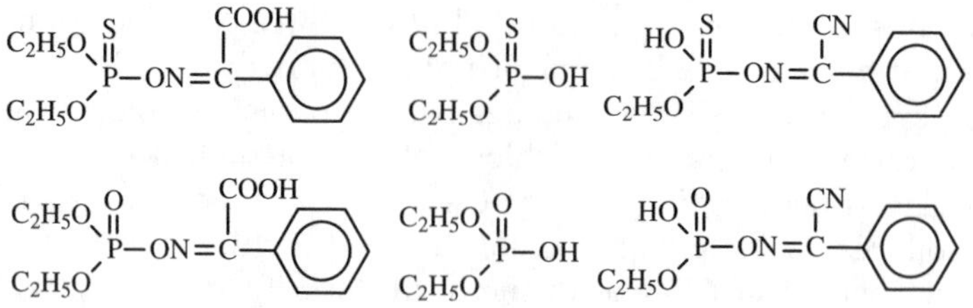

inactive biotransformation products formed in mammals

5.4.4.2 Aliphatic organophosphates

Many organophosphorus insecticides with aliphatic leaving groups have been developed. We shall describe trichlorfon, dichlorvos, dimethoate, and malathion. Trichlorfon has to be activated to dichlorvos by an intramolecular rearrangement combined with hydrolysis:

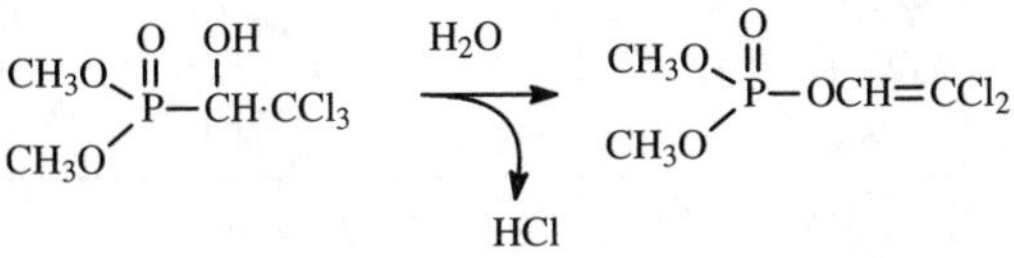

$$\underset{CH_3O}{\overset{CH_3O}{\diagdown}}\underset{}{\overset{O\ \ OH}{\underset{}{\|\ \ \ |}}}P-CH\cdot CCl_3 \quad\xrightarrow{\ H_2O\ }\quad \underset{CH_3O}{\overset{CH_3O}{\diagdown}}\underset{}{\overset{O}{\|}}P-OCH=CCl_2$$

HCl

Trichlorfon has a low mammalian toxicity. Dichlorvos is more toxic, but the difference in toxicity between insects and vertebrates is very high. It evaporates easily and may be formulated in plastic strips that slowly release the insecticide, killing the insects, but it is apparently harmless to humans. The two compounds have also been widely used to kill salmon ectoparasites, as well as parasites encountered in veterinary medicine. However, they are strong teratogens in some species, disturbing brain development (Mehl et al., 2000), and should therefore be used with care.

Dimethoate is one of the more popular systemic insecticides. It has a relatively low mammalian toxicity because it is easily hydrolyzed at the amide bond and demethylated by glutathione transferase. Its oxygen analogue, omethoate, is also used as an insecticide:

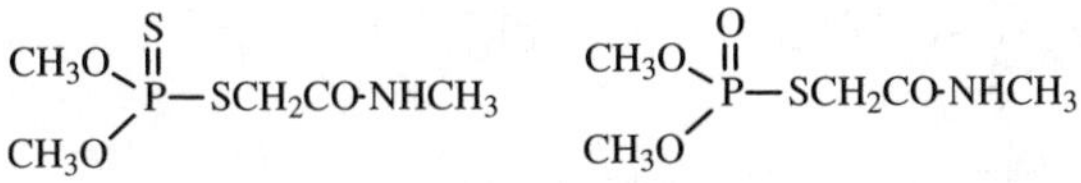

$$\underset{CH_3O}{\overset{CH_3O}{\diagdown}}\underset{}{\overset{S}{\|}}P-SCH_2CO\cdot NHCH_3 \qquad \underset{CH_3O}{\overset{CH_3O}{\diagdown}}\underset{}{\overset{O}{\|}}P-SCH_2CO\cdot NHCH_3$$

5.4.4.3 Examples of carbamates

The carbamates are sometimes divided into two groups — ordinary carbamates and oxime carbamates — but their biochemical modes of action are similar. The current edition of *The Pesticide Manual* describes 18 ordinary carbamates and 8 oximcarbamates.

We shall describe some properties of carbaryl and carbofuran, aldicarb and oxamyl, and pirimicarb and ethiofencarb because their properties are contrasting. Table 5.4 shows that their toxicities vary widely.

Table 5.4 The Difference in Toxicity of Various Carbamates

Name	Systemic	Bioactivation	LD50 mg/kg, Oral Rat	ADI	Year First Described
Carbaryl	No	No	850	0.003	1957
Carbofuran	Yes	No	8	0.002	1965
Carbosulfan	Yes	Yes	3820	0.01	1979
Aldicarb	Yes	Yes	0.93	0.003	1965
Oxamyl	Yes	Yes	8	0.03	1975
Pirimicarb	Yes (selective)	No	142	0.02	1969
Ethiofencarb	Yes	Yes	200	0.1	1974

Source: Data are collected from Tomlin, C., Ed. 2000. *The Pesticide Manual: A World Compendium.* British Crop Protection Council, Farnham, Surrey. 1250 pp.

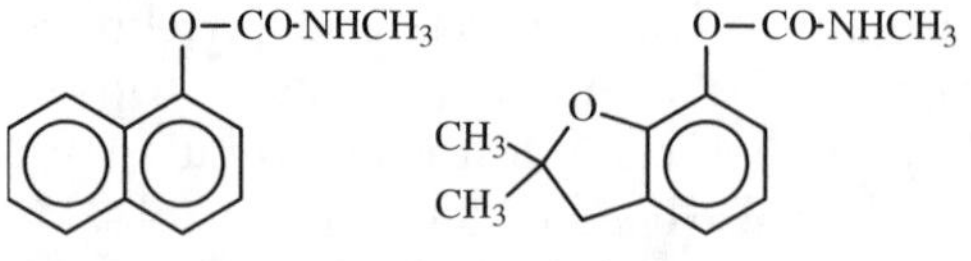

carbaryl and carbofuran

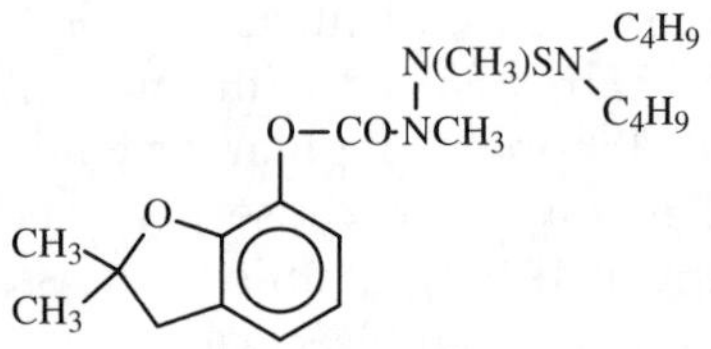

carbosulfan

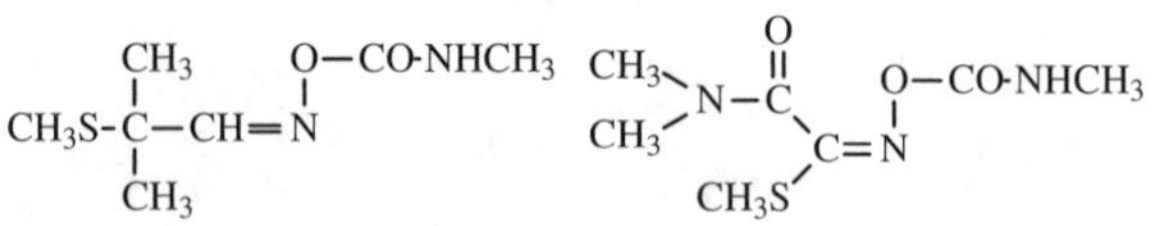

aldicarb and oxamyl

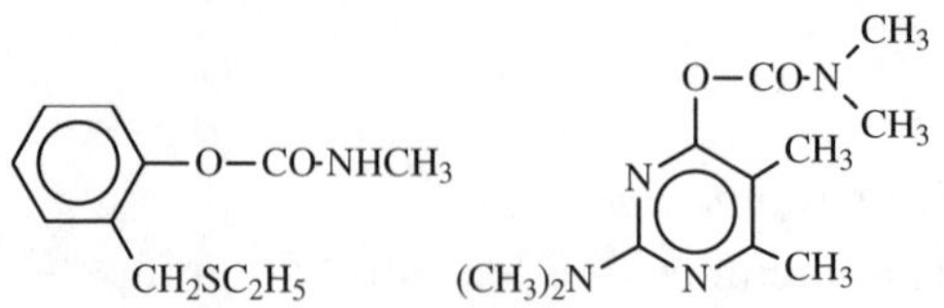

ethiofencarb and pirimicarb

The extreme difference in toxicity toward mammals for equally potent pesticidal carbamates is striking. The difference is at least partly due to the difference in cholinesterase sensitivity, but in some cases, metabolism is important. Carbosulfan must be transformed to carbofuran in order to become an active cholinesterase inhibitor — a conversion that does not occur in vertebrates. Aldicarb is extremely poisonous and stable. It can be activated through oxidation of the ether-thio group to a sulfoxide that is even more toxic. Besides being active against mites and aphids, many carbamates are very potent nematicides but are also highly toxic toward earthworms. This is particularly the case for carbaryl, carbofuran, and aldicarb (Stenersen, 1979, 1981; Stenersen et al., 1973). Carbamates are also often very active against slugs.

5.5 Other enzymes inhibited by organophosphates and carbamates

5.5.1 The butyrylcholinesterases

Acetylcholinesterase has a cousin that is called pseudocholinesterase, butyrylcholinesterase (BuChE), or just cholinesterase. There is a high degree of similarity between BuChE and AChE despite the fact that the studied enzymes are from species that are far apart in evolution. Human BuChE and Torpedo AChE have amino acid sequences that are 54% identical, whereas human BuChE and *Drosophila* AChE are 38% identical. Human BuChE and Torpedo AChE are more similar to each other than to *Drosophila* AChE. However, bovine AChE is closer to Torpedo AChE than to human BuChE (about 60 and 50%, respectively). The two enzyme types are inhibited by almost the same inhibitors. By definition, an esterase is either AChE or BuChE if inhibited by 10^{-5} M eserine. DFP is much more active toward BuChE, and both AChE and BuChE have inhibitors specific to either of them. The exact function of BuChE is not known (Chatonnet et al., 1999; Chatonnet and Lockridge, 1989). Cocaine is an example of a natural toxicant that is hydrolyzed by BuChE to pharmacological inactive compounds (Mattes et al., 1996). The enzyme is present in the blood plasma of humans and most other mammals. Many invertebrates have BuChE or enzymes similar to BuChE. It is sometimes claimed that the divergence between AChE and BuChE probably occurred in the deutorostomian lineage, but protostomian, like the earthworms, has at least two enzymes similar to both AChE and BuChE. The earthworm's enzymes are both highly affected by the typical cholinesterase inhibitors. In earthworms, both enzymes are important in the toxicodynamics of organophosphates and carbamates (Stenersen, 1981). Two different enzymes corresponding to AChE and BuChE have also been found in insects, but not in the fruit fly, *Drosophila melanogaster*, which has, besides AChE, an esterase with an amino acid sequence similar to that of BuChE but different enough to be regarded as another enzyme type.

Humans have several variants of BuChE, and as many as 11 forms of serum BuChEs have been described. One of these forms is associated with unusual clinical responses to succinylcholin, which is used as a muscle relaxant during surgery. Low-rate exposure of humans or other organisms to organophosphates or carbamates may cause an inhibition of the BuChE, without any sign of poisoning.

An example may illustrate how sensitive plasma BuChE is to inhibition. Rats dosed orally with bromophos needed only 10.1 mg/kg of body weight for the enzyme activity to be reduced to half, whereas the erythrocyte AChE I_{50} was 1938 mg/kg and the brain AChE I_{50} was 576 mg/kg. The plasma I_{50} dose did not give any symptoms of poisoning (oral LD50 for rats = 3750 to 7700 mg/kg). It is very clear that plasma enzymes may be inhibited at much lower concentrations than needed to give the more severe symptoms associated with AChE inhibition in the nervous system, and they can be used in early-warning monitoring (Shivanandappa et al., 1988).

5.5.2   The neurotoxic target esterase (NTE)

A disease called delayed neurotoxicity is generally associated with exposure to some organophosphorus esters. It has been described as a polyneuritis characterized by flabby muscles in the legs and arms. Muscular weakness is the primary symptom that begins in the feet and later extends to the legs and hands. The hind limbs are always more severely affected than the forelimbs. Clinical symptoms are not seen until 8 to 14 days after exposure. In mild cases patients can make a total recovery. The difference in sensitivity between species is great, and humans and chickens are considered to be among the most susceptible species, whereas animals often used in toxicity testing like rats, rabbits, mice, guinea pigs, partridges, and quails were not susceptible.

The disease was first observed in some tuberculosis patients that were treated with phosphocreosote, an uncharacterized mixture of esters derived from coal tar phenols and phosphoric acid. It was responsible for an epidemic outbreak of the delayed neurotoxicity disease in the U.S. in the 1930s because a mixture of cresyl phosphates was used to extract ginger and the extract was used to illicitly flavor distilled liquors. Early investigations revealed that the ortho isomer of tricresyl phosphate (TOCP) produced a toxic effect. The triarylphosphates are inert chemicals, but reactive cholinesterase inhibitors such as the warfare agent DFP and the insecticides mipafox, O-ethyl-O-(4-nitrophenyl)benzenethionophosphonate (EPN), trichlorfon, dichlorvos, and leptophos could give similar symptoms. It was shown that TOCP inhibited BuChE in the brain but that other organophosphates that not did result in delayed neurotoxicity also inhibited brain BuChE, and that some substances not inhibiting the BuChE were neurotoxic. The problem was solved mainly by the works of Eto et al. (1962) and M.K. Johnson, who published many papers on the subject during the late 1960s and during the 1970s (Johnson and Lotti, 1980; Johnson, 1982). To cut a long story short, TOCP was metabolically activated by being converted to saligenine phosphate, a strong esterase inhibitor acting exactly the same way as already shown for the reaction of organophosphates with acetylcholinesterase. But in this case, the active organophosphate inhibited an esterase not formerly described that had an unknown function in the nervous system.

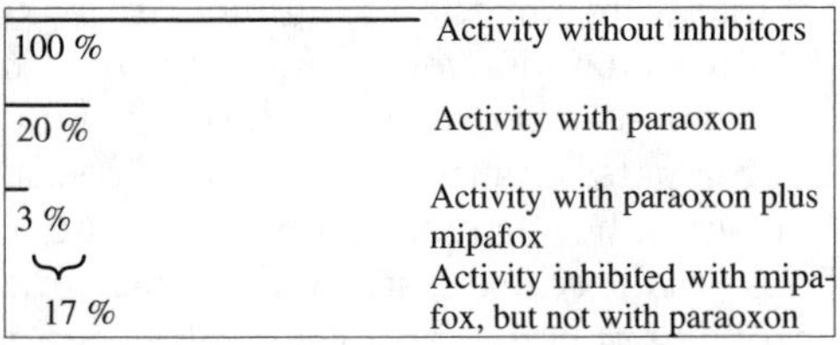

100 %	Activity without inhibitors
20 %	Activity with paraoxon
3 %	Activity with paraoxon plus mipafox
17 %	Activity inhibited with mipafox, but not with paraoxon

Figure 5.6 A diagram showing neuropathy target esterase as a percentage of total esterase activity with phenyl valerate as the substrate.

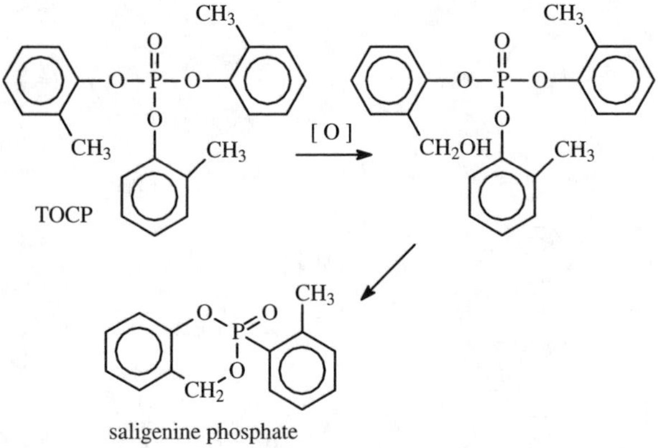

saligenine phosphate

The insecticides leptophos, EPN, cyanofenphos, trichloronate, and dioxabenzophos (salithion) cause irreversible ataxia not only to chickens but also to mice and sheep. It is believed that AChE inhibition is the reason for their acute toxicity, while NTE inhibition is responsible for causing paralytic ataxia.

The esterase got the names neurotoxic esterase or neuropathy target esterase, and the ester phenyl valerate (PV) was found to be a good substrate for this esterase. However, PV is also hydrolyzed by other esterases because paraoxon, which does not give symptoms of delayed neurotoxicity, inhibited the PV activity as much as 80%. The NTE is defined as the hydrolytic activity against PV that is not inhibited by paraoxon but by mipafox. About 3% of the activity is not inhibited by mipafox plus paraoxon. Thus, the part of PV activity due to neurotoxic esterase should be 17%.

The simple diagram in Figure 5.6 shows this for the phenyl valerate-hydrolyzing enzymes in a homogenate of a hen's brain.

But this is not the whole story. It was soon established that many compounds could inhibit this particular activity without causing delayed neurotoxicity. In fact, many inhibitors protected the animals against established neurotoxic inhibitors if administered as pretreatment.

Only inhibitors that were subjected to aging (e.g., loss of an alkyl group, etc.) when bound to the enzyme were neurotoxic. Insecticidal carbamates may also inhibit the enzyme but do not produce this kind of neurotoxicity. Exposure to some carbamates and dithiocarbamates results in neurotoxicity, but through mechanisms other than NTE inhibition. The organophosphorus insecticide leptophos was used extensively for some time as a very useful insecticide in the growing of cotton and vegetables. It caused neurotoxicity at doses of 1.0 to 20 mg/kg/day for 60 days in birds. Leptophos was implicated in the mass poisoning of more than a thousand water buffaloes in Egypt in 1971 and was suspected of causing neurotoxicity in humans.

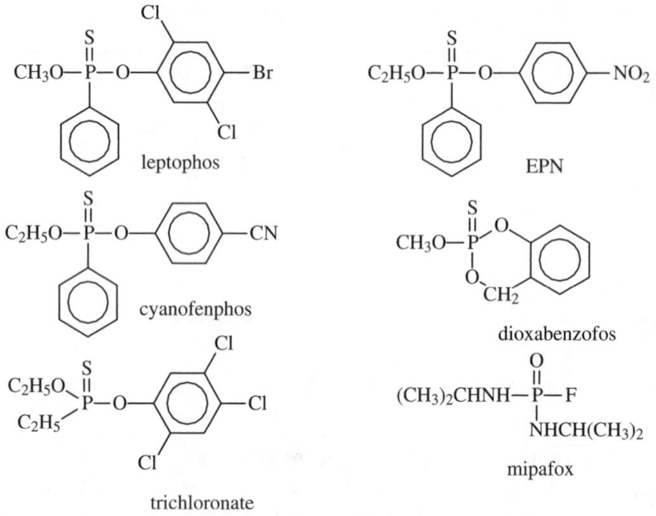

Note that the phosphorus atoms in leptophos and EPN are connected to four different groups and are therefore a mixture of two different stereoisomers, which may have different biological effects. With EPN only, the L(–) isomer is neurotoxic, whereas the L(+) is not. In fact when treated first with the L(+)isomer and thereafter with the L(–) isomer, the symptoms are less severe. Note also that an impurity, desbromo leptophos, in the technical product is 10 times more potent than leptophos as a neurotoxic agent.

The functions of neuropathy target esterase (NTE) and its natural substrate are not known, but NTE is now sequenced and the sequence can be compared with other proteins. It has a structural similarity (41%) with a protein in the *Drosophila* brain, suggesting a function in brain development, through involvement of a cell-signaling pathway. Similar proteins are found in many other organisms (Glynn, 1999). Neuropathy target esterase is an integral membrane protein present in all neurons and in some types of nonneural cells of vertebrates. Recent data indicate that NTE is involved in a cell-signaling pathway controlling interactions between neurons and accessory glial cells in the developing nervous system.

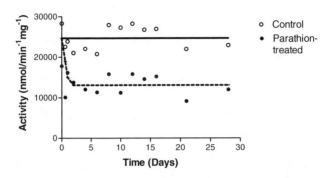

Figure 5.7 Carboxylesterase activity as measured with paranitrophenyl butyrate in homogenates of earthworms (*Eisenia veneta*). The worms were dipped in a solution of parathion (25 μg) in tap water (100 ml) for half an hour, rinsed in pure water, and then placed in pure soil. The carboxylesterase activity was approximately half that of the control for at least 1 month after treatment. There were no signs of toxicity or any other easily observable differences between the control (-○-) and treated (-●-) worms. Each point is the average of six worms (Source: Hoel, 1999).

5.5.3 Carboxylesterases

A considerable number of enzymes belong to a group called carboxy-lesterases (CBEs). Together, they are able to hydrolyze a wide range of endogenous and xenobiotic esters. Typical experimental substrates are 4-nitrophenyl acetate and butyrate; however, each of them has wide-ranging, yet characteristic, substrate specificities. Organophosphates and carbamates are typically strong inhibitors of CBEs. There is a wide variation between individuals and species. Their inhibition does not seem to cause any effect on the organism. When an organophosphate molecule has inhibited a CBE enzyme molecule, it cannot inhibit a more essential AChE molecule. The presence of carboxylesterase is therefore important to reduce the toxicity of organophosphates. In some animals, CBEs may remain inhibited a long time after exposure to organophosphates without any sign of poisoning; low activity may therefore be used as a biomarker for exposure (see Figure 5.7).

chapter six

Interference with signal transduction in the nerves

6.1 Potency of nerve poisons

Nerve poisons are the most biologically active substances known. Some naturally occurring toxins from bacteria such as the botulinum toxins have an LD50 (lethal dose in 50% of the population) in mice of 0.0003 µg/kg. They prevent the release of the transmitter substance acetylcholine from the nerve endings. Symptoms include respiratory problems, nausea, muscle paralysis, and visual impairments. Humans and animals may become seriously ill after consumption of spoiled food that has been kept anaerobically, due to growth of *Clostridium botulinum*. Fortunately, the toxin is heat labile and is destroyed by cooking. Poisoning by "sausage poison" was very common in the 19th century (Otto, 1838). Today poisoning by the mussel toxin, saxitoxin, originating from dinoflagellates of the gender *Gonyaulax*, is more common. Saxitoxin has an LD50 of 10 µg/kg in mouse. The poison blocks the sodium channels in the nerves leading to paralysis. Batrachotoxin, a toxin from frog, has a mice LD50 of 2 µg/kg. This extremely potent nerve poison also blocks the voltage-activated sodium channels. The poison from the black widow spider is also extremely strong, but is still not well characterized. Mushrooms, algae, green plants, animal venoms, and bacteria may have substantial amounts of nerve poisons. It seems that all phyla have species able to produce such nerve poisons.

Many heavy metals such as lead and mercury may also harm the nervous system. It is not surprising that most insecticides are nerve poisons. They have LD50 values in mammals between 1 and 1000 mg/kg.

6.2 Selectivity

It is evident that a typical nerve poison is selective because it only affects organisms having a nervous system, i.e., animals.

The structure and functional elements of the nervous system in different animals are quite different, and some selectivity between animals from different groups is expected. On the contrary, various types of nerve cells have a rather

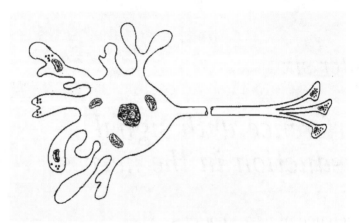

Figure 6.1 A diagrammatic representation of the structure of the neuron showing the cell body with dendrites, nucleus, mitochondria, and the axon terminating in synaptic knobs with mitochondria and vesicles.

similar general anatomy and chemical organization in widely different animals. We may therefore expect that nerve poisons used as pesticides seldom are very selective between animals, and may harm nontarget insects, earthworms, vertebrates, and birds as much as the pest itself. In spite of this, selective poisons are developed. The selectivity is often based on differences between organisms in uptake, distribution, and detoxication or bioactivation, but finer structural differences in the receptor sites for the poisons may make a big difference in sensitivity between various animal taxa. Cartap owes much of its selectivity to difference in bioactivation to the toxic agent, nereistoxin, whereas the neonicotinoids are selective due to differences in the nicotinic acetylcholine receptor sites in insects and mammals. The pyrethroids are selective mainly due to differences in uptake and distribution.

6.3   The nerve and the nerve cell

A nervous system may be composed of billions of nerve cells (*neurons*) connected by hundreds of contact points (*synapses*) in a complicated, but organized way. Neurons have a wide variety of shapes and sizes, but they have certain important features in common (Figure 6.1). There is a cell body that contains the nucleus and certain thin fibers extending from it. There is one long fiber, the *axon*, which in large animals may be several meters long, and a larger number of shorter fibers (*dendrites*), which are branched and usually less than 1 mm long. An integral part of the whole cell, including the fibers, is the nerve cell membrane. A nerve consists of hundreds of thousand of neurons (i.e., cells). The cell bodies of these neurons are aggregated in small organs known as *ganglia*. The axons transmit impulses to other cells through junctions called synapses. The synapse is essentially comprised of three parts: the presynaptic swelling of the axon terminal, the postsynaptic membrane of the receiving dendrite or cell, and a narrow space of the magnitude of 5 to 30 nm in between — the synaptic cleft. Through these synapses a single nerve cell might be connected to hundreds of other neurons, to muscle cells, or to glandular cells. The whole structure is called the *synaptosome*. The synapses may be excitatory or inhibitory; i.e., they may either aid transmission of an impulse to the contact cell (the postsynaptic cell) or inhibit the transmission of impulses coming from other (excitatory) synapses. The signal molecules that transfer the impulse across the synaptic cleft are called transmitter substances or neurotransmitters.

Many of the same transmitter substances are found in the housefly and man, but not always in analogous parts of the nervous system. The structure and function of the nerve cell and the nervous system are described in all textbooks of biochemistry, cell biology, and neurobiology (e.g., Alberts et al., 2002; Breidbach and Kutsch, 1995; Gullan and Cranston, 2000; Levitan and Kaczmarek, 2002; Nelson and Cox, 2000; Rockstein, 1978; Wilkinson, 1976); therefore, only short descriptions are given here. The general textbooks do not refer to the targets and mode of action of the insecticides, but some other poisons are mentioned if they have been used as tools in neurochemical research. Some very inspiring reviews are available (e.g., Bloomquist, 1996; Casida and Quistad, 1998; Keyserlingk and Willis, 1992; Zlotkin, 1999).

6.4 Pesticides that act on the axon

6.4.1 Impulse transmission along the axon

A nerve impulse propagated along the axon must be transmitted across the synaptic cleft to be further propagated. An impulse does not come alone, but in a train of impulses. Because impulse transmission in an axon is an all-or-nothing phenomenon, it is the frequency and not the amplitude of each impulse that determines the strength of the signal. The mechanism, now fairly well understood, is described briefly.

Ions cannot freely pass the cell membrane because it is made of a double layer of lipids. This makes it possible to have different concentrations of the same ion on the inside and outside of the neuron membrane. Typical inside values for some important ions are 400 to 140 mM K^+, 5 to 20 mM Na^+, 0.04 to 0.1×10^{-3} mM Ca^{+2}, and 20 mM Cl^-, while the outside concentrations may be 20 mM K^+, 450 mM Na^+, 1 to 2 mM Ca^{+2}, and 160 mM Cl^-. The nerve cell membrane is relatively impermeable to sodium and more or less open to chloride ions, but has a regulated permeability to potassium ions when at rest. The diffusion out, through so-called leakage channels down the concentration gradient, of some of the positively charged potassium ions leads to a difference in the electrical potential between the outside and inside. The voltage difference, approximately –70 mV, is called the *resting potential* (inside negative), which represents a very high field strength because the membrane is very thin. The high concentration of K^+ inside is sustained by special proteins, which pump K^+ back into the cell.

There are pores or channels in the membrane that may let various ions pass when open. These channels are gated and the gates are of two main types. One type is opened by the binding of various signal molecules, which function as keys. These channels are said to be *ligand gated*. Other channels are opened when the voltage difference falls below a threshold. These channels are said to be *voltage gated*.

The events occurring when an impulse travels along the axon and is then transmitted to the receiving cell at the synapse are not extremely complicated, and some knowledge about this mechanism is of importance in understanding the mode of action of some pesticides. The passage of a nerve impulse at a point on the axon is associated with a sudden drop, and even reversal, of the voltage difference of −70 to +30 mV at that point. This leads first to the opening of the voltage-gated sodium channels, allowing positive sodium ions to enter the cell, which enhances the voltage drop. Next, this leads to opening of voltage-activated potassium channels. This counteracts the voltage drop, because the potassium ions gush out. The opening of the sodium channels does not only lead to a decrease and reversal of the electrical potential difference at the site of the open channel, but also to a voltage drop a little further down the axon, causing the sodium channels at this point to open, with sodium influx at this point the result; the signal impulse has thus propagated a little further down the axon. The sodium channels are automatically closed after a very short time. The opening of the potassium ion channels occurs with a short delay, and they close a little more slowly. The efflux of K^+ ions compensates for the influx of Na^+ ions reestablishing the resting potential. Furthermore, sodium ions are continually pumped out and potassium in at the expense of adenosine triphosphate (ATP) by a so-called ion pump, so that resting potential and the concentration difference of ions between the inside and outside are maintained. Two K^+ ions are taken up for every three Na^+ ions that are kicked out by the pump. There are several thousand such pumps per square micrometer cell membrane. As isolated proteins, these pumps act as an ATP-hydrolyzing enzyme (Na^+/K^+-ATPase or Na^+/K^+-pump) that needs potassium, sodium, and magnesium as co-factors. The nerve poisons DDT and ouabain (a cardiac glycoside) are strong inhibitors of the enzyme (Koch, 1969), but the ion pump is not believed to be the major target of DDT. However, organisms that are very dependent on active transport of salt out of its cells may be very sensitive to DDT (Janicki and Kinter, 1971).

The impulse (drop in voltage difference) associated with the opening and closing of the gates proceeds along the axon until it reaches the synapse where it dies out. When the nerve impulse arrives at the presynaptic membrane, the drop in membrane potential briefly allows calcium ions to flow into the terminal through voltage-gated calcium channels. These channels are usually closed, but open in response to the drop in voltage. Remember that the calcium concentration is 10,000 times or more higher on the outside than on the inside of the cell, and calcium ions will therefore gush into the cell if the possibility is available. The rise in free calcium concentration inside the cell is extremely brief because the synaptic knob can remove calcium from the cytoplasm by pumping it out of the cell or taking it up into intracellular bodies. How this brief rise in intracellular calcium concentration results in the propagation of the nerve impulses is discussed later in this chapter.

6.4.2   Pesticides

The pyrethroids and DDT are by far the most important insecticides in this category. According to their modes of action, they are sometimes classified into two types. Type 1 includes DDT, its analogues, and pyrethroids without a cyano group, whereas type 2 compounds include the pyrethroids with an α-cyano-3-phenoxybenzyl alcohol. In mammals, they give slightly different symptoms by poisoning. Type 1 causes whole-body tremors, whereas type 2 causes salivation and choreoathetosis. Insects also show different symptoms, but not so distinct.

Several lines of evidence suggest that DDT and the pyrethroids react with the voltage-gated sodium channels. Pyrethroids prolong the period that the sodium channels are in the open state. Opening and closing should normally occur in less than a millisecond when an impulse passes. However, when poisoned by a pyrethroid, the closing is delayed and sodium leaks out when the channel should be closed. This tail current is much more distinct for the type 2 pyrethroids and may last for minutes. The resting potential is not achieved and the impulse does not pass distinctly, but comes as a train of action potentials because a lower potential rise is necessary to reach the threshold for the action potential.

The sodium channels probably have many binding sites, maybe six, for various toxicants. Besides DDT and its analogues, poisons from plants, scorpions, sea anemones, amphibians, and others may have their modes of action by binding to one of the sites. It is also important to know that cross-resistance between all the DDT analogues and the pyrethroids often occurs. This type of resistance is called knockdown resistance (kdr). The low sensitivity is caused by one point mutation. We believe that this is caused by a variant of the binding site protein, giving less sensitivity. The amino acid leucin[993] in the Na$^+$ channel protein changed to phenylalanine in houseflies. The super-kdr flies have in addition another mutation in the same gene — an exchange of methionin[918] with threonine (Ingles et al., 1997).

6.4.3   Pyrethroids

Pyrethroids form a uniform group of pesticides, some of which are naturally occurring, and many are synthetic analogues of these. The natural pyrethroids are obtained from pyrethrum, a substance that is extracted from the flowers of certain species of chrysanthemum. Pyrethrum is made up of six naturally occurring esters, two of which are sometimes referred to as pyrethrins, the other being known as cinerins and jasmolins.

Pyrethrum (from Chrysanthemum)

Pyrethrum I

Pyrethrum II

Cinerin I

Cinerin II

Jasmolin 1

Jasmolin II

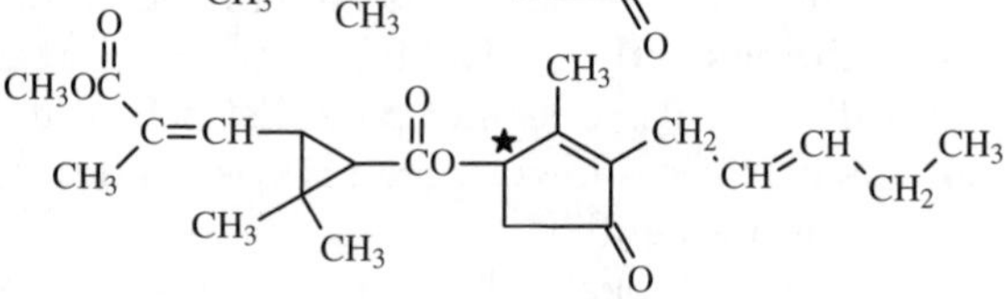

Originally, pyrethrum was manufactured by drying and pulverizing whole flowers. Today extracts from the plants that contain the active ingredients are usually used. Although pyrethrum is very toxic to mammals when injected, its toxicity, when injected or at skin exposure, is relatively low. The same is not true for arthropods, to which pyrethrum is highly toxic even when exposure is through their surface layer or through ingestion.

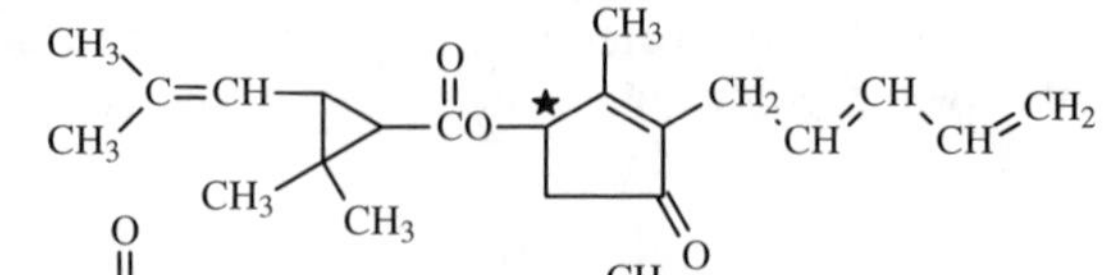

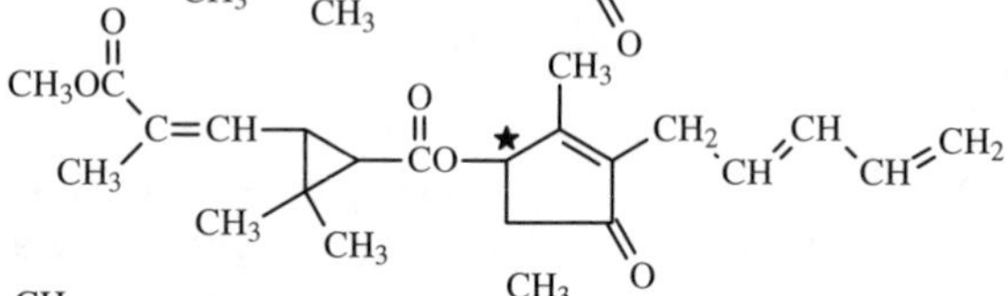

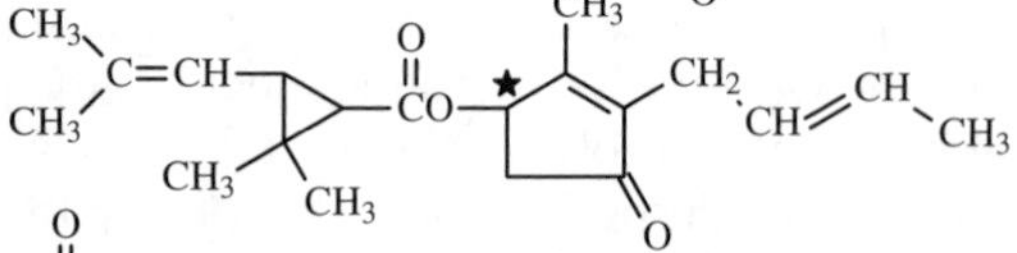

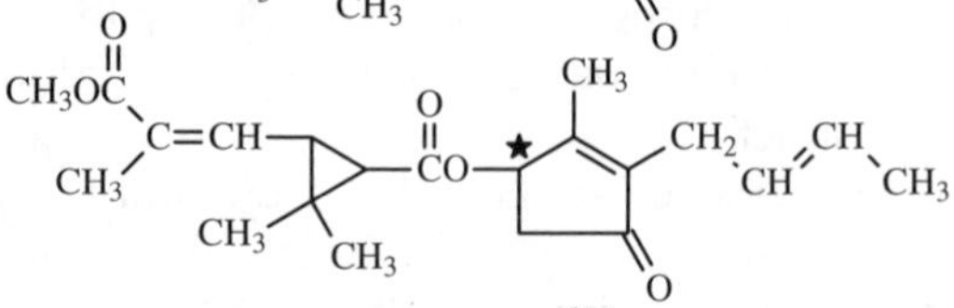

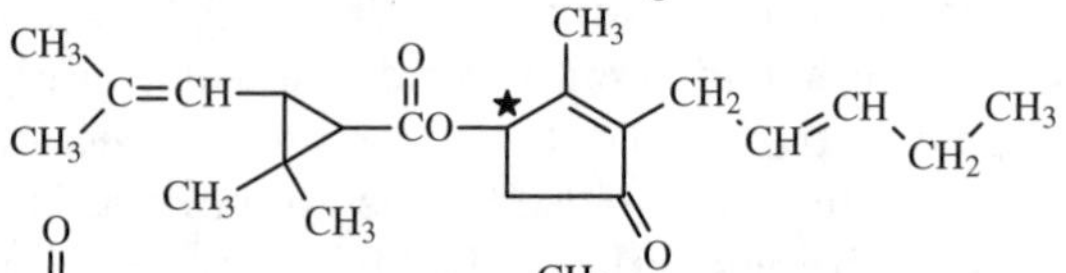

Pyrethrum was recognized as early as 1820 and used as a fast-acting insecticide. The chemical structure of the pyrethrins was elucidated in 1924. Pyrethrum is a very successful pesticide, but there are a number of problems associated with its use. The naturally occurring esters are easily degraded by light and the compounds are unstable, leading to easy and rapid oxidation when exposed to air and sunshine. Oxidation results in detoxication of the compounds. The natural pyrethroids also contain structures that make them vulnerable to fast detoxication in the target organism. As a result of these characteristics, pyrethrum was and is sold in an oily emulsion and stabilizers are added. The potential of the natural pyrethroids as models for the development of synthetic analogues, with the same or better effects, but without the problematic instability, became clear at an early stage. The development of the synthetic analogues took off in the 1960s when M. Elliott and his colleagues at Rothamstead Experimental Station, U.K., began an extensive study of the mechanisms of action and the relationship between the structures and activities of the natural pyrethroids and various synthetic analogues. Elliott et al. (1973, 1978) are central publications from their work. (A more recent review of pyrethroid research is that of Soderlund et al. (2002).) The Japanese company Sumitomo Chemical Company was also very active in this research. The goal of this effort was clear and is summarized by Casida and Quigstad (1998):

1. Photo stability without compromising biodegradability
2. Selective toxicity conferred by target site specificity (e.g., bioresmethrin) or metabolic degradation (lower toxicity for trans- than for cis-cyclopropanecarboxylates)
3. Modification of every part of the molecule with retention of activity
4. Maintenance of high insecticidal potency while minimizing fish toxicity (e.g., the non-ester silafluofen)
5. Development of compounds effective as fumigants and soil insecticides (e.g., tefluthrin)
6. Optimization of potency to allow corresponding reduction in environmental contamination

The development was very successful, and most of them are therefore extremely toxic for insects and many other invertebrates. Table 6.1 demonstrates their increasing efficiency against insects. Some examples of the development of pyrethroids are also shown.

The most remarkable compound listed is probably permethrin, a rebuilt chemical with much higher stability and insecticidal activity than the natural pyrethroid. Not much later the difference in activity between the various stereoisomers was taken into account. Permethrin is a racemic mixture, but in the products called *bio-*, as in bioallethrin and bioresmethrin, as well as in deltamethrin and several other newer pyrethroids, the inactive stereoisomers have been removed. Deltamethrin has a cyano group, making mirror-image isomerism possible. The one shown is the most potent. Substances without the cyclopropane moiety were also found. Fenvalerate was developed by Sumitomo Chemical Co. Ltd. and described in 1974, whereas its most active isomer was found and described in 1979.

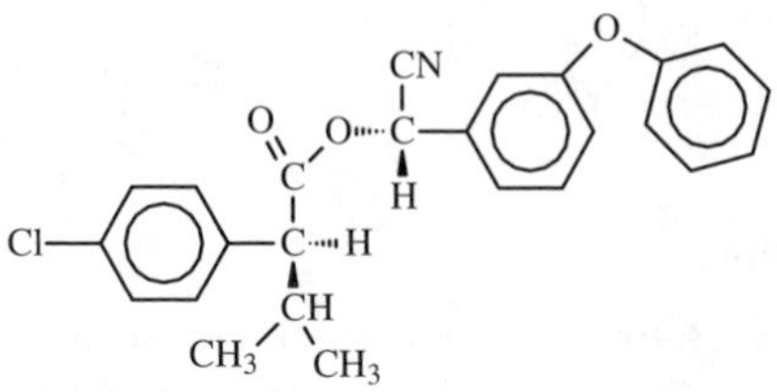

esfenvalerate:

Table 6.1 Examples Illustrating the Development of Pyrethroids with Increasing Potency

Name and Year of Publication	LD50, μg/fly	Structure
Pyrethrin I 1820	0.33	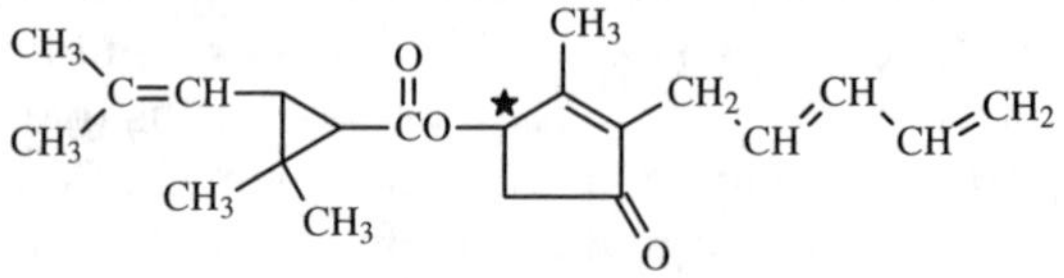
Allethrin 1949	0.1	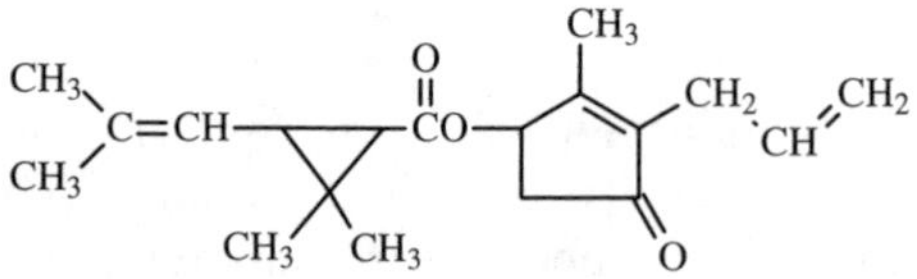
4-Allylbenzyl-chrysanthemate 1965	0.02	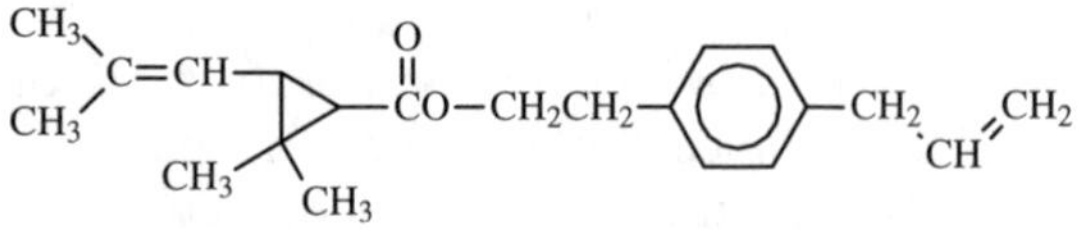
Bioresmethrin 1967	0.005	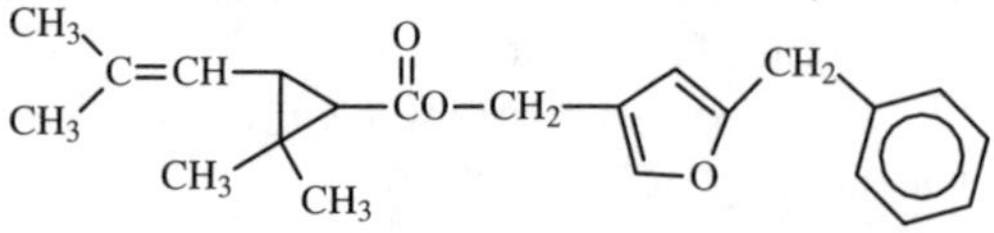
Permethrin 1973	0.002	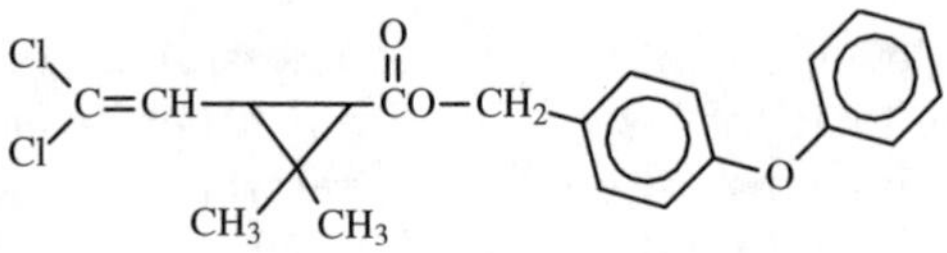
Deltamethrin 1974	0.0003	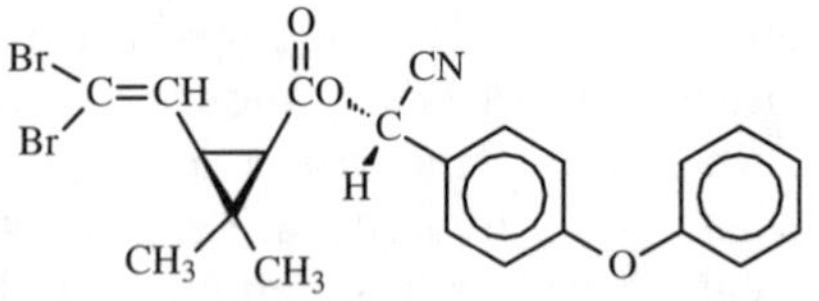

The structural similarity to the other pyrethroids is not striking. Even more different is the silicium-containing silafluofen lacking both the cyclopropane ring and the ester bond. This compound is remarkable for its very low fish toxicity, combined with good effects against insects.

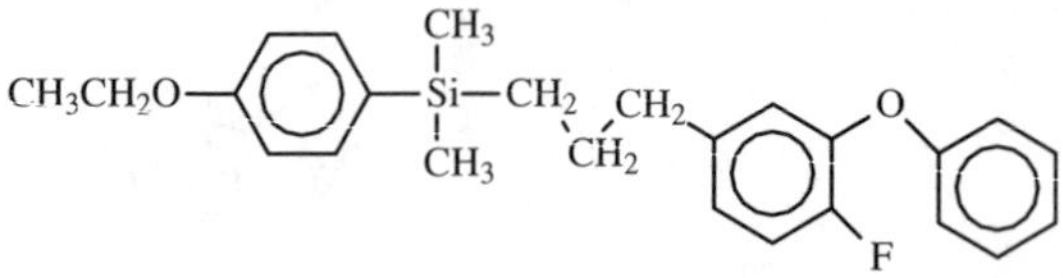

silafluofen

The toxicity to vertebrates does not increase at the same rate as the toxicity to invertebrates, making the synthetic pyrethroids generally better and more selective pesticides than the natural pyrethroids.

One of the first substances to be developed was permethrin. This substance differs from the natural pyrethroids in that two methyl groups have been replaced by chorine atoms, and an unstable side chain has been altered so that the substance is not so easily degraded by photooxidation or by enzymes in the insects.

pyrethrin I (from Chrysanthemum

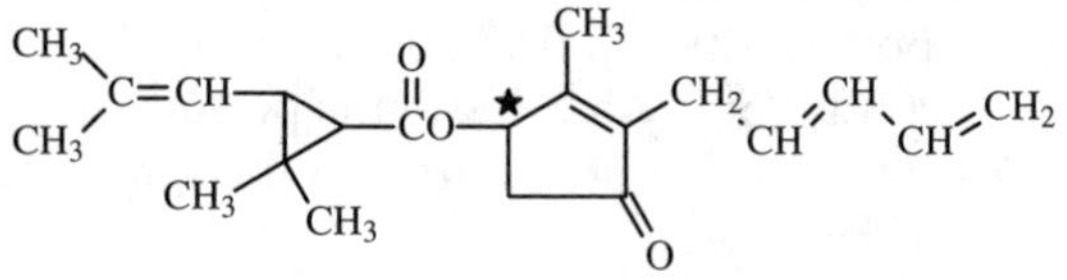

permethrin (Synthetic)

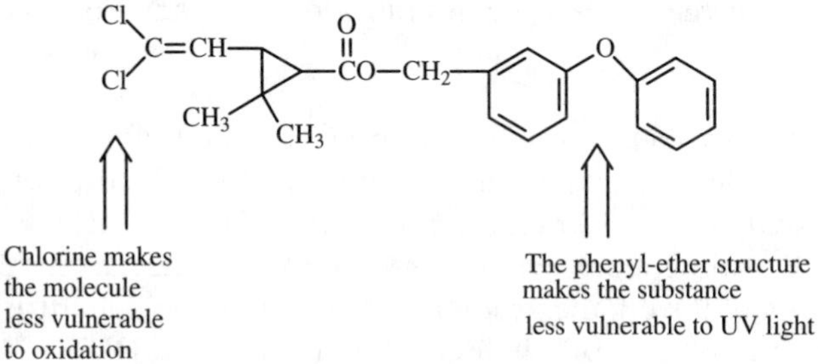

Chlorine makes
the molecule
less vulnerable
to oxidation

The phenyl-ether structure
makes the substance
less vulnerable to UV light

Next to the organophosphorus insecticides, the pyrethroids have been the most expanding group of pesticides. Although their structures and chemical names are very complicated, they are rather easy to recognize by name or structure. Most of them have a cyclopropane group substituted with an esterified carboxyl group in its 1 position, with two methyl groups in the 2 position, and with an isobutenyl group in the 3 position. Instead of an isobutenyl group, there may be a group of approximately similar shape. The alcoholic part contains a ring structure, oxygen, and double bonds or an aromatic structure. The alcoholic part may also have a chiral center, as in pyrethrins and deltamethrin (but not in permethrin). The chemical names are long and complicated. For instance, a pyrethroid with the simple common name *bioallethrin* has the name (*RS*)-3-allyl-2-methyl-4-oxocyclopent-2-enyl(1*R*,3*R*)-2,2-dimethyl-3-(2-methylprop-1-enyl)cyclopropanecarboxylate from the International Union of Pure and Applied Chemistry (IUPAC). To make it even more complicated, Chemical Abstracts has a slightly different system for naming, and Rothamstead Experimental Station, U.K., has its own system in order to make it easier to show the relationship between substances with similar stereoisomeries.

The Pesticide Manual from 1994 describes 33 pyrethroids. All but five have the suffix -*thrin* in their names. The five exceptions have name endings like -*thrinate* or -*valerate*, or are still named by a number (RU 15525). The current issue has 41, with a similar consistency in the names (Tomlin, 1994, 2000). It is very important to keep in mind the great differences in biological activity between the various stereoisomers.

6.4.4   DDT and its analogues

DDT was synthesized first by Zeidler (1874). He was only interested in organic synthesis and did not recognize its fantastic properties as an insecticide, which is described with enthusiasm by West and Campbell (1950). Dr. Müller and his colleagues at J.R. Geigy S.A. of Basle through systematic testing detected its insecticidal activity. We confine ourselves to mention the great uncertainty about the mechanism behind the toxicity during and immediately after the Second World War. One of the popular hypotheses promoted by Dr. Hubert Martin was that DDT satisfied three requirements:

1.   Ability to penetrate and concentrate at the site of action
2.   Adequate stability to reach this site
3.   Ability to release hydrogen chloride when adsorbed at the site of action

HCl release was believed to be essential. The first two points are important and even today correct, but the last point, although supported by many structure–activity considerations, is not correct. The HCl release hypothesis had even at that time many weaknesses. If, for instance, a chlorine and the hydrogen atom at the ethane group are exchanged, the dehydrochlorination in alkaline solutions is approximately similar to that of DDT, but its toxicity is much lower. The p,p'-chlorine substituents are important for toxicity and cannot be removed or moved to ortho positions without loss of activity. It therefore became clear early in the golden age of DDT that shape, size, and electronic configuration were the important parameters and not the reactivity. Although its exact site of action (in the sodium channels of the axon) could not be postulated at that time — because the mechanism of impulse propagation was not known — it was recognized that DDT is a nerve poison. West and Campbell (1950) quoted Smith, who did some studies on aphids (F.F. Smith, *J. Econ. Entomol.*, 39, 383, 1946): "the symptoms observed in the treated aphids are in agreement with other evidence that at least a part of the action of DDT is that of a nerve poison." Today it is established that DDT acts at the same site as most pyrethroids, but there may still be some uncertainty about a possible effect at the sodium pump. It has been found that brine shrimps (*Artemia salina*) as well as sea birds and eels are rather sensitive to DDT. These organisms have active sodium pumps that reduce the intracellular salt concentration by pumping Na^+ ions out at the expense of ATP. However, inhibition of the ATPases involved is also observed for other chlorinated hydrocarbons (see Janicki and Kinter, 1971; Koch, 1969).

Müller himself and other entomologists tested a wide array of compounds similar to DDT in order to reveal the relationship between structure and activity. The most important outcome besides DDT was methoxychlor having p,p'-methoxy groups instead of p,p'-chloro groups. The methoxy groups have approximately the same size and shape as the chloro groups.

Methoxychlor is much less stable and became popular when the environmental contamination caused by DDT was recognized. The methoxy groups are easily attacked by oxidative enzymes (CYP enzymes). Ethyl groups in the para position are also possible, as in Perthane. Another DDT analogue, more active as a miticide, is dicofol. The structures of DDT and some of the more important derivatives are shown.

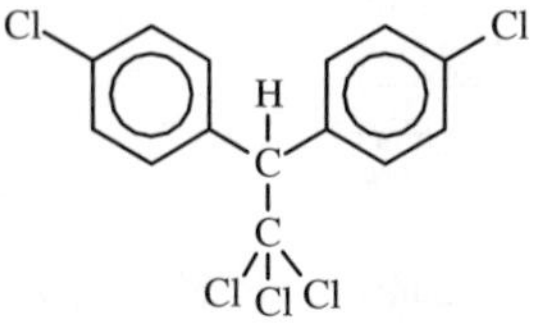

p,p'-DDT

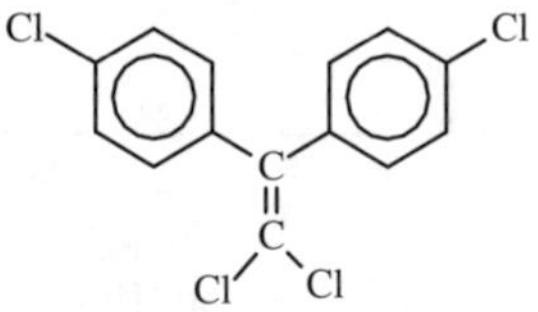

p,p'-DDE

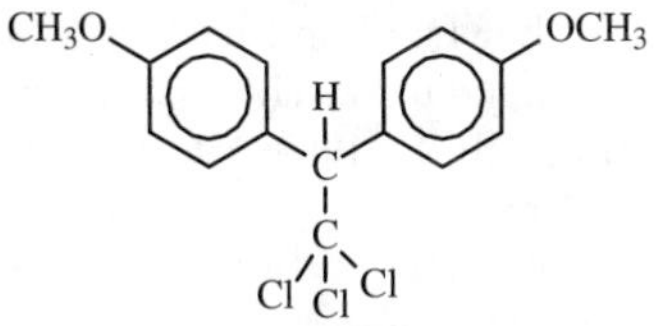

methoxychlor

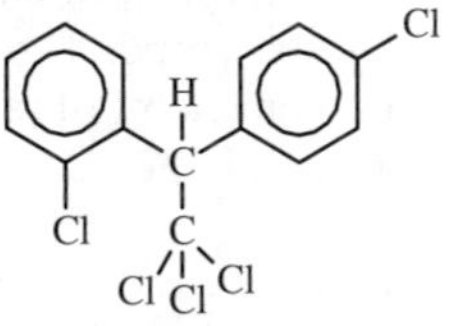

o,p'-DDT

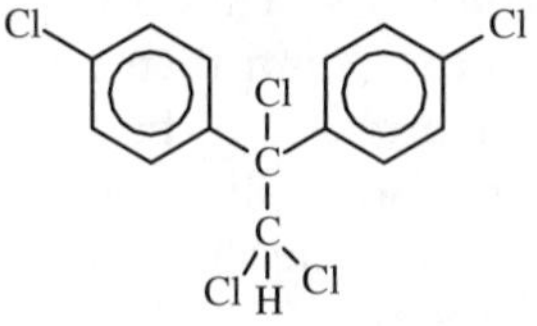

nonactive DDT analogue

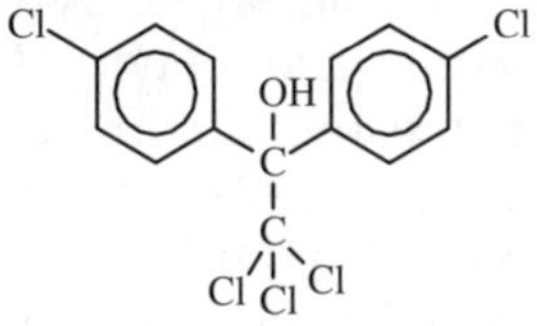

dicofol

It should be noted that DDT, analogues, and the pyrethroids have a negative temperature gradient for their toxicity (increasing LD50 with temperature). The diagram (Figure 6.2) is based on Holan's data (1969). He determined the toxicity of several halocyclopropane analogues of DDT in his efforts to relate toxicity to molecular shapes. DCC (1,1-di-(p-chlorophenyl-2,2-dichlorocyclopropane)), is rather toxic. A newer study of structure and activity of DDT analogues is that of Nishimura and Okimoto (1997).

6.5   Pesticides acting on synaptic transmission

The chemical used to transmit the signal to the next cell is packed in small vesicles in the nerve terminal knob. Calcium ions at a concentration of 1 to 10 μM reduce the energy barrier between the membranes of the cell and the vesicle membranes, allowing the membranes to fuse. The transmitter substance, stored in the vesicles, is discharged into the synaptic cleft. It has been calculated that one impulse to a neuromuscular junction releases 300 vesicles. The transmitter acetylcholine is stored in vesicles containing 5000 to 10,000 acetylcholine molecules. It takes much less than a millisecond for the released

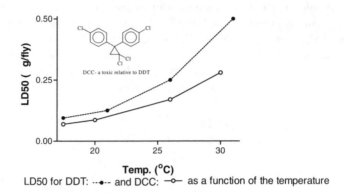

LD50 for DDT: --•-- and DCC: --o-- as a function of the temperature

Figure 6.2 Toxicity of halocyclopropane analogues of DDT. LD50 values for DDT and DCC as functions of the temperature. Note that DDT and its analogues and pyrethroids do not depend on chemical reactivity to be toxic. They can therefore be quite stable, as is the case with DDT. (Data from Holan, G. 1969. *Nature*, 221, 1025–1029.)

acetylcholine to diffuse across the synaptic cleft where it binds to specific receptor proteins located in the postsynaptic membrane. The receptors are proteins that form channels across the membrane. They are normally closed, but open in response to acetylcholine binding, and then permit sodium to flow in and potassium out. The channels are said to be chemically gated, and acetylcholine is the key, as opposed to the voltage-gated channels for sodium, potassium, and calcium mentioned earlier. Each channel molecule requires two acetylcholine molecules to open.

The electrical potential at the postsynaptic membrane falls because of the influx of sodium. The fall depends on how many gates are opened and how long they are kept in that position. If a satisfactory number of gates are open long enough, the voltage difference across the postsynaptic membrane is decreased sufficiently to open the voltage-gated sodium channels so that the voltage difference is further decreased and an action potential is achieved.

6.5.1   Inhibitory synapses

Some synapses deliver transmitter substances that do not decrease the membrane potential at the postsynaptic membrane but, on the contrary, increase it by binding to receptor sites at other specific channel proteins. These synapses are said to be inhibitory because, when activated, they inhibit the transfer of signals from the excitatory synapses, like the cholinergic ones. Most channels for chloride are of this type. Although chloride ions cannot flux freely across the membrane, the outside and inside concentrations of chloride are as if they could do this. Because of the voltage difference, the outside–inside concentration difference may be substantial (e.g., 570 μM outside and 40 μM inside). The concentrations are said to be at equilibrium at the resting electrical potential. Opening of the chloride channels makes it possible for chloride to enter because the concentration is much higher on the outside. Chloride influx reduces the effect of sodium influx caused by the opening of the sodium channels. The potential may even become more negative than the resting potential by influx of chloride. Stronger signals (e.g., more excitatory transmitter substances, like acetylcholine) are needed to create an action potential when inhibitory signals have been received. The most important transmitter substance acting at the chloride channels is probably gamma-aminobytyric acid (GABA).

Much research on the GABA receptors, and the chloride channels they regulate, has been done because many important drugs modulate their activity. Sleeping medicine like barbiturates and sedatives such as benzodiazepins are important examples.

GABA-ergic inhibitory synapses are also present in insects and other invertebrates and are targets for many pesticides — some of them are extremely potent.

The toxic properties of most of these pesticides reside in their inhibitory actions at one or more binding sites on insect GABA-gated chlorine ion channels. The chemical structures of these substances are different, and it is difficult to relate structure to activity. Chlorinated cyclodiene insecticides, lindane, and gamma-butyrolactones such as the plant toxin picrotoxinin possess a number of structural features in common, and a minimum requirement for the insecticidal activity of many of these convulsants is the existence of at least two electronegative centers and one region of steric bulkiness or hydrophobicity. This theory explains how a range of such structurally unrelated compounds have similar modes of action. The chloride channel blockers are often divided into four groups: A, B, C, and D, according to their exact binding site. Most insecticides (lindane, toxafen, the cyclodienes) are in group A. Type C includes fipronil, and group D the avermectins. Group B does not include any insecticides so far, and we shall restrict ourselves to give a short description of the more important insecticides. Figure 6.3 summarizes the events as an impulse passing along the axon to the synapse.

6.5.2   Pesticides

6.5.2.1   Lindane

The insecticidal properties of lindane were discovered at Imperial Chemical Industries (ICI), England, in 1942, but the hexachlorocyclohexanes (HCHs) had already been synthesized by Faraday as early as 1825. The research leading to the pesticide lindane started much later at ICI, where scientists looked for a chemical that could kill turnip beetles. They tried HCH, and many other synthetic compounds. HCH is very easy to make. It is just to bubble chlorine gas through benzene and at the same time illuminate with UV light. Chlorine will then add to benzene to give the nonaromatic HCH. Without UV light, chlorine is substituted with hydrogen to give the aromatic hexachlorobenzene (BHC). However, the synthesis of HCH always gives a mixture of many stereoisomers because the chlorine atoms can be in an

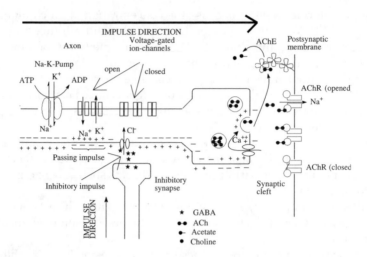

Figure 6.3 A simplified diagram of some of the events that happen in an axon and at the synapse when an impulse is passing. At one location (P) of the axon the following events occur when an impulse passes:

1. Before the impulse arrives, the voltage difference is at its resting potential. The Na+ channels and K+ channels are closed.
2. Immediately before the impulse has reached the location (P), but has reached a location very close to it, the opening of Na+ channels there decreases the voltage difference also at (P). The voltage drop leads to a brief opening of the Na+ channels and influx of Na+.
3. The influx of Na+ ions reduces the voltage potential difference even further and causes a reduced potential at a point further down the axon.
4. The Na+ channels close but the K+ channels are kept open for a while.
5. This leads to the restoring of the resting potential.
6. The ion pump throws Na+ out and K+ in at the expense of energy from ATP, restoring the concentration difference of ions at the inside and outside of the cell.

These events happen in a very rapid sequence. However, there is inertia in the system that maintains every impulse as a discrete event. When an impulse reaches the synapse, Ca++ channels open, vesicles with the transmitter substance fuse with the membrane, and the transmitter substance is released into the synaptic cleft. Transmitter substances act as keys for the gates on ion channels at the postsynaptic membrane. The transmitter substance from inhibitory synapses (e.g., GABA) opens Cl- channels that suppress or stop the action of Na+ influx on the voltage difference, and the positive feedback this has on the Na+ channels.

equatorial or axial position. Professor Hassel at the University of Oslo was awarded the Nobel Prize in chemistry for his work on the conformations of HCHs and other cyclohexane derivatives. As many as nine isomers are formed, but only one, the gamma-isomer, is useful as an insecticide. Four of the isomers, alpha-, beta-, gamma-, and delta-isomers, had been described by Van der Linden in 1912 (*Berichte*, 45, 236, 1912). The gamma-isomer is isolated and called lindane or gamma-HCH. The synthesis gives 10 to 45% alpha-, 5 to 12% beta-, 3 to 4% delta-, and only 10 to 14% of the useful gamma-isomer. It was supposed that because the structure of lindane is similar to that of inositol, its toxicity was due to interference with the inositol metabolism, although the importance of inositolphosphates in the internal signaling system of the cell was not known. Lindane was shown to inhibit the growth of yeast (*Saccharomyces serevisie*), and addition of i-inositol to the medium reversed the inhibition. Some toxic effects of lindane and other isomers may perhaps be explained by interference with inositol in signal transduction, because phosphorylated inositol plays an important role in signal transduction from the so-called metabotrophic receptors. The muscarinic receptors to be described soon are a good example. However, it now seems well established that the main reason for lindane's toxicity is blockade of the GABA-gated chlorine channels, inducing convulsions in insects as well as in mammals.

The symptoms in poisoning are in accordance with this theory and were well known in the earlier days. West and Campbell (1950) wrote (p. 511):

> When symptoms appear, the end is usually in sight and little can be done to save the animal. Symptoms of acute poisoning with Gammexan develop rapidly and include the following in this order:
>
>   i. Increased respiratory rate, sometimes very considerable.
>   ii. Restlessness, accompanied by frequency of micturition.
> iii. Intermittent muscular spasms of the whole body.
>   iv. Salivation, grinding of teeth, bleeding from the mouth and tongue resulting.
>   v. Backward movement, with loss of balance and somersaulting.
>   vi. Head-retraction, convulsions, gasping and biting.
> vii. Collapse and death.
>
> In hyper-acute cases, this train of events lasts 40–120 minutes; more resistant animals survive 12–20 hours.

Newer research on the mode of action of lindane may be found in Pajuelo et al. (1997).

The other convulsing GABA-blocking poisons give similar symptoms. Lorazepam or diazepam has the opposite action on the chloride channels and may be given intravenously as an antidote (0.1 mg/kg of body weight according to *Cassarett and Doull's Toxicology* (Klaassen, 2001)).

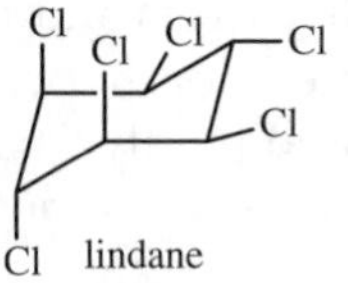

lindane 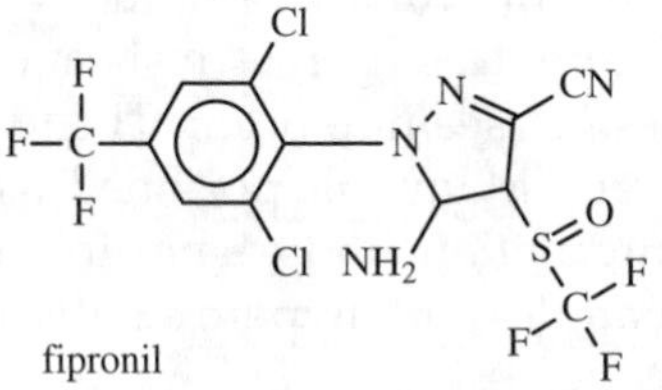   fipronil

6.5.2.2 Fipronil

Fipronil is a new (described first in 1992) and superefficient insecticide that blocks the chloride channel by binding to an allosteric site, or binding irreversibly. There is a good correlation between binding to the receptor *in vitro* and insecticidal activity for fipronil and various related substances (Ozoe et al., 2000). Because of its high stability and long residual insecticidal activity, it is not approved in all countries. Photooxidation leads to other very active compounds. Fipronil was a promising alternative in locust control, but because of its high persistence, it may harm the endemic part of the desert fauna.

6.5.2.3 Cyclodiene insecticides

The older cyclodiene insecticides like aldrin, dieldrin, heptachlor, chlordane, endrin, and endosulfan act also as antagonists on the GABA channels. These substances still represent some problems as environmental pollutants because many of them are very stable in organisms, soil, and sediments. They all have a characteristic "clumsy" structure. Endosulfan was introduced in 1956 and is still in use, whereas the other compounds were introduced between 1948 and 1950.

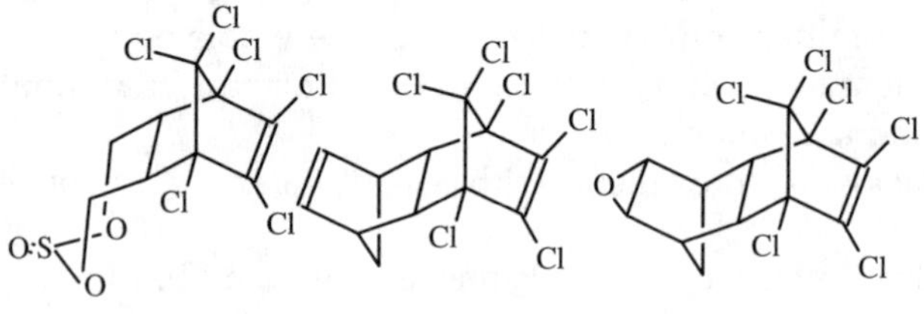

endosulfan   aldrin   dieldrin

Note that dieldrin is an oxidation product of aldrin. Epoxides are usually rather unstable, being hydrolyzed to diols or split up to the enols (double bond and hydroxyl group). Dieldrin may be formed in soil and organisms from aldrin and is very stable.

6.5.2.4 Avermectins

One important group of insecticides, the avermectins, works differently by being agonists and not antagonists as the other, acting on the chloride channels. The avermectins are produced by *Streptomyces avermitilies*. The binding site is different, and cross-resistance to fipronil, the cyclodienes, and lindane does not seem to occur. The toxic symptoms in insects and mammals are different. Mammals poisoned with avermectins exhibit hyperexcitability, incoordination, and tremor followed by ataxia and paralysis. In insects and nematodes, the hyperexcitation phase is absent. Their symptoms are therefore more in accordance with the postulated mode of action at the molecular level.

It is important to note that the interaction with regulatory binding sites at the chloride channels does not involve making or breaking covalent bonds. Pesticides of this class may therefore be rather stable.

6.5.3   The cholinergic synapses

The transmitter substances do not bind covalently to the receptor site and will diffuse off. Binding and dissociation will follow the law of mass action so that high concentrations of transmitters in the synaptic cleft will lead to more molecules binding to the receptor and stronger signals. Before the next impulse arrives, the concentration of the transmitter substance in the synaptic cleft must be reduced, either by diffusion out, uptake in the cells involved, or enzymatic degradation. Most important is the *acetylcholinesterase*, which degrades acetylcholine, described in the previous chapter.

The synapses using acetylcholine (ACh) as the transmitter substance are the target for a wide variety of pesticides and therefore need a more detailed description. Acetylcholine is used as a transmitter substance in nearly all animal phyla, but at different parts of the nervous system. It is also present in single-cell animals and even in plants. Enzymes that catalyze the hydrolysis of acetylcholine, the cholinesterases, are also present in various organisms not having a nervous system. In insects and other arthropods, ACh is the transmitter of messages from sensory neurons to the central nervous system (CNS) and within the CNS, but not from motor neurons to skeletal muscles, where the transmitter is glutamate. In annelids, the excitatory transmitter for the body wall muscles is acetylcholine, as at the neuromuscular junctions in vertebrates.

There are two types of cholinergic synapses. They are called nicotinic and muscarinic synapses, respectively, because the postsynaptic membranes have receptors that are sensitive to either nicotine or muscarine, although they are both sensitive to acetylcholine.

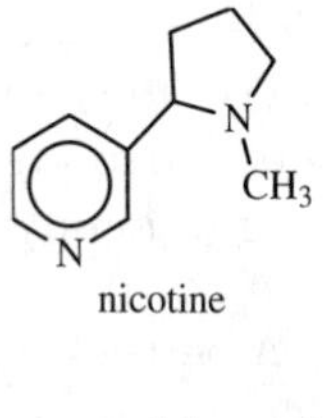

nicotine 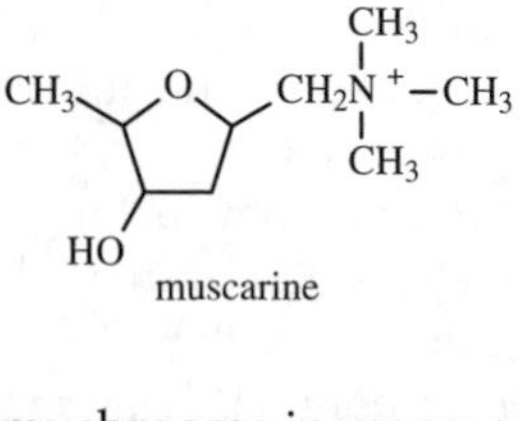   muscarine

Muscarine is present in certain mushrooms in many genera, notably the *Inocybe* and *Clitocybe* genera, but small amounts are also present in the fly agaric, *Amanita muscaria*. The ecological function of this nerve poison in the fungus is not understood. In mammals, the typical symptoms are unrest, irritability, excitement, sweating, salivation, respiratory trouble, feeble pulse, and small pupils. The symptoms are in accordance with its agonistic action at the cholinergic synapses in the peripheral parasympathetic nervous system. Nicotine is present in many plants, notably tobacco plants, where it probably has a function to protect against insect attack. Extracts from tobacco are used as a contact insecticide and fumigant. Nicotine acts in the ganglial synapses in insects' central nervous systems and in the nicotinic synapses in the autonomous systems of vertebrates, as well as in their neuromuscular junctions. Symptoms in humans include salivation, muscular weakness, fibrillation, chronic convulsions, and cessation of respiration. Large daily intake by humans is quite common because of nicotine's stimulatory action. It causes serious addiction problems, and nicotine itself and substances associated with it cause illnesses and early deaths in millions of people worldwide. The nicotinic receptors are placed in the sodium channels in the postsynaptic membranes in certain parts of the nervous systems. Binding of two acetylcholine molecules opens the channels, leading to influx of sodium and transmittance of the impulse. Symptoms of poisoning are therefore referred to as nicotinic and muscarinic according to the parts of the nervous system that are affected.

The muscarinic symptoms are miosis (constriction of the pupil of the eye), vomiting, diarrhea, bradychardia (slow heartbeat frequency), and cardiovascular collapse. Muscarinic symptoms are attributable to peripheral parasympathetic stimulation.

The nicotinic symptoms are salivation, vomiting, muscular weakness, fibrillation (fast, irregular muscle constrictions), chronic convulsion, and cessation of respiration. The symptoms are caused by overstimulation of the autonomic ganglia and the neuromuscular junctions in the voluntary muscles.

There is a fundamental structural difference between the muscarinic and nicotinic receptor systems, which should be mentioned here and studied further in other textbooks. Whereas the nicotinic receptor is composed of five subunits designated, for instance, α, β, γ, δ, and ϵ, and may have a structure α_2, β, γ, δ, etc., the muscarinic receptor is only one peptide chain. This chain crisscrosses the cell membrane seven times. When acetylcholine (or muscarine or another agonist) binds at the receptor site that is located on a part of the receptor molecule at the outside of the membrane, a cascade of chemical reactions is started on the inside. There are several types of muscarinic receptors and they belong to a large family referred to as G protein-coupled receptors. They have been much studied, and the reader should consult a textbook in cell biology (e.g., Alberts et al., 2002). The study of the muscarinic receptors has been facilitated by the availability of radiolabeled ligands that bind specifically and with high affinity to them. The benzilic acid ester of 3-quinuclidinol (QNB) is a powerful muscarinic receptor antagonist that can be radioactively labeled. It binds specifically and exclusively to all types of muscarinic receptors and has therefore been used as an incapacitating chemical warfare agent, but is also an excellent tool in neurochemical research.

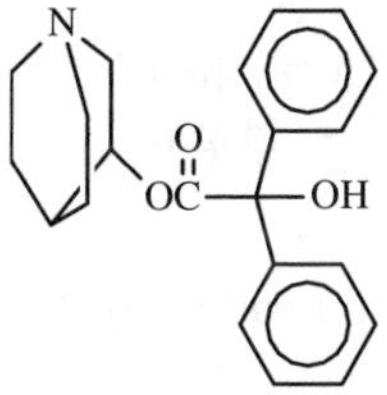

QNB

The venom of the Southeast Asian banded krait (*Bungarus multicinctus*) contains α-bungarotoxin, which binds exclusively and by high affinity to the nicotinic receptors. By means of these and many other substances, it is unveiled that insects and other invertebrates like the vertebrates have both types of acetylcholine receptors.

6.5.3.1   Atropine

Of direct relevance for pesticide science is the antagonist atropine. This toxicant also binds specifically to the muscarinic receptors where it blocks the effect of ACh. The symptoms are therefore the opposite of those caused by muscarine or acetylcholine (pupil dilation, dry mouth, inhibition of sweating, tachycardia, palpitations, hallucinations, delirium, etc.). Atropine is an important antidote when one is poisoned with a cholinesterase-inhibiting insecticide.

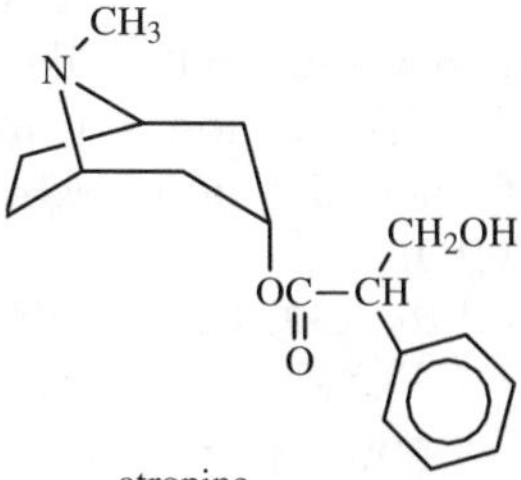

atropine 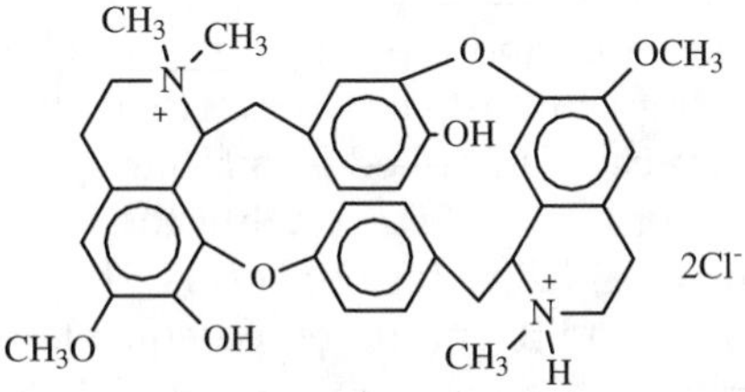   tubocurarine chloride

Tubocurarine is another important natural antagonist. It blocks the nicotinic receptors, but because it does not penetrate into the brain, it acts mainly at the neuromuscular junctions, causing paralysis without disturbing the consciousness or acting as an anesthetic. It has been used as an arrow poison, but is also quite useful as a muscle relaxant during surgery. This compound or others with similar modes of action are extremely unpleasant if not given together with a general anesthetic. The subject can see, hear, and feel but cannot move a finger and needs help with breathing. Atropine and tubocurarine are present in various plants (*Atropa belladonna* and *Chondodendron tomentosum*) where they probably protect the plant from grazing animals. Succinylcholin is a synthetic substance used as a muscle relaxant with the same physiological properties as tubocurarine. It is better to use at surgery because it is degraded very fast to nontoxic substances by butyryl-cholinesterases in the blood of most patients.

$$
\begin{array}{c}
\text{O} \qquad\qquad \text{CH}_3 \\
\parallel \qquad\qquad \mid \\
\text{C}-\text{OCH}_2\text{CH}_2\text{N}^+\text{-CH}_3 \\
\text{CH}_2 \qquad\qquad \mid \\
\mid \qquad\qquad\quad \text{CH}_3 \\
\text{CH}_2 \qquad\qquad \text{CH}_3 \\
\qquad\qquad\qquad \mid \\
\text{C}-\text{OCH}_2\text{CH}_2\text{N}^+\text{-CH}_3 \\
\qquad \text{O} \qquad\qquad \text{CH}_3
\end{array}
$$

succinylcholin

6.5.3.2 *Nicotinoids and neonicotinoids*

Nicotine and certain nicotine analogues have been used a long time as insecticides, but the nicotinic acetylcholine receptor (nAChR) is now used as the target for a new class of synthetic compounds, the neonicotinoids. Imidacloprid was the first commercialized member of this new class of insecticides. Nicotine and its analogues have a basic nitrogen that even at physiological pH picks up a proton to form a positive ion, whereas the neonicotinoids contain a chlorinated pyridyl group, or another heterocyclic group, that withdraws electrons from an imido group and thus makes it partially positive without being protonized. Imidacloprid and the other neo-nicotinoids bind selectively to the nicotinic acetylcholine receptors in insects. Because they are not ions, they penetrate easily into the nervous systems of insects. Many of them have a very low toxicity to vertebrates, nematodes, and crustaceans, and are frequently used against ectoparasites like lice on cats and dogs, but are also very efficient in plant protection. For instance, nitenpyram can be applied as a foliar spray against sucking insects on rice at a rate of only 15 to 75 g/ha, but the LC50 (lethal concentration in 50% of the population) (24 h) for *Daphnia* is >10 g/l and the LD50 for (male) rats is 1680 mg/kg, and the no-observed-effect level (NOEL) (2 years) for male and female rats was determined to be 129 and 54 mg/kg, respectively. The efficiency is therefore not very different from that of deltamethrin. (Compare it, for instance, with the organophosphate fenthion, or other organophos-phates, which have recommended application rates of 60 to 1200 g/ha, depending on the crop, pest, pest stage, and application method.) The vet-erinary use of neonicotinoids is described extensively in Krämer and Mencke (2001), who also give a good introduction to the basic toxicology and phar-macology of imidacloprid and other neonicotinoids.

The structures of some of them are shown:

$$
\begin{array}{c}
\text{CH}_3 \\
\mid \\
\text{Cl}-\!\!\!\bigcirc\!\!\!-\text{CH}_2-\text{N}\diagdown\;\diagup\text{CH}_3 \\
\quad\;\text{N} \qquad\quad \text{C} \\
\qquad\qquad\qquad \parallel \\
\qquad\qquad\qquad \text{N} \\
\qquad\qquad\qquad \diagdown\text{CN}
\end{array}
\qquad \text{acetamiprid}
$$

clothianidin

dinotefuran

imidacloprid

nitenpyram

thiamethoxam

nithiazin

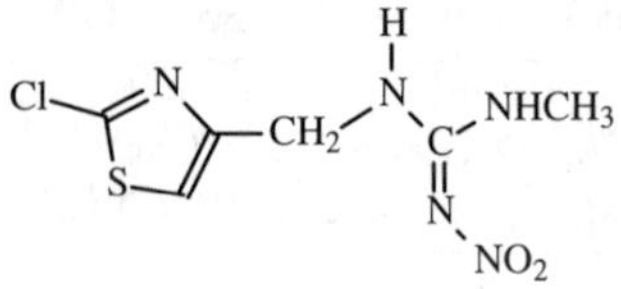

Besides an electron-withdrawing heterocyclic ring, they have a nitromethylene, nitroimine, or cyanoimine group that can distinguish the insect receptor from the vertebrate nicotinic acetylcholine receptor (Tomizawa et al., 1995a, 1995b; Yamamoto et al., 1995).

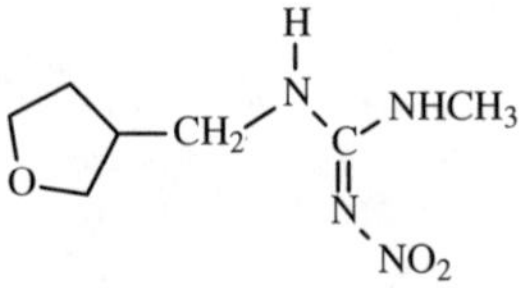

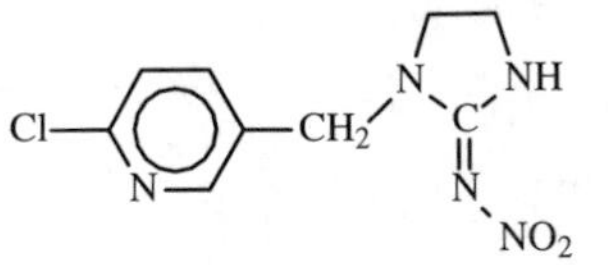

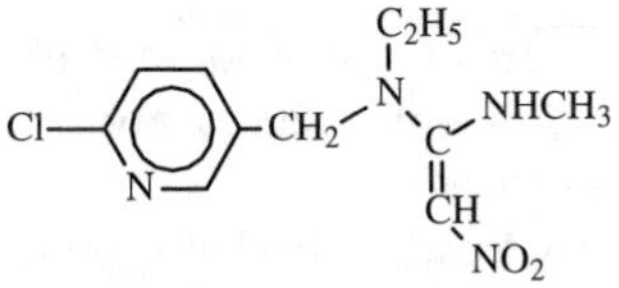

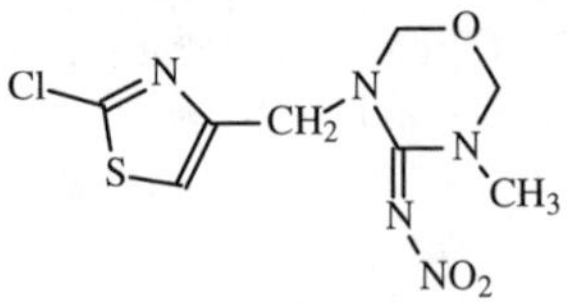

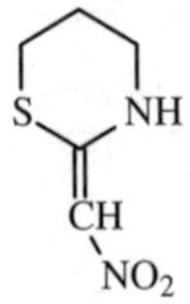

The neonicotinoids' many favorable properties may be summarized in these points:

- Broad spectrum of activity against insect pests
- Relative low toxicity against vertebrates
- New mode of action, with less likelihood for cross-resistance
- High NOEL and acceptable daily intake (ADI) values (when determined)

The neonicotinoids were mainly developed in Japan, but imidacloprid is sold by the German company Bayer AG.

Table 6.2 Toxicity of Neonicotinoids Compared to Some Other Nerve Poisons

Name	Rat (Male) Oral LD50 (mg/kg)	NOEL Rat (2 years) (mg/kg in food)	Toxicity Class (WHO)	Daphnia EC50 (mg/kg)	Application Rates (g/ha)
Acetamiprid	217	7.1	—	>200	75–700
Cartap	345	10	Xn	—	400–1000
Clothianidin	—	—	—	—	—
Dinotefuran	2804	100	—	>1000	100–200
Imidacloprid	450	—	II	85	—
Nicotine	55	—	Ib	0.24	—
Nitenpyran	1680	129	—	—	15–400
Thiamethoxam	1563	—	III	>100	10–200
Fipronil	97	0.02	II	0.19	10–80
Carbaryl	850	200	II	0.0016	250–2000

A few of the properties of cartap, nicotine, and the neonicotinoids are taken from *The Pesticide Manual* (Tomlin, 2000) and are given in Table 6.2 together with some other insecticides.

The neonicotinoids have very low fish toxicity, are not adsorbed through mammalian skin, and are not irritating or allergenic in the tests carried out so far. The pyrethroids (and DDT group) have a negative temperature correlation; these insecticides are more active in warm weather.

6.5.3.3 Cartap

Cartap is also an important insecticide acting at the nicotinic acetylcholine receptor site. It causes insects to stop feeding, is systemic, and has a low mammalian toxicity. It should therefore be a perfect insecticide. Cartap is not toxic per se, but is biologically converted to the cholinergic agonist nereistoxin, described later.

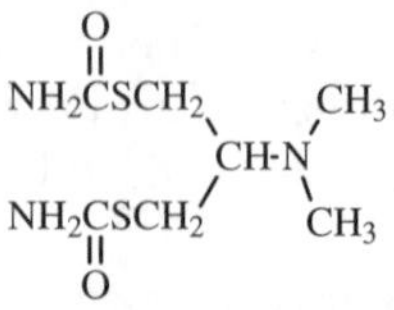

$$
\begin{array}{c}
\overset{\displaystyle O}{\underset{\displaystyle \|}{}} \\
NH_2\overset{O}{\overset{\|}{C}}SCH_2 \qquad CH_3 \\
\diagdown \qquad \diagup \\
CH\text{-}N \\
\diagup \qquad \diagdown \\
NH_2\underset{\|}{\overset{}{C}}SCH_2 \qquad CH_3 \\
\underset{\displaystyle O}{}
\end{array}
$$

cartap

6.5.4   Calcium channels as possible targets for insecticides

The calcium level inside the cells is under very strict control. Opening of the voltage-gated calcium channels in the synapse, caused by the nerve impulse, triggers the release of the transmitter substance. Calcium thus has a very important role in the transmittance of the impulse. The onset of necrosis is also triggered by increased calcium concentration because many hydrolytic

Table 6.3 Sites of Action on the Nerve Cells of Important Insecticides and the Antidote Atropine

Site	Substance	Action	Consequence
Na+ channels	DDT Pyrethroids	Inhibit proper closing of the channels	The resting potential is not fully restored and trains of false impulses are produced; tremors and other symptoms follow
GABA receptor	Fipronil Lindane Cyclodienes	Inhibits opening of Cl⁻ channels	Inhibits the signals from inhibitory synapses
ACh receptor	Atropine	Antagonistic block, inhibiting transmission	Paralysis, and reduces the effects of nicotinoids and ACh; useful as an antidote against organophosphorus and carbamate poisoning
ACh receptor	Nicotine Neonicotinoids	Causes false signals in cholinergic synapses	Overstimulation with tremor and paralysis
AChE	Organophosphorus insecticides Carbamates	Inhibits the hydrolysis of ACh, causing overstimulation of the cholinergic synapses	Nicotinic and muscarinic effects

enzymes are activated. The calcium channels should therefore be excellent targets for insecticides. A genus of tropical American shrubs and trees (Ryania) contains insecticidal compounds in it bark, which is ground and used as a commercial insecticide. Ryania was described in 1945 (Pepper and Carruth, *J. Econ. Entomol.*, 38, 59). The active ingredient, ryanodine, activates the calcium channels in the sarcoplasmic reticulum. The compound thus seems to have its main site of action inside the cell, but new insecticides with the calcium channels as the target are expected.

6.6 Summary

There is a myriad of venoms and poisons from animals and plants, as well as insecticides and warfare agents, that act on the described or other sites in the nervous system. Table 6.3 summarizes the modes of action of the main groups of insecticides acting on nerves.

chapter seven

Pesticides that act as signal molecules

In this chapter we concentrate on hormones, pheromones, and kairomons. Toxicants, such as the dioxins and the many poisons, act by interfering with the signal systems, causing cells to divide. The can also cause genes to transcribe RNA, as well as the nerve poisons and interfere with the action of signal molecules. This is described in other chapters.

7.1 Insect hormones

Much of the following description is based on Dhadialla et al. (1998), Eto and Kuwano (1992), Gullan and Cranston (2000), Rockstein (1978), Yu and Terriere (1974), and Zeleny et al. (1997).

7.1.1 Insect endocrinology

Many details of the anatomy and biochemistry of insect development from egg to adult are known. Growth and development are strictly regulated by hormones, and hormone concentration, relative and absolute, is vital for this to proceed correctly. The hormones control cuticle synthesis, molting, sexual maturation, color differentiation, reproduction, etc.

The endocrine organs of insects are of two types: neurosecretory cells within the nervous system and specialized endocrine glands, such as the corpora allata, the corpora cardiaca, and the prothoracic glands. *Corpora cardiaca* are a pair of organs closely associated with the main vessel (aorta). *Corpora allata* and the prothoracic glands are more diffuse glandular bodies.

Molting, which is defined as the shedding of an outer covering as part of a periodic process of growth, is central to the development of insects. Release of ecdysteroids, or molting hormones, from the prothoracic glands is controlled by a neuropeptide called the prothoracicotropic hormone (PTTH) produced by neurosecretory cells and released by corpora cardiaca.

The ecdysteroids are responsible for cellular programming in cooperation with the juvenile hormones secreted from the corpora allata. The juvenile hormones (JHs) regulate the result of a molt — whether a new larva or a pupa will be the next stage following the molt. When the amount of JH secreted is high, the epidermis is programmed for a larval molt. When the level declines, the cells are programmed for metamorphosis and pupa formations (in holometabolous insects).

The prothoracicotropic hormone and the eclosion hormone are very important in starting and terminating the process of ecdysis. PTTH is a small protein composed of two identical subunits of 109 amino acids. Although different in primary structure, it has a shape similar to that of nerve growth factors and other growth factors in vertebrates, and may be classified together with them in a superfamily of proteins. Eclosion hormone is a 62-amino acid neuropeptide that plays an integral role in triggering the behavior at the end of each molt. At least three populations of cells are thought to be targets for the hormone, each of which shows an increase in the intracellular messenger cyclic guanosine monophosphate (cGMP).

7.1.2 Juvenile hormone

The juvenile hormone, which is secreted from a little appendage under the brain (corporus allatum), determines the result of *ecdysis*, defined as molting of the outer cuticular layer of the body. Four or more chemically related sesquiterpene derivatives constitute the natural hormones. They are synthesized through many steps from acetyl coenzyme A (CoA) via mevalonate and are also found in crustaceans. The JH receptor belongs to the basic helix–loop–helix (bHLH)-PAS family of transcriptional regulators described in textbooks of molecular biology (Alberts et al., 2002).

In the late 1930s and in early 1940 Bouhiol and Piepho, working with *Bombyx* and *Galleria mellonella*, respectively, found that removal of the corpora allata from early instar larvae led to precocious metamorphosis and production of miniature pupae and adults, whereas implantation of corpora allata prevented metamorphosis and resulted in supernumerary giant larvae. Its concentration decides whether a new larval stage, a pupa, or an imago is formed. During the development, its concentration is reduced. When present in high concentrations, a molt is followed by a new larval stage. When the normal number of larval moltings has been carried out, the production of juvenile hormones is abolished and a pupa is produced. Juvenile hormones are virtually absent in pupae but are present in the adults in which they have some functions related to ovarian development.

The juvenile hormones are easily degraded by membrane-bound epoxide hydrolases and carboxyl esterases. The degradation products have no hormone activity and are excreted as sulfates. A carrier protein protects it from degradation, and its concentration is important for the regulation of the amount of active JH.

7.1.2.1 American paper towels

Substances in plants that resemble juvenile hormones were detected as a result of peculiar circumstances and may be of interest. Early in the 1960s Dr. K. Slama, an entomologist from Czechoslovakia, visited the University of Harvard in order to work with insect hormones at Dr. Williams' lab. He brought with him from Prague a laboratory strain of the European linden bug (*Pyrrhocoris apterus*), a Heteroptera, much used for studies of embryology, reproductive biology, theoretical and applied endocrinology, and other aspects of insect biochemistry. To his surprise, the bugs did not develop normally under the conditions at Harvard. Instead of normal, fully grown insects, the larvae developed into bugs with many larval characteristics, and they died before becoming mature. The cause was found to be the paper towels placed in the cages for the insects to hide in. American paper towels, the *New York Times*, the *Wall Street Journal*, and even *Science* had an unfortunate influence on the insects' growth and reproduction, but the *London Times* and *Nature* were excellent. The reason for this preference for European literature was that the American newspapers and journals were produced from *Abies balsamea*. The paper was shown to contain a natural chemical, subsequently called the paper factor, that interfered with the normal development of the larvae. The substance was successively isolated and was found to act as a juvenile hormone. This discovery led to a search for more synthetic and natural substances with juvenile hormone activity. The paper factor is the methyl-ester of a monocyclic sesquiterpene called juvabione. Other natural terpene derivatives also have juvenile hormone activity to variable degrees. The synthesized pesticides that were soon found and marketed are not terpenoids, but some (methoprene and hydropene) resemble the juvenile hormone in structure.

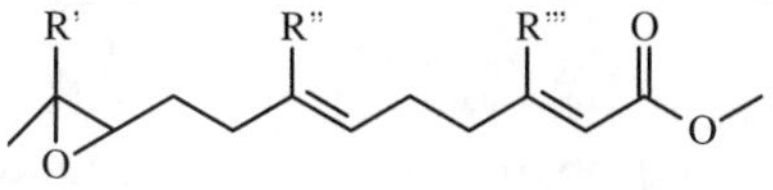

juvenile hormone

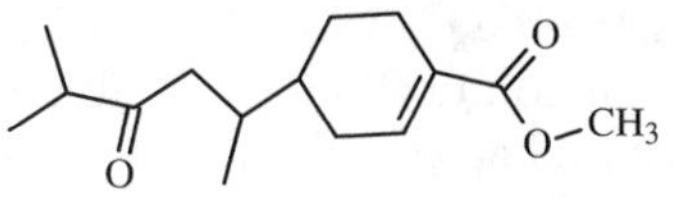

paper factor

R', R", and R''': Either ethyl or methyl groups, and there are four analogues
JH O: All three R groups are ethyl
JH I: R' and R" are ethyl groups while R''' is a methyl group
JH II: R' is an ethyl group, while R" and R''' are methyl groups
JH III: All three are methyl groups

7.1.2.2 Juvenile hormone agonists as pesticides

Methoprene is a juvenile hormone mimic preventing metamorphosis and is used to control a wide variety of insects, especially Diptera and Pharaoh's ant, but also beetles and various Homoptera. It has a very low mammalian toxicity and a low toxicity to adult insects like bees. Methoprene was the first synthetic pesticide marketed with JH activity and has a structure related to that of the juvenile hormones, whereas the newer pesticides such as fenoxycarb, pyriproxyfen, and diofenolan deviate more in structure from the true hormones:

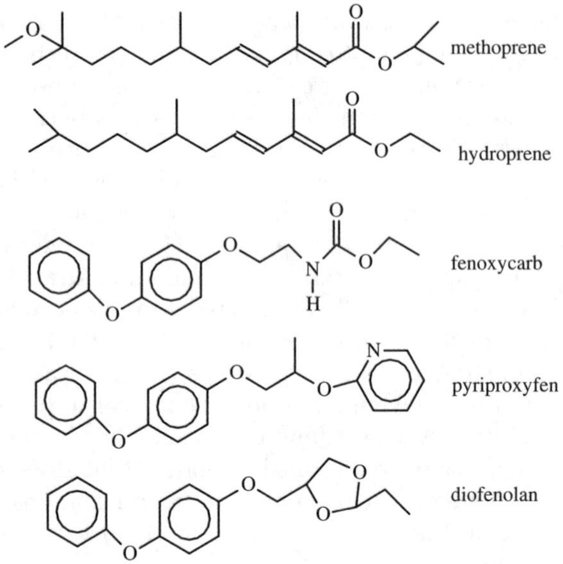

Ecdysone induces the larval–pupal metamorphosis in the absence of JH or a JH-active insecticide, but the presence of active compounds leads to a new larval stage at ecdysis, or to the development of larval–nymphal, larval–pupal, or larval–adult intermediates that are unable to give rise to normal adults. Treated pupae (e.g., tobacco cutworms treated with pyriproxyfen) may develop into normal adults. However, the females are unable to deposit eggs because a substance that induces oviposition behavior after mating is not released in the hemolymph. Other physiological and behavioral effects of JH-active insecticides are also observed.

The toxicity to mammals of the JH-active insecticides is very low. In the World Health Organization (WHO) classification system, they are all in class III (Table V). The rat oral LD50 (lethal dose in 50% of the population) is >5000 mg/kg, and the toxicity to fish and birds is also low. Indications for mutagenic, carcinogenic, or teratogenic effects are not found. To bee larvae, JH-active insecticides are, of course, rather toxic (e.g., 0.1 μg/bee for hydroprene), while for adult bees they are essentially nontoxic.

7.1.2.3   Antagonists

A priori there is good reason to believe that inhibitors of CYP enzymes may influence the synthesis or degradation of ecdysteroids and juvenile hormones, and thus interfere with the normal development of insects: this is the case. Piperonyl butoxide, a well-known CYP enzyme inhibitor that is widely used as a synergist for pyrethrins, retards the development of houseflies and other insects by a few days. The mechanism is probably an interference with the metabolism of JH and not JH activity.

A wide variety of fungicides act by inhibiting a CYP enzyme important to ergosterol synthesis. These fungicides have an imidazole ring structure or another ring structure with nitrogen, which binds to the iron atom of the CYP enzyme. It could therefore be possible to find some similar inhibitors acting on juvenile hormone synthesis. Eto and Kuwano (1992) describe how imidazoles that are structurally similar to JH inhibit JH synthesis and lead to premature pupation in silkworms. The last epoxidation step is probably the target for the action of imidazoles. It is interesting to note that methoprene counteracted the effect.

One of the most active imidazole derivatives with JH antagonist activity was 1-benzyl-5-[(E)-2,6-dimethyl-1,5-heptadienyl]imidazole:

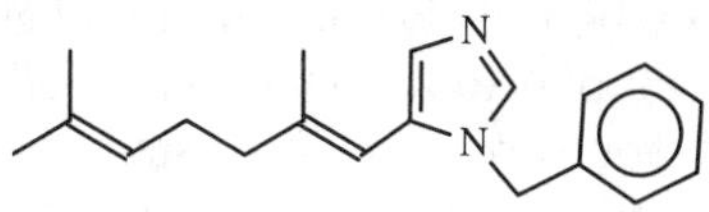

7.1.3   Ecdysone

An interesting peculiarity of insects is the requirement of sterols in the diet. Insects are not able to convert mevalonic acid to squalene, and the step from squalene epoxide to lanosterol seems absent. The absence of enzymes for transforming squalene epoxide into a cyclic structure may have evolutionary significance in the development of juvenile hormones by protecting the hormone molecule, which possesses a 2,3-epoxide linkage, from possible cyclization. Lanosterol, or many other sterols, can completely or partially substitute for cholesterol in the diet, but some species differences are observed in this ability. Plant sterols are often adequate substitutes for cholesterol, and many insects can rely on fungal symbionts in the gut to produce ergosterol. Besides being the precursor to ecdysone, the sterols are important constituents of surface wax of insect cuticles, lipoprotein carrier molecules, and subcellular membranes.

Cholesterol or other sterols available from the diet or gut microflora are converted to the active hormone 20-hydroxyecdysone (β-ecdysone or ecdysterone) through several oxidative steps involving CYP enzymes. The last step is oxidation of α-ecdysone to β-ecdysone, which can occur in many tissues, such as in the organ called the fat body.

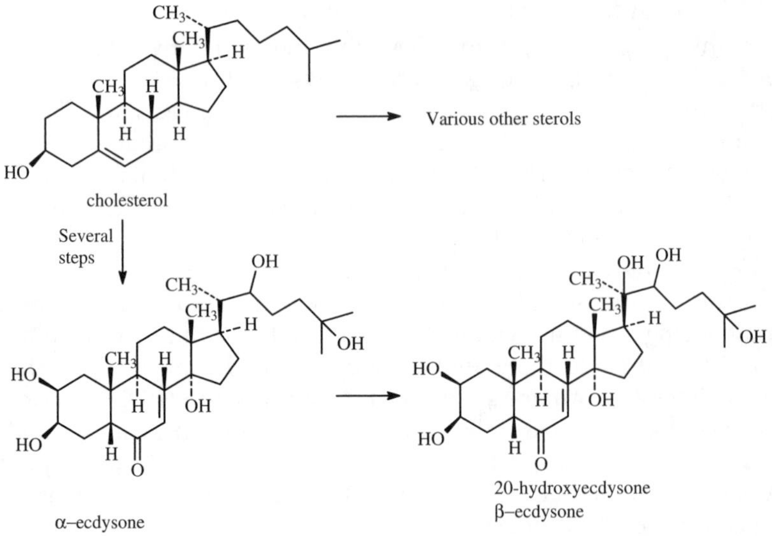

cholesterol → Various other sterols

Several steps ↓

α–ecdysone

20-hydroxyecdysone
β–ecdysone

Secretion of ecdysone starts the many biochemical processes that are necessary for the molting. The cells in the epidermis are stimulated to produce a new cuticle, and when ready, the insect will creep out of its old skin. The molecular mechanism of ecdysone has been studied in some detail. The molecular target of ecdysone and other ecdysteroids consists of at least two proteins, the ecdysteroid receptor (EcR) and ultraspiracle (USP). Both EcR and USP are members of the steroid hormone receptor superfamily with characteristic ligand-binding domains. An EcR–USP–ecdysteroid complex is formed, which activates several genes that code for transcription factors, i.e., proteins that activate or repress the activity of other genes, and the appropriate amounts of proteases and other enzymes necessary to degrade old structures and rebuild new ones are formed in a time-controlled sequence.

7.1.3.1 Phyto-ecdysones

Obviously, ecdysone or analogues that have similar or antagonistic modes of action could disturb the normal development of the insects and so can be used as insecticides. Inhibitors or inducers of the ecdysone biosynthesis may also be developed as insecticides.

It is not too surprising that many plants produce ecdysteroids to defend themselves against potential insect pests. The so-called phyto-ecdysones have a potent molt-inducing effect, and some of them may be even more potent than β-ecdysone. Gymnosperms and ferns often contain much of phyto-ecdysones. Bracken (*Pteridium aquilium*) and the rhizomes of the well-known liquorice fern (*Polypodium vulgare*), for instance, have up to 1% of their dry weight of β-ecdysone, and the rhizomes of liquorice can be used as a commercial source of β-ecdysones. A Siberian medicinal plant (*Leuzea carthamoides*) was recently studied by Zeleny et al. (1997). The plant has as much as 300 to 1000 ppm 20-hydroxyecdysone equivalents in the leaves, but in spite of this, it had a rich arthropod fauna, with a total of 126 species, 74 of which were feeding on the leaves. Thirty-three of these could complete their development on the plants without apparent difficulties. The insects and mites were from groups where the action of ecdysteroids is little known, such as spider mites, collembolans, thysanopterans, and procopterans, or from taxa with sucking mouthparts. Possible mechanisms of insensitivity may be a result of elimination of the ecdysteroids by excretion, or by the use of other substances as the endogenous hormone. Juvenile hormone-active compounds are also present in many other plants besides *Abies*.

7.1.3.2 Synthetic ecdysteroids used as insecticides

Many ecdysteroid agonists with structures different from those of the sterols have been found and are used as valuable insecticides, notably the diacyl-hydrazine derivatives, including tebufenozide, halofenozide, methoxy-fenozide, and RH-5849. RH-5849 is a code number for Rohm & Haas, a firm that spends much effort in developing such nontoxic pesticides. Tebufenozide and methoxyfenozide seem to be exclusively toxic to lepi-dopteran larvae and nontoxic to other insect species, whereas halofenozide is also active against Coleoptera larvae. These synthetic ecdysteroids are much more potent than β-ecdysone in inducing lethal molts in all larval stages in many Lepidoptera.

The toxicity of these substances to other animals, including man, is very low. They are not mutagenic and are regarded as safe.

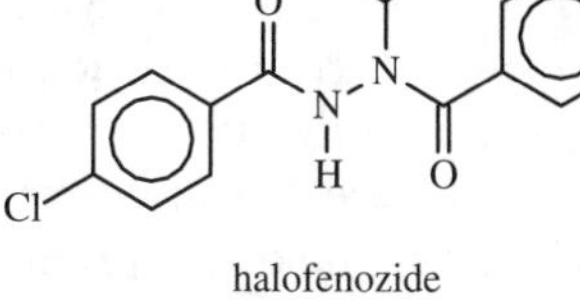

RH-5849

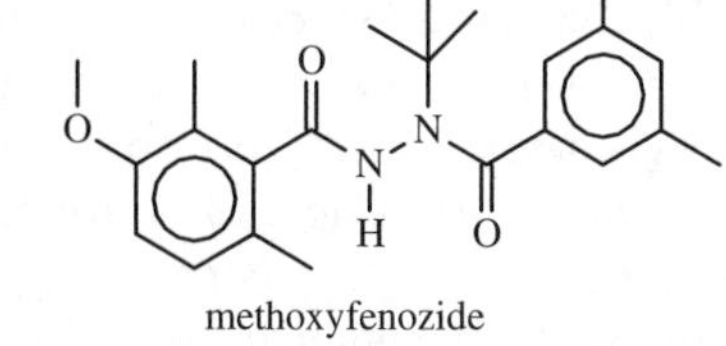

tebufenozide

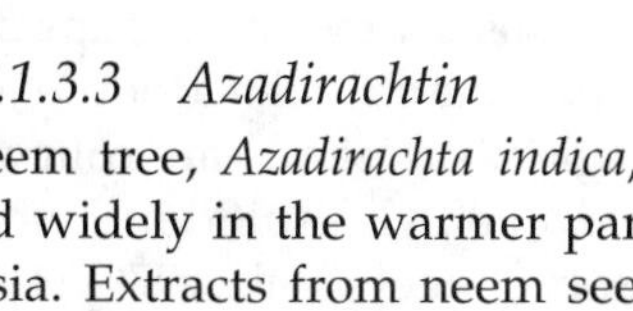

halofenozide

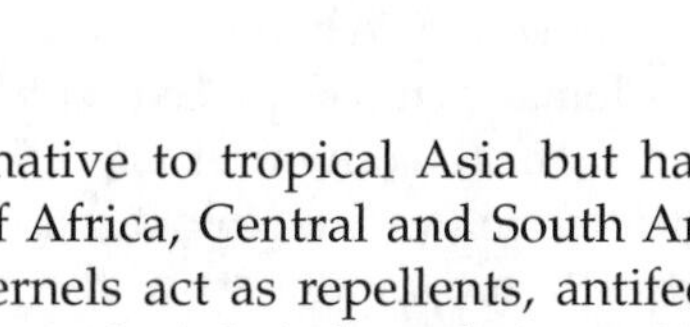

methoxyfenozide

7.1.3.3 Azadirachtin

The neem tree, *Azadirachta indica*, is native to tropical Asia but has been planted widely in the warmer parts of Africa, Central and South America, and Asia. Extracts from neem seed kernels act as repellents, antifeedants, and growth disruptants. The main active principle in kernels is azadirachtin (AZ), a limonoid with a very complicated structure. A range of other compounds is also present. These neem substances can repel insects, prevent settling, inhibit oviposition, inhibit or reduce food intake, interfere with the regulation of growth, and reduce the life span of adults. The antifeedant action of neem apparently has a nongustatory component because injection or topically applied neem derivatives can reduce feeding even though the mouthparts are not directly affected. The biochemical mode of action is not fully understood, but azadirachtin probably functions as an antiecdysteroid by blocking the binding sites of ecdysteroids. Molting inhibition or abnormal development is one of the more distinct symptoms at low concentrations. The extract also has fungicidal properties and the mammalian toxicity is very low.

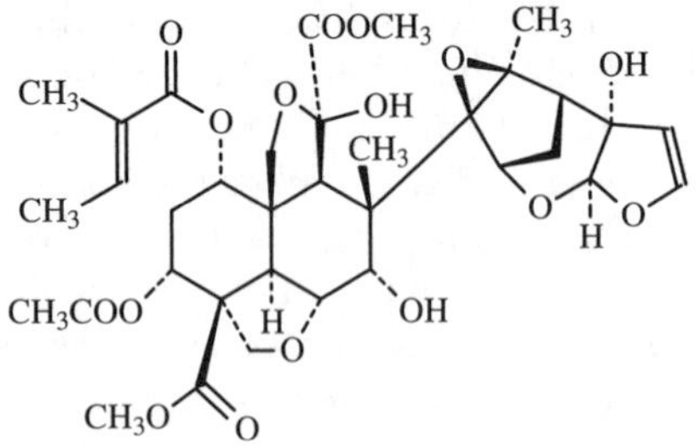

azadiracthin

7.2   Behavior-modifying pesticides

"An insect is like an old-fashion soldier: it does what it is ordered to do, and when it has no orders it does nothing" (a citation from Wright, 1963). The orders are very often given as chemical signals that induce certain behavior or change the course of development in the insect. Pheromones are the most important and the most active of such substances, but food contains substances that induce or suppress feeding, oviposition, etc. Houseflies, for instance, are very fond of certain amino acids (e.g., leucin) and nucleotides (e.g., GMP) that induce feeding behavior (Robbins et al., 1965). Mosquitoes are attracted by CO_2 — a reaction believed to be inhibited by N,N-diethyl-m-toluamide and some other compounds used as mosquito repellents. Methyl-iso-eugenol, a volatile chemical present in carrot leaves, induces oviposition in the carrot rust fly. Mammals also use pheromones. Male dogs are, for instance, sexually stimulated by the smell of methyl 4-hydroxybenzoate. When small amounts of this compound are sprayed on anestrous females, males placed with them become sexually aroused and try to mount them (Goodwin et al., 1979). Vaginal secretions are found to be highly attractive to male golden hamsters. Dimethyl disulfide is one of the active attractive components of the secretion (Singer et al., 1976). The fungicide thiram has been approved as a repellent that prevents browsing in orchards by deer, and may also be used in forestry against moose or mice (e.g., Christiansen, 1979; Nolte and Barnett, 2000). Hundreds of other such compounds have been found, and thousands still wait to be identified. Their potential in pest control is certainly not fully exploited. During the decades from 1960 to 1980 it was a serious belief that lures and repellents would more or less replace the other insecticides. Interesting reviews from the earlier period of pheromone research include Silverstein (1981, 1984), Wilson (1963), Wright (1963), and Karlson and Butenandt (1959). Rockstein's (1978) book *Biochemistry of Insects* has two informative chapters, "Chemical Control of Behaviour — Intraspecific" (pp. 359–389) by N. Weaver and "Chemical Control of Insects — Interspecific" (pp. 391–418), and a chapter about "Chemical Control of Insects by Pheromones" (pp. 419–464) by W.L. Roelfs. A book that gives details on recent developments is *Insect Pheromone Research* (Carde and Minks, 1997).

The growing public concern about the negative aspects of the uncritical use of insecticides is the driving force for much of the research. New analytical tools, like the gas–liquid chromatograph coupled to a mass spectrometer, developed by Professor Stina Stenhagen, University of Gothenburg, and Professor Ragnar Ryhage, Karolinska Institutet, Solna, Sweden, made it possible to detect and determine the structures of the extremely small amounts of substances present in scents. The pioneers Butenandt et al. (1959) had to use kilograms of silkworms in order to produce the minute amounts of the pheromone bombykol necessary to determine the chemical's identity.

Bombykol — the pheromone from female silkworms,
isolated by Butenandt et al. after 25 years of research

7.2.1 Definitions

Behavior-modifying compounds are divided into two broad categories — pheromones and allelochemicals. A pheromone is a substance secreted by an animal that influences the behavior or development of other animals of the same species. The term was suggested by Butenandt, Karlson, and Lüscher in 1959 (Butenandt et al., 1959; Karlson and Butenandt, 1959; Karlson and Lüscher, 1959). Allelochemicals act between different species and are often called kairomones, provided that the recipient has an advantage by sensing the substance. If the substance only benefits the emitter species, the allelochemical is called an allomone, and if both emitter and receiver are benefited, the allelochemical is known as a synomone. The scent from flowers that attracts bees may therefore be called synomone.

Pheromones are divided into primer and releaser pheromones. They are also classified according to what kind of reaction they induce in the receiver, e.g., sexual pheromones, aggregation pheromones, or alarm pheromones. Substances used in baits to attract animals are called *lures*. They may be pheromones, synthetic analogues of pheromones, or other attractants.

7.2.2   Pheromones

As explained, we make a distinction between primer pheromones that influence the physiology of the recipient and the releaser hormones that influence its behavior. Primer pheromones are used by social insects to suppress the production of sexual individuals, and by locusts to stimulate the production of migratory winged types when the population density becomes high. The releaser pheromones or analogues are most relevant as insecticides.

Releaser pheromones are quite common in many insect orders. Good examples are alarm pheromones released by angry bees that tell other bees of the same hive to attack an intruder. Attractants released by female insects of many species are among the most active biological compounds known. R.H. Wright (1964) wrote the following in one of his papers with the optimistic title "After Pesticides: What?":

> The nature of the olfactory process is such that a very few molecules can initiate a nerve impulse that sets off an amount of activity out of all proportion to the size of the original stimulus. Let me make a comparison. The largest hydrogen bomb releases energy equivalent to 100 megatons or 10^8 tons of TNT. I do not know exactly how much chemical explosive is needed to trigger the bomb, but let us suppose it is 200 lb., or evidently the magnification factor between the trigger and the full force is 10^9 or 1,000 million. Now consider that the sex attractant of American cockroach. If the reports are correct, 30 molecules of the pure chemical are enough to excite a male cockroach [referring to Jacobsen et al., *Science*, 139, 48, 1963]. The kinds of energy involved are difficult to determine, but let us consider the ordinary chemical energies. Thirty molecules of pure stuff weigh 10^{-20} g, and at room temperature their kinetic energy of translation is about 10^{-11} ergs (10^{-18} joule). The male insect weighs 2 g and is induced by the scent to run with a speed of 4 cm/sec, his kinetic energy is something more than 10^4 or ten thousand ergs. Thus the magnification factor — the disparity in energy between cause and effect — is 10^{15}, which is a million times greater than the disparity between the energy of the biggest hydrogen bomb and the trigger which sets it off.

The sexual pheromones are usually species specific or are only active when released together in a mixture. They are often more sophisticated chemicals than the aggregation or alarm pheromones.

7.2.3   Structure–activity relationships

Is it possible to say something about the structure and activity of pheromones?

Most of them are esters. They may also be alcohols, carboxylic acids, lactones, aldehydes, ketones, and hydrocarbons (Silverstein, 1984). Crucial properties of releaser pheromones are their volatility, their stability, and, of course, the degree of specificity possible to build into a relatively small molecule. The molecular weight tends to be between 80 and 300. On the basis of evaporation and diffusion rates, it can be predicted that long-distance sex pheromones would have a molecular weight between 200 and 300. A pheromone may be one chemical, but usually it is a mixture of chemicals, each of which is a component of the pheromone. Mixtures increase the specificity, a property important especially for the sex pheromones.

7.2.3.1 Alarm and trail pheromones
Most social insects use alarm and trail pheromones. Their specificity is not so important, but fast release and dispersal are important. They tend to be small volatile substances; for instance, (S)-4-methyl-3-heptanone is an alarm pheromone in the leaf-cutting ant, *Atta textata*.

Alarm pheromone in   Aggregation pheromone
leaf cutter ant   in elm bark beetle

Note that the natural pheromone is the S-enantiomer, and the (synthetic) R-enantiomer is inactive or has a very low activity. The parent alcohol, 4-methyl-3-heptanol, exists in four stereoisomers. Only one of these stereoisomers (3S, 4S) is used as an aggregation pheromone by the elm bark beetle (*Scotylutus multistriatus*), the beetle responsible as the vector of the Dutch elm disease fungus (*Ceratocystis ulmi*).

7.2.3.2 Aggregation pheromones
4-Methyl-3-heptanol is only part of the elm bark beetle pheromone. It is released together with α-multistriatin and α-cubebene. The latter compound is to be regarded as a kairomone released from the tree. The three substances work synergistically (see Silverstein, 1981 and references therein for further details).

Aggregation pheromones of spruce bark beetles, e.g., from the genus *Ips*, were widely studied in order to fight back their destructive actions in the forests of Canada and Scandinavia. They tend to be simple metabolites of terpenes, present in the resin of the trees they live on, but a rather complicated system of chemical communication has evolved. A great amount of research was necessary to clarify the complicated systems of chemical communication within and between the species, and some points of interest are described here. The bark beetle attack from the genus *Ips* starts with a few males. If the tree is suitable for them, the pioneer beetles secrete pheromones

Table 7.1 Pheromones Found in Three *Ips* Species

	(S)+ Ipsdienol	(R)- Ipsdienol	Ipsenol	cis- Verbenol	Methyl- Butenol[a]
Ips duplicatus		Present			
Ips acuminatus	Present		Present		
Ips typographus	Present			Present	Present

Note:　It may be interesting to note that 3-methyl-3-buten-1-ol may be present in bark beetle pheromones as well as in pheromones from mammals.

[a]　2-methyl-3-buten-2-ol　　3-methyl-3-buten-1-ol

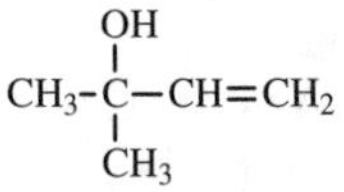

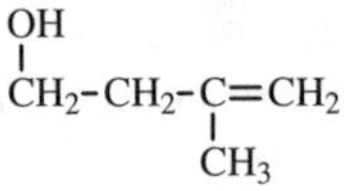

$$
\begin{array}{cc}
\text{OH} & \text{OH} \\
| & | \\
CH_3\text{-}C\text{-}CH=CH_2 & CH_2\text{-}CH_2\text{-}C=CH_2 \\
| & | \\
CH_3 & CH_3
\end{array}
$$

Source: Data from Bakke, A. 1977. *Naturwissenschaften*, 64, 98. Bakke, A. 1978. *Naturen*, 31–37.

that attract more individuals, which also start eating and secreting phero-mones. Their combined effort leads to the death of the tree, which then becomes an excellent substrate for the beetles and their larvae. From an evolutionary point, their pheromones are interesting because they are oxi-dation products of terpenes present in great amounts in the tree, and are simple biotransformation products of CYP enzymes. A quite straightforward biotransformation or detoxication product has thus become a pheromone (White et al., 1979). (Rat liver microsomes have CYP enzymes able to perform the same oxidation of α-pinene, but less efficiently.) The tree secretes resin as a protective response, but instead of being killed, the beetles use it to attract other beetles so that together they can kill the tree and render it suitable as a resource. The importance of the various oxidation products of natural terpenes such as α-pinene and myrcene is different among *Ips* species (see Table 7.1).

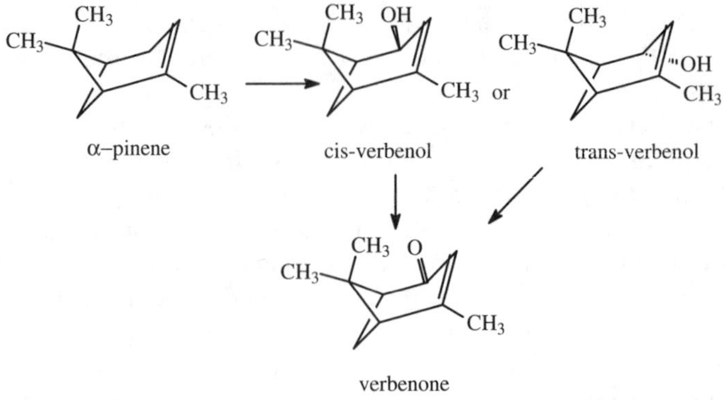

α–pinene　　　　cis-verbenol　　or　　trans-verbenol

verbenone

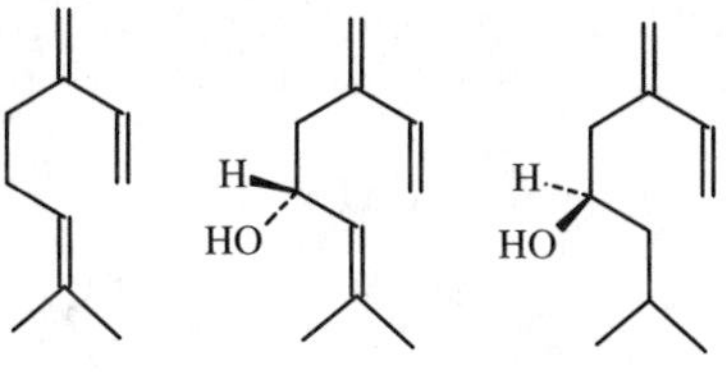

myrcene   (R)-ipsdienol   (S)-ipsenol

It may be of interest to note that 3-methyl-3-buten-1-ol also may be present in bark beetle pheromones as well as in pheromones from mammals.

Ips typographus is the most aggressive of beetles. It is very actively attracted if three pheromones are present. When *Ips typographus* has attacked a tree, *Ips duplicatus* may be attracted by the (R)-ipsdienol made by the other species. However, *Ips duplicatus* prefers the finer upper part of the tree and does not compete with *Ips typographus*, which prefers the coarser part. They thus cooperate to kill the tree.

7.2.4   Pheromones used as pesticides and lures

Aggregation and sexual pheromones, as well as food attractants and deterrents, can be used as pesticides. Their practical use can be as lures in traps combined with a poison, or they may be used to disturb the normal mating process when sprayed in great amounts. Males become confused and do not find the females for mating. Pheromones in traps are also useful for monitoring increase or decrease in population density. Active pheromones from the bark beetles described above were synthesized, and special traps were made and placed in the thousands in the forests of Scandinavia during the 1980s. Billions of beetles were caught, but the effect was not powerful enough to have a dramatic effect on the damage the beetles caused. On the other hand, the damage was formidable, and it was argued that even a small reduction of the damage would have paid back the effort and money put forth.

Twenty-two pheromones and kairomones and two synthetic analogues are now listed in *The Pesticide Manual* (Tomlin, 2000). Most of them are long aliphatic unsaturated compounds, with substituents containing oxygen, e.g., esters, alcohols, ketones, and aldehydes.

7.2.4.1   Coleoptera

Some esters of long-chain saturated fatty acids are aggregation pheromone beetles. The ethanol-ester of 4-methyloctanoic acid is, for instance, a commercialized aggregation pheromone (Oryctalure®) used to control the coconut rhinoceros beetle (*Oryctes rhinoceros*) (Morin et al., 1996). American palm weevil (*Rhynchphorus palmarum*) is attracted by the aggregation pheromone 6-methyl-2-hepten-4-ol. The pheromone has been tried for control of the weevil and is sold under the names Rhyncopherol® and Rhynkolure®.

Other chemically related long-chain branched alcohols are commercially available for other weevils.

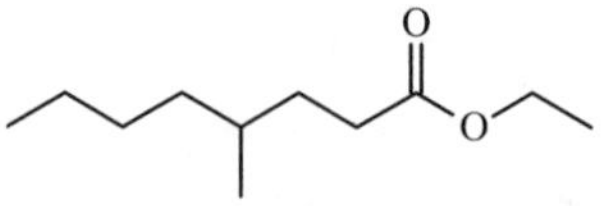

ethanol-ester of 4-methyloctanoic acid 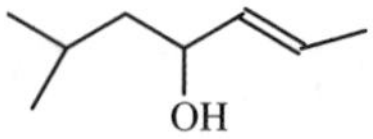   6-methyl-2-hepten-4-ol

7.2.4.2 Lepidoptera

Long-chain aliphatic aldehydes or esters, such as hexadecenal and hexade-cenyl acetate or chemical-related analogues, are mating pheromones in several species of Lepidoptera — either as single compounds or in mixtures. The natural mating hormone of rice stem borer (*Chilo suppressalis*) is, for instance, a mixture of (Z)-hexadec-11-enal, (Z)-hexadec-9-enal, and (Z)-octa-dec-13-enal in the proportion 50:5:6. This mixture is formulated in slow-release capsules for monitoring the population density or for mass trapping under the trade name Fersex ChS®. Dispensers for mating disruption may be placed in rice fields.

(Z)-hexadec-11-enal

A related substance with a triple bond, the acetyl-ester of hexadecenynol, is the sole female sex pheromone of the pine processionary moth (*Thaume-topoea pityocampa*) (Quero et al., 1997) and may be used for mass trapping.

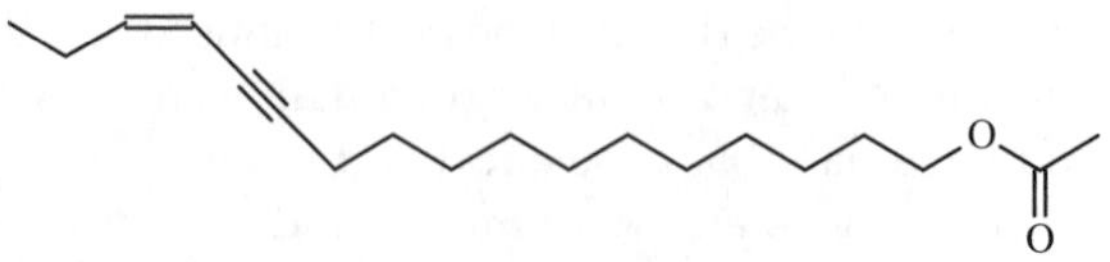

(Z)-hexadec-13-en-11-yn-1-yl acetate

It may be of interest to note that in order to give a response in the male moth, the ester must be hydrolyzed in its antennae. Some fluorinated ana-logues of the pheromone inhibit the esterase involved in the hydrolysis and may therefore block the response (Duran et al., 1993; Parrilla and Guerrero, 1994).

Pheromones are used to a large extent to monitor the population density of the gypsy moth (*Lymantria dispar*, *Porthetia dispar*), a forest pest in the U.S. introduced by accident. The story of the gypsy moth as a pest in America is interesting. The following is taken from the U.S. Department of Agriculture (USDA) website (http://www.fs.fed.us/ne/morgantown/4557/gmoth/, last modified on September 15, 1998):

The gypsy moth, *Lymantria dispar*, is one of North America's most devastating forest pests. The species originally evolved in Europe and Asia and has existed there for thousands of years. In either 1868 or 1869, the gypsy moth was accidentally introduced near Boston, MA by E. Leopold Trouvelot. About 10 years after this introduction, the first outbreaks began in Trouvelot's neighborhood and in 1890 the State and Federal Government began their attempts to eradicate the gypsy moth. These attempts ultimately failed and since that time, the range of gypsy moth has continued to spread. Every year, isolated populations are discovered beyond the contiguous range of the gypsy moth but these populations are eradicated or they disappear without intervention. It is inevitable that gypsy moth will continue to expand its range in the future. Over the last 20 years, several millions of acres of forestland have been aerially sprayed with pesticides in order to suppress outbreak gypsy moth populations.

For understandable reasons, Trouvelot lost his interest in entomology, but became a clever astronomist. Materials used to fight back gypsy moth include the chemical pesticide Dimilin®, an inhibitor of chitin synthesis. The biological pesticides, *Bacillus thuringiensis*, or naturally occurring gypsy moth virus is also quite useful.

The pheromone lure disparlure is a synthetic gypsy moth sex pheromone introduced in 1998 under the trade name Disrupt II GM:

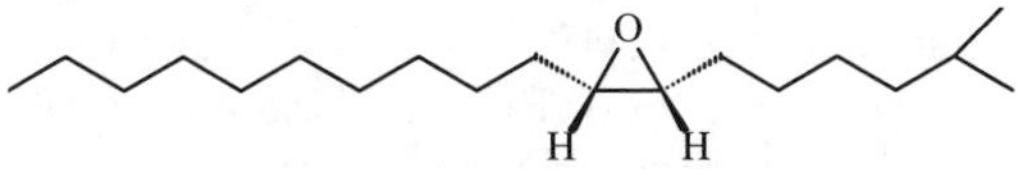

cis-disparlure

The cis-isomer that is shown is the natural sex pheromone of the gypsy moth. Disparlure is an aliphatic hydrocarbon (methyl-octadecan) with an epoxy group. It is used as an attractant and is formulated in plastic flakes. The racemic mixture may also be used.

7.2.4.3 Fruit flies

To control the Mediterranean fruit fly, a synthetic sex attractant pheromone mixture, trimedlure, is efficient. Chemically this lure is a mixture of substituted chlorinated cyclohexanecarboxyl-esters of tertiary butanol:

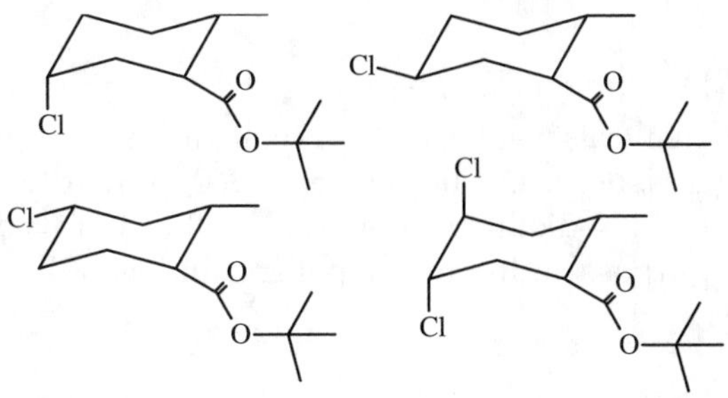

Methyl-eugenol and cue-lure are highly active commercialized synthetic kairomone lures to the oriental fruit fly, *Bactrocera dorsalis*, and the melon fly, *Bactrocera cucurbitae*, respectively (e.g., Vargas et al., 2000):

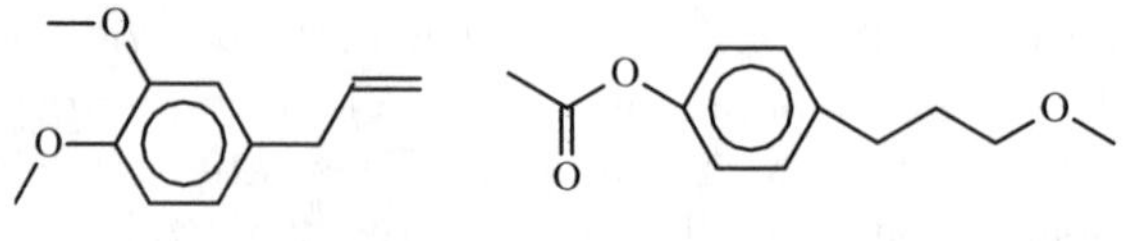

methyl-eugenol and cue-lure

Traps with these substances, together with an organophosphate, such as malathion, are efficient in the control of these flies in Hawaii.

7.2.4.4 Aphid food deterrent

Substances that remove the appetite from the insects may be efficient insecticides. A good example is pymetrozine, although its biochemical mode of action seems to still be unknown. This compound, produced by CIBA-Geigy (Novartis Crop Protection Corp.), is remarkable. Its selective action makes it very useful. The insecticide is selective against Homoptera, affecting their behavior and causing them to stop feeding before they die. It is rapidly degraded in soil and has very low toxicity for birds and other vertebrates, as well as for Daphnia and insects other than aphids (Flueckiger et al., 1992). Insects often have very exact demands about flavoring substances. As mentioned earlier, houseflies love GMP or leucine. Ticks use glutathione as a flavor or signal substance for food. Pymetrozine may probably mask some feeding stimulants for the aphids, as N,N'-diethyltoluamide masks the scent of CO_2 for mosquitoes.

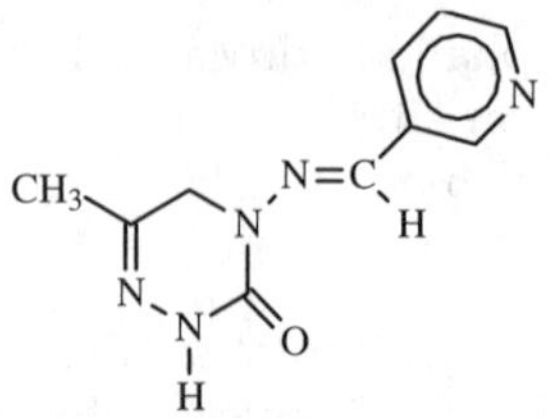

pymetrozine

7.2.4.5 Mosquito repellents

More than 9000 chemicals were tested as repellents for mosquitoes and other biting flies, chiggers, fleas, and ticks before 1960, primarily for military use (Smith et al., 1960). N,N-diethyl-m-toluamide (DEET) proved to be the most outstanding all-purpose individual repellent and is still the most useful substance today.

N,N'-diethyltoluamide

As a curiosity, DEET has been isolated from female pink bollworm and can therefore not be regarded as a xenobiotic (Jones and Jacobson, 1968). Many factors are involved in the host-seeking behavior of blood-sucking insects. Laboratory and field experiments have shown that CO_2, lactic acid, and other kairomones are natural cues for host location by biting flies. In Muscidae, CO_2 is initially involved as an activator before the insect becomes oriented to the source (see Nicolas and Sillans, 1989). Lactic acid and other substances, as well as heat, are also involved. Despite the widespread use of DEET in insect repellent products, nothing is known about the molecular basis for the mode of action of the repellent function of DEET (Reeder et al., 2001). Interestingly, the authors made mutant fruit flies that were not repelled by DEET. However, other repellents such as benzaldehyde and citrotellal were still active. The mutation is recessive and located on the X chromosome. It has been suggested that it blocks the CO_2 receptors, i.e., that it is not a real repellent but an antikairomone. Interference with lactic acid sensing has also been suggested (Kuthiala et al., 1992). Interestingly, DEET seems to have activity outside the biting insects, ticks, and mites. It was recently found to also be active against *Schistosoma mansoni*, the parasite worm that causes schistosomiasis (Secor et al., 1999). Another repellent, 1-(3-cylo-hexen-1-yl-carbonyl)-2-methylpiperidine, had a similar effect on preventing *S. mansoni* from infecting mice. Other substances, of different structures, such as phthalates, terpenes, etc., may also be active.

DEET is absorbed through the skin, and much research has been carried out to determine possible harmful effects on humans. In a recent trial, no harmful effect was found on survival, growth, or development at birth and on 1-year-old children whose mothers had used DEET quite extensively during pregnancy (McGready et al., 2001).

7.5   Plant hormones

In 1926 Frits Went, a student in the Netherlands, detected that the tips of wheat seedlings contained a substance that caused the seedlings to bend toward the light. The identity of the substance, which was given the name auxin from the Greek *auxein* (meaning "to increase"), was unknown. A few years later it was shown that auxin was indole-3-acetic acid (IAA). Today, four natural auxins are known. Besides IAA there are indole-3-butyric acid, phenylacetic acid, and 4-chloroindole-3-acetic acid. (It is quite interesting to find a chlorinated aromatic substance as a natural substance. It is synthesized from 4-chlorotryptophane in extracts from young seeds of *Vicia faba* (Fock et al., 1992).) The auxins regulate the balance between root and culm growth. Low concentrations stimulate growth, whereas higher concentrations inhibit the growth of the root. The inhibiting effect may be caused by a stimulation of synthesis of ethene, a hormone that inhibits root growth. Auxins are also involved in the regulation of bud sprouting.

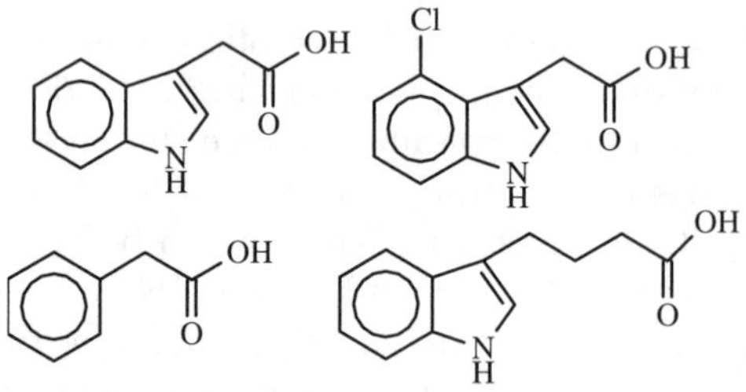

The natural auxins: indole-3-acetic acid,
4-chloroindole-3-acetic acid, phenylacetic acid,
and indole-3-butyric acid

In spite of the importance of auxin, the versatility of synthetic auxins as herbicides, and the long period they have been known, little is still known about how they work (see Leyser, 2002). However, new receptor proteins with very high affinity to indole-3-acetic acid have recently been found (e.g., Kim et al., 1998). Peterson (1967) has written an excellent review of the history of 2,4-(dichlorophenoxy) acetic acid (2,4-D), and the book by Bovey and Young (1980) describes most aspects of the development, use, and mis-use of 2,4-D, (2,4,5-T), and related herbicides, 2,4,5-(trichlorophenoxy) acetic acid such as (4-chloro-2-methylphenoxy) acetic acid (MCPA), as well as the dioxin problem. The current knowledge of the mode of action of auxin hormones is reviewed by Leyser (2002), whereas the book by Kearney and Kaufman (1975) is very instructive. Newer aspects are found in *Molecular Biology of Weed Control* (Gressel, 2002). During the Second World War different structural analogues of auxin were tested as possible herbicides. The intention was to develop herbicides that could be used to destroy the German wheat harvest, and Pokorny (1941) already described the synthesis of 2,4-D and 2,4-T. At low concentration these synthetic auxins stimulated growth, but at higher concentrations the plants died. Before the weed dies, we can observe a curiously distorted appearance of the plants.

The synthetic auxin analogues have a much stronger effect on broad-leaved plants than on grasses, which gives a high degree of valuable selec-tivity, but made them useless for their intended application: to destroy the wheat harvest in Germany.

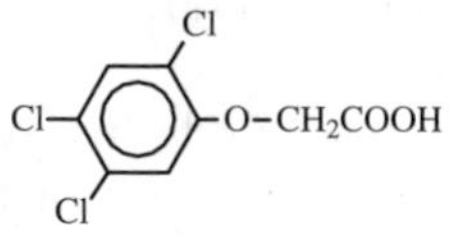

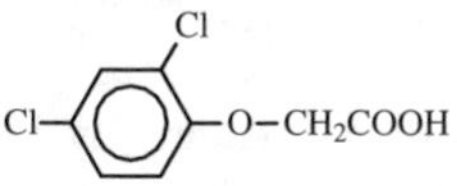

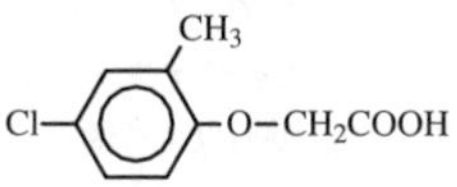

2,4,5-T   2,4-D   MCPA

As with the organophosphorus compounds, some structure–activity relationships were found, although this could not be related to affinity to specific binding sites on one receptor protein or enzyme. Newer research (e.g., Kim et al., 1998) has established that it is more than one receptor protein and each can have several binding sites for auxins. The best-characterized auxin-binding protein (ABP1), with $K_D = 5 \times 10^{-8} M$, was first described in maize in 1972 (Hertel et al., 1972). Simple test methods with wheat coleoptiles were developed soon after the first detection of auxin, and it was shown that auxins cause growth of sprouts and inhibit growth of root tips at very low concentrations (10^{-11} to $10^{-7} M$). The use of wheat coleoptiles may sound contradictory, but the insensitivity of wheat and other grasses is caused by low uptake and translocation to the sensitive sites and not by low sensitivity at the site of action. Synthetic auxins also have an intrinsic effect on grasses. The auxins affect cell division as well as cell elongation, and the antiauxins (competitive inhibitors to auxin) work in the opposite direction. Very often one of two enantiomers can have an effect as an auxin, and the other enantiomer can be an antiauxin. The antiauxins bind to the receptors in competition with the natural or synthetic auxins and thus block their action. Antiauxins may themselves have a weak auxin-like effect.

The simple test system with wheat coleoptiles made it possible to extensively study how chemical structure influences activity. It was found that an auxin (or antiauxin) had to have:

1. An acid group — a carboxyl, thiocarboxyl, sulfono, sulfate, phophono, or tetrazole group.
2. A ring structure with one or more double bonds, or another planar structure.
3. More often than not, one carbon atom between the acid group and the planar structure.
4. Isometry: 2-phenozypropinoic acids and others may have an asymmetric carbon atom — the D-enantiomer is often an auxin, whereas the L-enantiomer is an antiauxin.
5. Of the alkanoic acids with a phenoxy group in the 2 position, D-enantiomers of propanoic and butanoic acid have higher activities than phenoxy acetic acids. If the phenoxy group is further away, there must be a metabolic activation, as shown for 4-(4-chloro-2-methylphenoxybutanoic acid (MCPB).

6. The carbon atom closest to the carboxy group must have at least one hydrogen atom.
7. Ring substituents influence the activity. Unsubstituted phenoxyacetic acid has low activity. Introductions of halogens increase the activity in the following sequence: 4 > 3 > 2.

There is a close relationship between the effect of various substances as auxins and their herbicidal effects, with some easily explained exceptions. 4-Phenoxybutanoic acids (e.g., MCPB) have no auxin effect, yet they are efficient herbicides in many plants, providing they are metabolically transformed to the corresponding acetic acid. Although MCPA is selective with toxicity mainly against broadleaves, MCPB has an even more narrow selectivity because broadleaved plants, which do not metabolize it to MCPA, are not killed.

Metabolic activation of MCPB
that occurs in sensitive weeds

The Pesticide Manual describes 5 pyridine carboxylic acid derivatives, 3 benzoic acids, and 10 aryloxyalkanoic acids in current use as herbicides. 2,4,5-T seems to be banned worldwide and is not included in the current issue (Tomlin, 2000) but is described in earlier issues (e.g., Worthing, 1979). It was first registered in 1948 by Amchem Products, Inc., Pennsylvania, and the Dow Chemical Company, and was mainly used to control shrubs and trees, for example, in forests and along railways. Although the content of dioxin was eventually controlled, with a limit of <0.05 mg/kg in the preparation, it was banned soon after the termination of its use in Vietnam. Other products with 2,4,5-trichlorophenol were also banned or restricted after the Seveso accident (see Hay, 1978a, 1978b). 2,4-D is also produced from a chlorophenol and may contain dioxins, but of the much less toxic congeners. It is widely used as esters or amine salts for weed control in cereals. Its persistence in soil is approximately 1 month. MCPA is also formulated as salts or esters and has the same use as 2,4-D, whereas MCPB is more selective and can therefore be used in some cultures such as clover, peas, peanuts, and grassland. Mecoprop has an asymmetric carbon and the D-enantiomer, (R)-2-(4-chloro-o-tolyloxy)propionic acid, is the active isomer, sold under the standard name mecoprop-P. The other isomer is an antiauxin. Its applications are not very different from 2,4-D and MCPA.

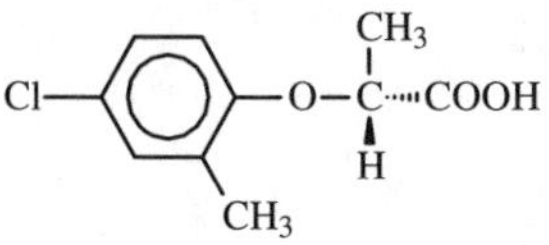

mecoprop-P
(active enantiomer)

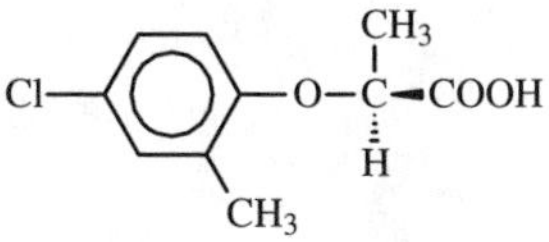

inactive enantiomer

Chloramben is a synthetic benzoic acid derivative acting as an auxin. It is selective because some plants (soya beans) detoxicate it by making stable N-glucosides of it. It has low adsorbtion in soil and may be active in the soil for several weeks. Picloram is an example of a pyridinecarboxylic acid, which has been widely used alone or mixed with other herbicides. It is taken up by the roots, has a long action time, and leakage from the site of application may occur. It was marketed since 1963 by the Dow Chemical Company.

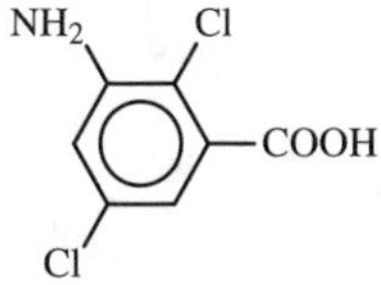

chloramben

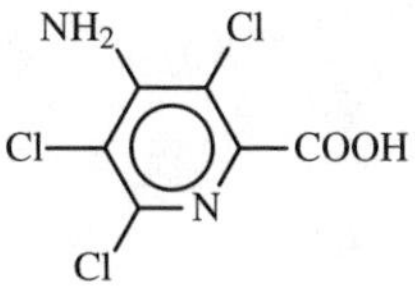

picloram

Some synthetic auxins are not used as herbicides, but are used for such purposes as preventing loss of unripe fruit, in plant cell cultures, and in root formation on cuttings. Naphthyl acetic acid is an example.

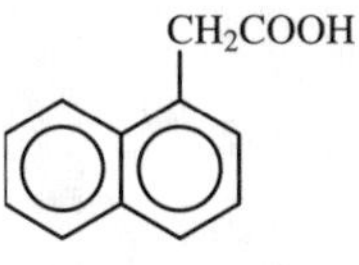

naphthyl acetic acid

chapter eight

Translocation and degradation of pesticides

8.1 The compartment model

For all systems, an animal, a piece of soil, a lake, or a landscape, we may use the same mathematical models to describe the uptake and elimination of a substance or contamination. The substance may enter the system in one event or at a more or less fixed rate (R — amount of substance per unit time). It distributes to different compartments in the system. A compartment is defined as a hypothetical volume of a system wherein a chemical acts homogeneously in transport and transformation (Hodgson et al., 1998). The substance disappears through excretion (animals), leakage (soil), or evaporation (from soil or water, or together with respiratory air), or the chemical is transformed to other substances through the influence of sunlight, the biotransformation enzymes of microorganisms, etc.

$$\text{Chemical} \xrightarrow{\text{k}} \text{Products}$$

The disappearance rate,

$$\left(-\frac{dC}{dt}\right)$$

by one process may nearly always be described by first-order or pseudo first-order kinetics; i.e., the rate is proportional with the concentration

$$-\frac{dC}{dt} = kC \text{ or integrated: } C = C_{START} \times e^{-kt}$$

where C is the concentration in the compartment, k is a constant, and t the time. The start concentration is C_{START}. An animal can often be described as

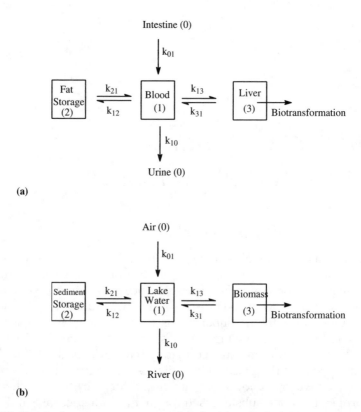

Figure 8.1 Simple three-compartment models for (a) oral administration to an animal and (b) a lake receiving pollution from aerial fallout.

a three-compartment system or sometimes, more appropriately, as a four-compartment system (Figure 8.1). One central compartment (blood) takes up a substance (from the intestine) at a certain rate. The substance is translocated to the liver, to fat, and other organs. Because the transfer back to the blood increases when the concentration in these peripheral compartments increases, a dynamic equilibrium will sooner or later be reached. The substance may be eliminated from the blood to the urine and by biotransformation in the liver. A lake can also be described as a three- or four-compartment model (see Figure 8.2).

The change of concentration in a compartment, e.g., blood or lake water, may be described by a differential equation,

$$\frac{dC_1}{dt} = k_{01}C_0 + k_{21}C_2 + K_{31}C_3 - k_{10}C_1 - k_{12}C_1 - k_{13}C_1$$

with concentrations and velocity constants. It is simpler to put up the equation than to do the integration because all the concentrations, with C_0 as a possible exception, change over time. Simpler models may be used in many cases.

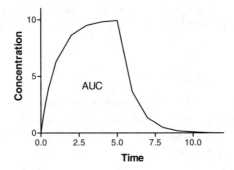

Figure 8.2 A situation with exposure during five time units, followed by an elimination time with no further uptake.

A fish swimming in a lake with uptake through the gills and elimination through just one biotransformation system, for instance, by an enzyme in the liver, may be regarded as a one-compartment system:

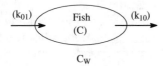

The concentration in the water (C_W) may be constant for some time and a steady-state equilibrium concentration in the fish (C_{MAX}) will be reached. The uptake rate (R), defined as change in fish concentration due to uptake, is proportional to the concentration in the water ($R = k_{01} \times C_W$) and is therefore approximately constant. The elimination due to liver metabolism or other first-order processes is proportional to the concentration in the fish. The total change of concentration by time will be the difference of uptake rates and elimination rates. In our simple case,

$$\frac{dC}{dt} = k_{01} \times C_w - k_{10} \times C$$

where C is the concentration in the fish:

$$\int_0^{C_t} \frac{dC}{k_{01} \times C_w - k_{10} \times C} = \int_0^t dt$$

By integration and rearranging, remembering that the term $k_{01} \times C_W$ is constant, we get

$$C_t = \frac{k_{01} \times C_W}{k_{10}} \left(1 - e^{-k_{10}t}\right)$$

where C_t is the concentration in the fish at a specified time. The exponential term approaches zero, and C_t becomes constant (C_∞), but is proportional with C_W.

8.1.1 The bioconcentration factor

The exponential term (e^{-kt}) will approach zero and the concentration will reach a level (C_∞) where uptake rate and elimination rate are the same; when this happens, there is equilibrium. This is the philosophy behind introducing the term *bioconcentration factor* (BCF):

$$BCF = \frac{\text{Concentration in fish}}{\text{Concentration in water}} = \frac{C_\infty}{C_W} = \frac{k_{01}}{k_{10}}$$

For lipophilic substances this factor can be quite high, but theoretically, there will always be an equilibrium concentration, where no net uptake takes place. The factor is quite versatile as a simple parameter that describes the tendency of a substance to accumulate.

Organic chemicals are often much more soluble in organic solvents and fats than in water and are said to be lipophilic. The BCF and also to a great extent the binding to soil are dependent on the lipophilic nature of the compound. In principle, this is simple to measure experimentally by shaking a small amount of the substance in a separating funnel with n-octanol and water. The two solvents separate into two phases, and the substance distributes between them. The distribution constant at equilibrium (KOW) is defined as

$$KOW = \frac{\text{Concentration in octanol}}{\text{Concentration in water}}$$

Chromatographic methods using separation columns in which substances separate according to their lipophilicity are often used to determine the KOW. The retention times of the substances are compared to known standards.

8.1.2 The half-life

The first-order disappearance rate is also the theoretical basis for the half-life $(t_{1/2})$ concept given by some simple calculation:

$$C = C_{START} \times e^{-kt} \quad \text{or} \quad \frac{C}{C_{START}} = e^{-kt}$$

where

$$C = 0.5 \times C_{START} \text{ then } \ln(0.5) = -kt_{1/2} \text{ and } t_{1/2} = \frac{0.69}{k}$$

The half-life is therefore independent of the initial concentration (C_{START}). When multicompartment models have to be used, the half-life is not independent of the start concentration, but is still a useful parameter. Toxins may have extremely different half-lives. Dioxin (2,3,7,8-TCDD) and DDT have half-lives of several years in the human body, whereas the hydroxyl radical probably has a half-life of less than a microsecond.

8.1.3 The area under the curve

The toxic effect is often a function of the concentration of the toxicant multiplied by time of exposure, or the concentration in a tissue multiplied by the time. The integral of the concentration–time function is called AUC, the acronym for the area under the curve (Figure 8.2). AUC is easily determined by measuring the area under the concentration vs. time, either by mathematical integration if the function is known or by some more pragmatic methods. AUC is useful to determine the uptake or bioavailability of substances. AUC in blood can be determined after intravenous injection and compared with the AUC after oral administration.

$$\frac{AUC(\text{oral administration})}{AUC(\text{injection})} = \text{Bioavailability}$$

The uptake rate is R = 10 and k_{01} = 1; i.e., the concentration is

$$C_t = \frac{10}{1}\left(1 - e^{-1t}\right)$$

for the first five time units, and then $C_t = e^{-1(t-5)}$ for the last five time units. AUC is the area under this curve.

Very often, and always when we use a two-compartment model, the disappearance is better described by a function with two exponential terms (e.g., $C = Ae^{-k't} + Be^{-k''t}$)

8.1.4 Example

8.1.4.1 Disappearance of dieldrin in sheep

The disappearance of organochlorines in mammals very often follows a two-compartment model. The following may apply to dieldrin in a sheep:

$$C = 0.054 \times e^{-0.54t} + 0.030 \times e^{-0.051t}$$

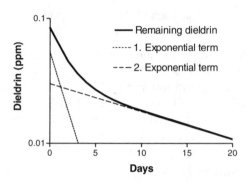

Figure 8.3 Disappearance of dieldrin (in ppm) from blood in sheep during the first 20 days after administration.

Figure 8.3 shows the disappearance of dieldrin from blood during the first 20 days after administration. When plotted in a semilogarithmic diagram, the two branches of the curve are seen and may be resolved into two straight lines. Dieldrin that has been taken up through the food disappears, for instance, via urinary excretion and via metabolism in the liver.

8.1.4.2 Dieldrin uptake in sheep

Uptake may also be described by two-compartment models: $C = C_{MAX} - A \times e^{-k'xt} - B \times e^{-k''t}$. Figure 8.4 shows that this is the case for sheep and dieldrin, which can be described by

$$C = 700 - 230 \times e^{-0.4xt} - 470 \times e^{-0.0077xt}$$

8.2 Degradation of pesticides by microorganisms

The majority of pesticides eventually find their way into soil and aquatic environments, where attack from microorganisms is an important mechanism of degradation. Low-level environmental contamination as a result of the normal use of the pesticides, as well as large-scale accidental spillage and the illegal disposal of unused heeltaps or outdated products, will reach soil and water sooner or later. An extensive and valuable book is *Pesticide Microbiology*, which covers most of the subjects described here (Hill and Wright, 1978). Most of the examples and concepts described here are taken from the book, and a few additional references are provided.

8.2.1 Degradation by adaption

A well-known postulate says that under the right condition there will always be one or more microorganisms that are able to degrade any organic compound. But, unlike an animal, a single microorganism does not have enzymes that are specialized in degrading lipophilic secondary metabolites of plants

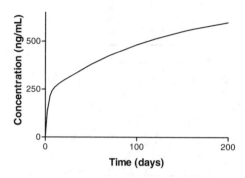

Figure 8.4 Increase of dieldrin concentration in blood of sheep on a diet containing dieldrin, resulting in an exposure of 2 mg of dieldrin/kg of body weight/day.

and xenobiotics. However, the enormous number of species, the high frequency of mutations per unit time in reproductive cells, and the high selection pressure that may occur in a population of microorganisms will sooner or later create an organism that may degrade the compound. This new biotype will increase in number if it has an advantage over the other microorganisms present. Simple Darwinian selection of organisms that can utilize the substance will cause an increase in their number. The rate of degradation will therefore increase in time. Genes that code for degradation enzymes are often located in plastids, and so-called horizontal gene transfer can occur. This mechanism also speeds up the evolution of microflora that can degrade a particular compound. A paper on the degradation of the aromatic hydrocarbon toluene (Roch and Alexander, 1997) illustrates the typical progress of degradation by adaptation. A low concentration of toluene is not degraded.

8.2.2 Degradation by co-metabolism

A substance can also be degraded by co-metabolism. In this case no biotype gains any particular advantage by being able to degrade a substance, but the substance is degraded because some of the thousands of enzymes present in the microflora may use it as a substrate. Such degradation can go on rather slowly. Many chlorinated hydrocarbons can have a half-life of many years in the soil, and even some natural organic substances (e.g., humic acids) can take thousands of years to be degraded.

8.2.3 Kinetics of degradation

Degradation by co-metabolism starts immediately and follows a first-order kinetic progress. A plot of log concentration against time will follow a straight line, whereas degradation when adaptation must first occur follows a somewhat more complicated pattern. There will be a lag period with slow degradation, followed by a more or less logarithmic phase. At very low

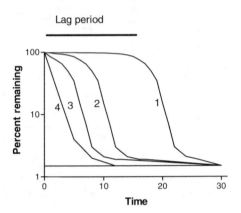

Figure 8.5 Disappearance of a substance when there is an adaption phase (1) with a long lag period, for instance, when MCPA or another herbicide that can support some microorganisms is applied for the first time. After having been used for several years, the degradation starts almost immediately (2 and 3). A small residual amount (e.g., 1.5%) is not degraded, due to strong binding to the soil or because the concentration is too low to be of interest to the microorganisms. Co-metabolism or complete adaption is shown by (4).

concentration the degradation may slow down. The remaining residue may be too strongly adsorbed to soil particles and it does not pay for the microorganisms to concentrate and degrade it (see Figure 8.5).

8.2.4   Importance of chemical structure for degradation

It is not necessary to be a microbiologist or chemist to get an idea about the degradability of a substance just by looking at the chemical structure.

- Chemicals that are likely to be strongly adsorbed to soil and sediment particles will have a reduced microbial degradation. Therefore, polar, water-soluble substances degrade faster than nonpolar, insoluble substances. Anionic substances are more easily degraded than cationic ones because positive ions are adsorbed strongly to the soil particles. Good examples of chemicals that are strongly adsorbed in the soil are DDT, dioxin, and paraquat, whereas trichloroacetic acid, malathion, and dalapon are not absorbed and are therefore more easily degraded.
- Aliphatic molecules, or aliphatic parts of molecules, are degraded faster than aromatic ones. Toluene is attacked at the aliphatic part.
- Esters are likely to be hydrolyzed. Examples are malathion and pyrethroids. Ester bonds between polar groups hydrolyze easier than bonds between nonpolar groups. Microorganisms, like animals, have unspecific carboxylesterases that may facilitate hydrolysis.

- Compounds with a high oxidation state such as those with a lot of chlorine are recalcitrant to further oxidation. These compounds must therefore be degraded anaerobically. Chlorine is substituted by hydrogen or HCl is removed and a double bond is introduced, for example, DDT that may be dechlorinated to 4,4′-dichlorodiphenyl-dichloroethane (DDD) by anaerobic processes or slowly converted to 4,4′-dichlorodiphenyldichloroethylene (DDE) (Stenersen, 1965). Compounds such as mirex and hexachlorobenzene are extremely recalcitrant to degradation, and microorganisms do not attack the highly fluorinated or chlorinated polymers, such as Teflon and PVC.
- The pattern of substitution of aromatic compounds strongly influences the degradation rate. The degradation of the 12 different chlorobenzenes is dependent on where the chlorines are placed. If there are two hydrogen atoms in vicinal positions, the degradation is much faster because oxygen is added as an epoxide bridge.
- Substances that are highly toxic to microorganisms are not easily degraded. Such compounds may also delay degradation of other compounds in the same mixture. Examples are pentachlorophenol and some corrosion inhibitors, which are very toxic to microorganisms.

Some rather obvious environmental factors should also be borne in mind:

- High temperatures increase the degradation rate because the substances become more soluble and adsorb less to the soil colloids and will be more available for the microorganisms, and because the number of and metabolic activity of microorganisms increase.
- Moisture strongly influences degradation. Substances that need anaerobic conditions will be more easily degraded at very high soil moisture because increased water combined with microbial activity will remove oxygen. An intermediate amount of moisture will stimulate aerobic microbial growth.
- Rich soil, with high microbial activity, will usually increase the degradation due to co-metabolism.
- A higher pH seems to be favorable for degradation.

8.2.5   Examples

8.2.5.1   Co-metabolism and adaptation

Good examples of co-metabolism are polychlorinated biphenyls (PCBs), which have been studied quite extensively. PCBs with few chlorine atoms may be degraded through co-metabolism by organisms that can live on biphenyl. By adding biphenyl to the soil, the organisms that can use biphenyl as a nutrient will increase in number. The degradation rate of PCBs with few chlorine atoms will also increase. The chlorinated analogues to biphenyl cannot support growth, but are degraded by the same enzymes as biphenyl (see Quensen et al., 1998a, 1998b).

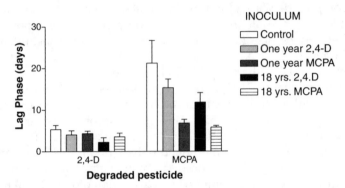

Figure 8.6 The first five columns show the lag period in the degradation of 2,4-D with inoculums of control soil, and soils from 2,4-D- and MCPA-treated soil, respectively. There is a significantly shorter lag period in soils from lots treated for 18 years with 2,4-D. Pretreatment with MCPA has a dramatic influence on the degradation of MCPA, as the second set of columns illustrate. Inoculums with soil treated for 18 years with MCPA gave a much shorter lag period than when untreated soil was used. Treatment with 2,4-D reduced the lag period.

Highly chlorinated PCBs and pesticides (e.g., mirex and DDT) may be dechlorinated anaerobically by functioning as an electron acceptor. DDT is dechlorinated to DDD by many facultative anaerobic microorganisms under anoxic conditions. DDD can be further dechlorinated or degraded aerobically.

Reaction 1 is mainly due to facultative bacteria grown anaerobically, whereas reaction 2 occurs often in various animals, notably some DDT-resistant flies. Alkali UV light and some metal salts also catalyze dehydrochlorination.

Torstenson et al. (1975) published a nice example of adaptation of microflora to (2-methyl-4-chlorophenoxy) acetic acid (MCPA) and (2,4-dichlorophenoxy) acetic acid (2,4-D). Soils from lots that had been treated with herbicides for 18, 1, and 0 (controls) years were used to inoculate a salt medium where 2,4-D or MCPA had been added as the carbon source (100 μM). The lag time before the degradation started was very much influenced by the type of inoculate, as shown by Figure 8.6 with data from Torstensson's work.

8.2.5.2   Parathion and other pesticides with nitro groups

Parathion provides an excellent example for describing important differences in transformation between microorganisms, animals, and sunlight:

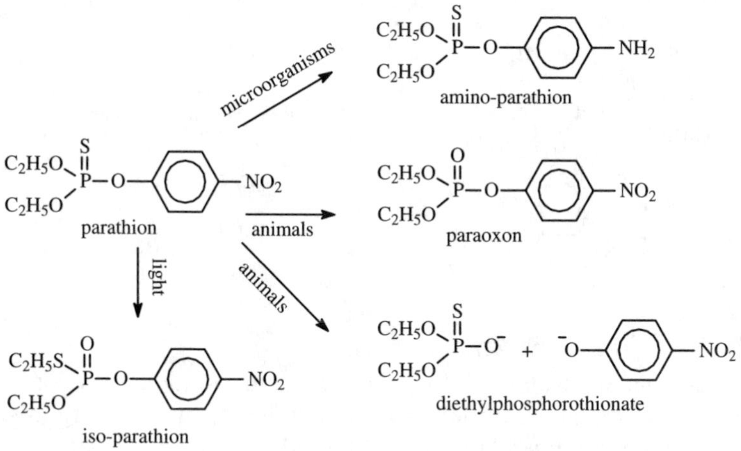

Paraoxon is the toxic metabolite produced in animals, whereas diethyl-phosphorothionate is a detoxication product. Microorganisms may, under anaerobic conditions, produce amino-parathion, which has a much lower animal toxicity than parathion. Therefore, parathion by oral administration is less toxic to ruminants than to other mammals. The highly active rumen flora detoxicates parathion by reducing it to amino-parathion. However, parathion deposited on leaves or dust particles can absorb light energy and be isomerized to iso-parathion, which has a high animal toxicity.

8.2.5.3   Ester hydrolysis of carbaryl

Microorganisms hydrolyze carbaryl to 1-naphthol, whereas in mammals different oxidation products are formed.

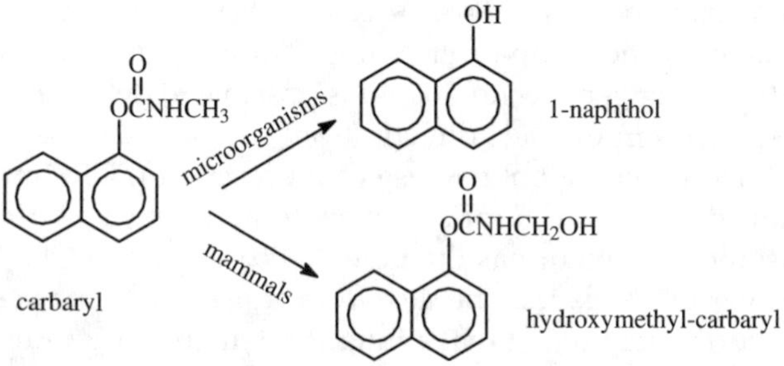

Ester hydrolysis is the most important microbial degradation of diazinon, whereas mammals produce glutathione conjugates and oxidation products.

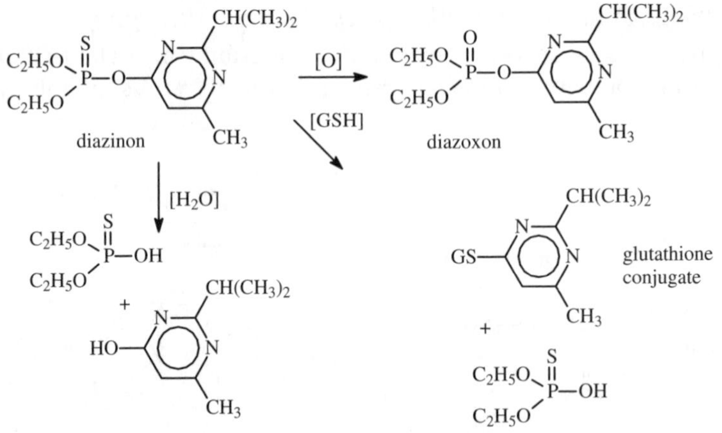

8.2.5.4   Mineralization of dalapon

Dalapon is a small, water-soluble aliphatic compound that is completely degraded to inorganic compounds (is mineralized) by microorganisms in soil, probably after some adaptation of the microflora.

$$CH_3CCl_2COOH \rightarrow CO_2 + HCl$$

dalapon

Trichloroacetic acid (CCl_3COOH) is also mineralized, but at a slower rate. Note that such compounds may contaminate groundwater. The number of microorganisms in groundwater is low, and thus after-leakage degradation will be much lower.

8.2.6   The degraders

The most important pesticide degraders in soil are within the genuses *Alcaligenes, Arthrobacter, Aspergillus, Bacillus, Corynebacterium, Flavobacterium, Fusarium, Nocardia, Penicillium, Pseudomonas*, and *Trichoderma*. Of great interest are strains of *Alcaligenes* and *Pseudomonas* that are very good at degrading PCBs. They have a gene complex encoding four enzymes necessary for the degradation. By using gene technology it is possible to detect extremely low levels of these genes in extracts directly from soil and sediment. Extracted DNA is amplified by using polymerase chain reaction (PCR), and the presence of organisms with appropriate gene complexes can be detected. If present, the soil or sediment has the potential to degrade PCBs (Hoostal et al., 2002). Similar methods may be developed for recalcitrant pesticides.

It is also worth mentioning that a fungus often grown commercially and sold as a delicacy, the lignin-degrading white rot fungus (*Phanerochaete chrysosporium*), is an exceptionally efficient degrader of recalcitrant contaminations

(PCBs, dioxin, lindane, DDT, etc.) because of its ability to produce hydroxyl radicals. Oyster mushroom farmers may therefore have a potential by-product in used growth substrate, which can be mixed into contaminated soil to aid degradation of the contaminant.

8.3   Soil adsorption

Adsorption is very important for the biological properties of the chemicals. Many soil pesticides may be applied in higher quantities when the soil has strong adsorption properties. Adsorption inactivates and makes toxicants less harmful and reduces leakage, but on the other hand, it can make the pesticides more recalcitrant to microbial degradation. The adsorption process is quite fast, and often less than an hour is needed to produce equilibrium. The opposite process, desorption, takes longer and sometimes a low residue is bound irreversibly.

8.3.1   Why are chemicals adsorbed?

The adsorption of soluble substances to solids is a surface phenomenon. A solute does not distribute evenly in the liquid phase. The more lipophilic substances will have a much higher concentration at the liquid surface. This is very characteristic for detergents such as long-chain fatty acids. The molecules orient the carboxyl group into the solvent and the hydrophobic hydrocarbon outward. The solute concentration will be much higher in the surface layer than in the bulk solution. Nonpolar chemicals with higher solubility in organic solvents than in water will also have a higher concentration in the surface layer. Electrolytes, on the other hand, will have higher concentrations within the water solution. Good adsorbents like humus, clay, and active carbon have an extremely large surface area, which of course implies that the surface of the water in contact with the adsorbent will also be extremely large. Dissolved hydrophobic substances will therefore have plenty of space at the surface. Different weak chemical bonds contribute to keep the substances at the boundary between the liquid and solid phases.

The bonds may be electrostatic forces. Positive charges are kept by the negative charges in the soil matrix. Charge-transfer complexes may be formed, and van der Waal's bonds are important for keeping the chemicals at the interface. Humus is the most important soil fraction for binding, and clay particles may also contribute. Sandy soils without humus and clay do not contribute to the binding. There is a good correlation between the KOW values (see p. 164 for definition) and adsorption.

8.3.2   Examples

The triazine herbicides are useful to illustrate that higher water solubility does not necessarily lead to less binding. Simazin, atrazin, and propazin are

Table 8.1 Summary of Some of the Properties of a Selection of Triazine Herbicides

Pesticide	Simazine	Propazine	Atrazine	Prometon	Ametryn	Prometryn
Solubility (ppm)	6.2	5	33	750	200	33
Acidity (pK$_a$)	1.7	1.7	1.7	4.3	4.1	4.1
Adsorption (KOC)	160	152	172	300	380	400

Source: Data are taken from Tomlin, C., Ed. 2000. *The Pesticide Manual: A World Compendium.* British Crop Protection Council, Farnham, Surrey. 1250 pp.

used as soil herbicides. They have very low water solubility and are therefore translocated very slowly in the soil. They act in the topsoil, but they are not kept tightly to the soil colloids and are easily taken up by the roots. Terbutryn, ametryn, and prometryn have much higher water solubility, but are more strongly adsorbed because of their base character — the molecules tend to pick up a H$^+$ ion and get a positive charge. They can take up H$^+$ and therefore tend to have a positive charge. On sandy soils these substances are taken up by the plants' root systems, but they bind more tightly in humus-rich soil. Table 8.1 shows some of the properties of triazine herbicides. They are mostly used as leaf herbicides. Atrazine has less base character, has a higher water solubility than simazine, and penetrates farther down into the soil. Plants with deeper roots will therefore take it up and be killed. Plants, animals, and microorganisms easily degrade atrazine. Nevertheless, it may penetrate to the subsoil water and cause problems where groundwater is used in waterworks for which a zero tolerance of pesticide residues in potable water has been enforced.

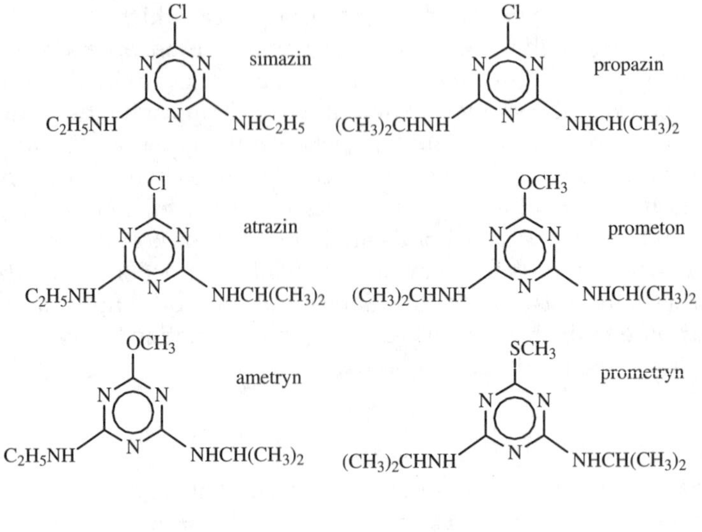

Comparison of the structures of some triazine herbicides

Ametryn reacts as a base (B),

$$B + H^+ \xrightleftharpoons{K_a} BH^+$$

$$K_a = \frac{[B][H^+]}{[BH^+]}$$

$$pk_a = -\log K_a$$

with $pK_a = 4.1$, and if pH is not very high, some of the molecules will become positively charged by binding H^+ ions. By applying the equation of Henderson–Hasselbach (see textbooks in chemistry), assuming a soil pH of 5.1, we get

$$\log \frac{[B]}{[BH^+]} = pH - pK_a = 5.1 - 4.1 = 1$$

which implies that the concentration of charged ametryn $[BH^+]$ is one tenth of the uncharged, and therefore constitutes 9.1% of the total ametryn dissolved in soil water at equilibrium. The charged form will be bound to negatively charged soil particles and more ametryn will be protonated to restore the equilibrium, which will then be bound until the final equilibrium is established.

8.3.2.1   Measurements of adsorption

Because adsorption is an important inactivation mechanism and determines leakage and biological activity, we need methods to describe adsorption properties in quantitative terms. One part of the substance is adsorbed to soil, and another is freely dissolved in the soil water. It is believed that these two compartments are in equilibrium. If the soil water (or total) concentration increases, the adsorbed amount will increase in a function bending downward.

Two functions, Freundlich's and Langmuir's adsorption isotherms, are used to describe adsorption (see Figure 8.7). The former is more often used:

$$\frac{x}{m} = k \cdot C^n$$

or

$$\log(x/m) = \log(k) + n \cdot \log(C)$$

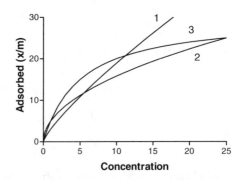

Figure 8.7 Two adsorption isotherms according to Freundlich's formula with (1) k = 3 and n = 0.8 and (2) k = 5 and n = 0.3. A Langmuir's isotherm is shown by line 3, with a = 0.2 and b = 30. The scales have arbitrary units. Experiments must be done in order to find the best relation.

where x is the amount of chemical adsorbed to m weight units of soil. C is the concentration, whereas n and k are constants that describe the relationship. The constant n is ≤1 and is often close to 1 for low concentrations. Note that sometimes the exponent n is replaced by a reciprocal value (1/n) in the formula.

Langmuir's adsorption isotherm may sometimes be more appropriate. It was developed to describe the adsorption of gases to solid phases and has a better theoretical foundation.

$$\frac{x}{m} = \frac{ab \cdot C}{1 + a \cdot C}$$

In this equation a and b are constants. At low concentrations the adsorption (x/m) is proportional to the concentration ($x/m \approx abC$) because $1 + aC \approx 1$, whereas at high concentrations the adsorption is independent of a further increase of concentration ($x/m \approx b$) because $1 + aC \approx aC$.

Very often adsorption is dependent on the humus content, which is proportional to the content of organic carbon. A constant called KOC is therefore very useful and more independent of soil type than the distribution coefficient (K_d). These parameters are defined by the equations

$$K_d = \frac{\text{Adsorbed on soil } (mg/kg)}{\text{Concentration in soil water } (mg/kg)} \qquad KOC = K_d \cdot \frac{100}{\%C}$$

KOC is determined by dividing K_d by the relative amount of total carbon. The distribution coefficient may be determined by making a water solution of low concentration, adding some soil, shaking for 3 to 5 hours, centrifuging, and analyzing the solid and liquid phases. Table 8.2 shows some of the properties of a selection of pesticides.

Table 8.2 Summary of Some Important Properties that Influence Pesticide Behavior in Soil

Pesticide	log (KOW)	KOC	Water Solubility (mg/l)	DT50
TCA	—	0	12,000	21–90
Chloramben	—	12.8	700	15–45
2,4-D	0.04–0.33	32	20,000	<7
Glyphosate	< –3.2	—	11,600	3–174
Propham	—	51	250	5–15
Bromacil	—	71	700	180
Propazine	—	152	5	80–100
Simazine	2.1	160	6.2	27–102
Dichlobenil	2.7	165	14.6	30–180
Atrazine	2.5	172	33	16–77
Chlorpropham	—	245	89	30–65
Prometon	—	300	750	360
Ametryn	2.63	380	200	11–120
Diuron	2.85	400	36.4	90–180
Prometryn	3.1	400	33	14–158
Paraquat	–4.5	20,000	620	180–360
DDT	6.2	243,000	0.0012	4–30 years

Note: The half-life in soil (DT50) is dependent on degradation, leakage, and evaporation, which in turn are dependent on soil type, moisture, and temperature.

Source: Data from Tomlin, C., Ed. 2000. *The Pesticide Manual: A World Compendium*. British Crop Protection Council, Farnham, Surrey. 1250 pp.

8.3.3   Desorption

Desorption is often a slower process than adsorption, and a phenomenon called aging prevents a total extraction by water or mild extraction solutions (methanol:water (1:1 v/v), ammonium acetate, etc.) to extract the pesticide completely from soil, even if it is sufficiently soluble in the solvent. The amount that is nonextractable increases over time. The reason for this has not yet been settled. One theory is that the soil has many extremely small pores (nanopores). In the course of time the adsorbed molecules diffuse into such pores, where they are not easily extracted and have a low and decreasing bioavailability. An alternative theory is that they are first adsorbed to some easily available low-affinity binding sites and over time jump over to high-affinity binding sites.

The movement of a pesticide through the soil can also be determined in the field by direct measurements of the changes in concentrations at different depths. Such measurements are consistent with results from laboratory experiments.

TCA, chloramben, and 2,4-D are carboxylic acids that dissociate to form negative ions that do not bind much to the soil matrix. Microorganisms degrade these pesticides easily. Glyphosate does not bind to humus, but makes insoluble salts with calcium and other metals in the soil. It does not

Table 8.3 Vapor Pressure and Water Solubility
of Some Persistent Insecticides

Insecticide	Vapor Pressure 20–25°C (mmHg)	Water Solubility 20–25°C (ppm)
p,p'-DDT	1.9×10^{-7}	0.0012
Dieldrin	1.0×10^{-7}	0.1
Endrin	2.0×10^{-7}	0.1
Aldrin	6.0×10^{-6}	0.05
Toxaphene	1.0×10^{-6}	3
Lindane	9.4×10^{-6}	10
Chlordane	1.0×10^{-5}	—
Heptachlor	3.0×10^{-4}	—

Note: Data are taken from various sources.

adsorb but can be trapped as insoluble salts. Microorganisms degrade it readily to give aminomethylphosphonic acid. Simazine has a low leaching potential because of its low solubility in water, whereas propazine is described as mobile. Note also that chlorpropham, prometon, diuron, and ametryn are water soluble in water to some extent but are adsorbed in soil due to their alkaline properties and ability to make positive ions. Microorganisms readily degrade these compounds. Paraquat is very strongly bound to the soil because it is an aromatic positive ion that can form charge-transfer complexes as well as ionic bonds. In spite of its very high water solubility, leakage is seldom or never observed. Only on very sharp sandy soil without humus may paraquat penetrate to the subsoil water. DDT is very slowly degraded, but some DDT disappears due to co-distillation with water. A small survey of DDT content in orchard soil showed that between 21 and 85% of all DDT applied between 1945 and 1968 remained in the soil as DDT (about 75%) and DDE (about 25%) (Stenersen and Friestad, 1969).

8.4 Evaporation

Volatilization and vapor phase transport are important processes in the dissipation of even the so-called nonvolatile pesticides, such as the chlorinated hydrocarbons. The vapor pressure or the vapor saturation density is therefore an important parameter to assess the persistence of the pesticide. Much work was done very early in the era of the (persistent) pesticides. Table 8.3 shows the vapor pressure and water solubility of some important chlorinated pesticides and contaminants in pesticide formulations.

Maximum vapor densities (weight of pesticide in the gas phase at equilibrium/volume of air (w/v)) are good indications of evaporation potency. It may be calculated from the vapor pressure by using the gas equation: $p \times v = (w/M) \times RT$ or $w/v = p \times (M/RT)$. w/v is the vapor density, p the vapor pressure, M the molecular weight of the substance, and R the universal gas constant ($R = 0.08205\ l \times atm \times °K^{-1} \times mol^{-1}$). T is the temperature in °K.

Pressure is often given in very different units (e.g., atm, psi, mmHg, or Pa), and the following equation may be useful for conversion to the appropriate unit: 1 mmHg = 133.3 Pa = 1.316×10^{-3} atm = 1.934×10^{-2} psi.

Vapor density of DDT (molecular weight = 354.5 g/mol) at 20°C (= 293°K) can easily be calculated:

$$w/v = \frac{1.9 \times 10^{-7}\,\text{mmHg} \cdot 1.316 \times 10^{-3}\,\text{atm} \cdot \text{mmHg}^{-1} \cdot 354.5\,\text{g} \cdot \text{mole}^{-1}}{0.08205\,\text{atm} \cdot \text{L} \cdot 293°\text{K}^{-1} \cdot \text{mole}^{-1}} = 3.7\,\text{ng/L}$$

Although this figure is very low, some vaporization will occur.

A similar calculation for lindane, using the figure in Table 8.3 (9.4×10^{-6} mmHg), gives a saturation vapor density of 150 ng/l, which indicates that lindane disappears by evaporation much more easily than DDT. It is important to note that the vapor density, and thus evaporation velocity, will be reduced by adsorption in the soil, but will be enhanced by higher moisture content in the soil due to co-distillation. A parameter called Henry's constant (H) is important to determine the volatilization of pesticides when dissolved in water. According to Henry's law there will be equilibrium of concentrations in water and air at a specified temperature:

$$H = \frac{\text{Concentration in air}}{\text{Concentration in water}}$$

Henry's constant depends on the temperature. By applying the law for ideal gases, the concentration may be replaced by the partial pressure. H is therefore often given as (unit of pressure)/(mole/unit of volume water) (e.g., atm $\times$ l $\times$ mol^{-1}, or mPa $\times$ m$^3 \times$ mol^{-1}). (Note that H sometimes is defined as the concentration in water divided by the concentration in air.)

8.4.1   Example

The fungicide cymoxanil has the vapor pressure 0.15 mPa at 20°C; the solubility in water is 890 mg/l, which is 4.495 mol $\times$ m^{-3} because the molecular weight is 198.2 g/mol.

Henry's constant is therefore

$$\frac{0.15}{4.495}\,\text{mPa} \times \text{m}^3 \times \text{mol}^{-1} = 3.33\,\text{Pa} \times \text{m}^3 \times \text{mol}^{-1}$$

The biodegradation is probably high because of cymoxanil's high solubility, its aliphatic structure, the absence of xenobiotic bonds, etc. The adsorption to soil is low (Freundlich's K is between 0.29 and 2.86 depending on soil type). Disappearance from soil through evaporation, leakage, and bio-

Table 8.4 Vapor Density and Vapor Pressure of DDT, Its Metabolites, and Analogues

Chemical	Vapor Density at 30°C (ng/l)	Vapor Pressure at 30°C (mmHg × 10⁻⁷)
p,p-DDT	13.6	7.26
o,p-DDT	104	55.3
p,p-DDE	109	64.9
p,p-DDD	17.2	10.2
o,p-DDE	104	61.6
o,p-DDD	31.9	18.9

Source: Data from Spencer, W.F. and Cliath, M.M. 1972. *Agric. Food Chem.*, 20, 645–649.

degradation is therefore likely to be fast. Typical DT50 (half-life) values are indeed low; values between 0.75 and 1.5 days have been recorded in laboratory experiments.

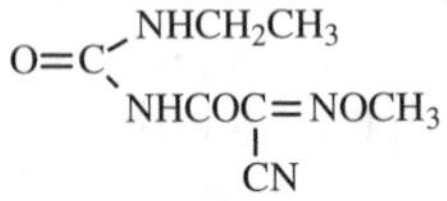

$$O=C\begin{array}{l} \diagup NHCH_2CH_3 \\ \diagdown NHCOC=NOCH_3 \\ \qquad\quad | \\ \qquad\quad CN \end{array}$$

cymoxanil

The highly persistent chlorinated hydrocarbons may also slowly disappear from soil through evaporation. Spencer and Cliath (1972) found that the vapor densities of DDT and DDE were approximately 21 times greater at 7.5% moisture content in soil than at 2.2% moisture content. They also found that increased temperature raised the vapor pressure according to the Clapeyron–Clausius equation ($\log p = A - B/T$, where A and B are constants). They measured the vapor densities of p,p-DDT on sand at 20, 30, and 40°C and found them to be 2.9, 13.6, and 60.2 ng/l. Furthermore, Spencer and Cliath found that DDE and other derivatives of DDT have a higher vapor density than DDT (Table 8.4). DDT therefore probably disappears from soil by first being converted to DDE by microorganisms and soil animals, which then slowly disappear by evaporation.

8.5 Biotransformation in animals

The biotransformation or metabolism of xenobiotics has been studied in detail but is seldom described in general textbooks on biochemistry. However, textbooks in toxicology usually have one or two chapters on the subject; for example, see Parkinson (2001), although they tend to concentrate on rats and other vertebrates. Valuable data on pesticide biotransformation are also found in Hayes and Laws (1991a and b), Miyamato et al. (1988), and Rockstein (1978). Chambers and Yardbrough's (1982) book is also inspiring reading, with an interesting chapter by Wilkinson and Denison on pesticide interactions with biotransformation enzymes. Also, for a short review of detoxication in earthworms, Stenersen (1992) is available. This chapter summarizes some basic ideas and describes some important enzymes.

Animals encounter poison in the plants they eat, from other animals that use poisons in attack and for protection, and from toxin-producing bacteria and fungi. Animals are exposed to toxic metals released from minerals, and even the oxygen that all animals depend on is a very poisonous gas. Leakage of mineral oil, with many toxic and lipohilic substances, is also an old challenge that organisms had to adapt to. Substances produced inside the animal (e.g., ammonia, epoxides, phenolic substances) may also be toxic. Animals must therefore have many protection mechanisms against toxic substances in order to survive and reproduce.

Plants are extremely inventive in making new poisons to protect themselves against attack from insects, mites, nematodes, fungi, grazing mammals, and even against other plants. Harborne (1978, 1993) addresses the interesting subjects of plant toxins in many of his publications. Besides producing all kinds of toxins for the purpose of deterring enemies, plants produce many carotenoids, steroids, higher alcohols, and hydrocarbons as part of their normal metabolism. As a consequence, terrestrial polyphagous herbivores, through natural selection, have a wide variety of enzymes that detoxicate and render such substances more soluble in water. The advantage of making the toxicants more soluble in water is to be able to excrete them in a small volume of water. Aquatic animals may dilute and excrete endogenous waste products and ingested toxins in plenty of water, whereas terrestrial animals must rely on much less water. A mammal or an insect therefore usually has more highly developed biotransformation enzymes than a fish or a shrimp. Being polyphagous, humans and rats have good detoxication enzymes compared to many carnivores. A human consumes more than a gram of natural pesticides such as alkaloids, mustard oils, toxic terpenoids, etc., each day. Peculiar to many humans is the consumption of fried barbecue meat that contains many carcinogenic and mutagenic substances. The voluntary exposure of lung tissue to toxic smoke containing alkaloids such as nicotine and carcinogens such as benzoe(a)pyrene is an even more dangerous challenge for the organism. The detoxication of synthetic pesticide residues in food, which seldom constitutes more than a milligram a day, proceeds via the same routes as those used for the vast number and amount of natural poisons. The main biotransformation reactions are oxidation, hydrolysis, and conjugation.

8.5.1 Oxidation

Oxygen is very toxic and is the source of many other reactive and toxic substances, such as the so-called ROS, or reactive oxygen species, which include the superoxide anion, the nitrogen oxides, the singlet oxygen, hydrogen peroxide, and organic peroxides. Animals have a wide variety of enzymes and low molecular compounds that protect them from the devastating actions of these compounds. Without antioxidants such as ascorbic acid, glutathione, vitamin C, vitamin E, carotenes, and uric acid, as well as the enzymes, including peroxidases, superoxide dismutases, epoxides hydrolases, and glutathione transferases, aerobic life would not be possible.

However, oxygen is very useful, not only for energy production in respiration, but also as a first-line reagent to transform unwanted lipophilic substances to more polar and water-soluble derivatives. A large family of enzymes called the CYP enzymes, or cytochrome P450, add one atom of oxygen from dioxygen (O_2) to the substrate and reduce the other to water. The co-substrate nicotineamide-adenine dinucleotidephosphate (NADPH) reduces the oxygen to water. The catalytic cycle is complicated and involves binding of the substrate to the enzyme, oxidation of Fe^{+++} to Fe^{++}, change of the valence electrons of Fe^{++} from low-spin to high-spin orbital, binding of oxygen to iron, and successive transfer of one oxygen atom to the substrate. The other oxygen atom is reduced to water:

$$R + O_2 + H^+ + NADPH = RO + H_2O + NADP^+$$

One individual fruit fly has 90 gene loci coding for CYP enzymes, and each locus may have many different allelic forms. Other organisms may have as many, and a general classification system is necessary. Today the natural and most convenient method is to classify the enzymes according to their amino acid sequence. They are named by the letters CYP accompanied with an Arabic number, a Latin letter, and a new Arabic number, e.g., CYP2E1. The shorter acronyms CYP2 and CYP2E are used for any enzymes in the superfamily CYP2, or the family CYP2E, whereas the last Arabic number designates the individual enzyme with a unique amino acid sequence. If two enzymes have more than 40% similarity in the amino acid sequence but less than 55%, they get the same Arabic number, but a different letter (CYP2B, CYP2E). If the similarity is more than 55%, but less than 98%, the letter is also the same (CYP2B1 and CYP2B2). If the similarity is greater than 98%, the two proteins are regarded as the same enzyme, even if they are from different sources (species or organ). The system is quite simple but requires a full sequence determination. It is also important to remember that CYP1 and CYP2 are not necessarily more closely related than, for instance, CYP1 and CYP4. In the years after CYP enzymes were detected they were called mixed-function oxidases or microsomal oxygenases. It was believed that there were only two enzymes. They got the names cytochrome P450 and cytochrome P448 due to a slight difference in their absorbance of light under specified laboratory conditions. Cytochrome P448 is similar to the enzymes now called CYP1A1 and CYP1A2. Certain substances, such as many dioxins, induce an increased concentration of CYP1 enzymes when animals are exposed. The toxicity of dioxins is related to their ability to increase (induce) the activity of CYP1 enzymes. Other toxicants such as DDT may induce other CYP enzymes. Plant secondary metabolites sometimes induce or inhibit various CYP enzymes, and therefore the diet of animals can modify the

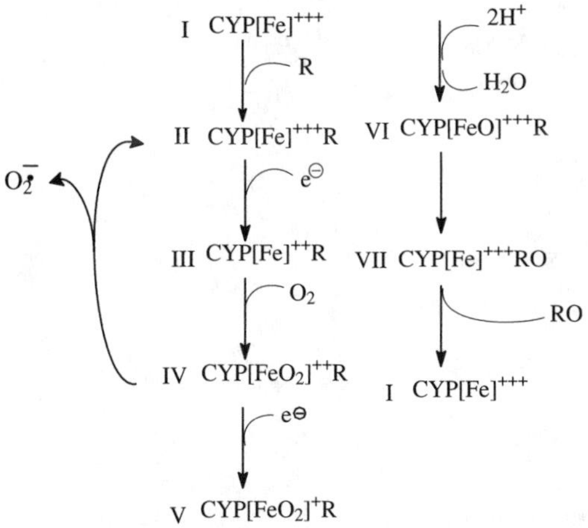

Figure 8.8 The catalytic cycle of the CYP enzymes. The substrate binds to a hydrophobic site of the enzyme (I → II). This leads to a shift in the valence electrons of iron, from low-spin to high-spin status, and an uptake of an electron from cytochrome P450 reductase (II → III). Oxygen is then added (III → IV) to the iron and another electron is added (IV _ V). Some of the intermediates may lose reactive oxygen species (IV → II), which has harmful consequences for the cell. There are some electronic rearrangements (V → VI → VII). The product (RO) leaves the enzyme and its ground state is restored (VII → I). RO is often more toxic than the substrate (R) and may be an alcohol, a phenol, or an epoxide that can be rendered harmless by other enzymes and made ready for excretion.

toxicity or effect of a pesticide or a drug. This applies to humans taking medicines, as well as to insects being exposed to insecticides.

The CYP enzymes play an important role in the endogenous metabolism and are very important in the synthesis and degradation of sterols. Some pesticides, e.g., the inhibitors of ergosterol synthesis described in Chapter 5, act by specifically inhibiting CYP51. Piperonyl butoxide, an important synergist used to increase the efficiency of pyrethrum preparates, also inhibits CYP enzymes. It must first be activated by oxidation and catalyzed by the CYP enzyme it inhibits. Enzymes in the families CYP2 and CYP3 play the greatest role in the metabolism of drugs and pesticides. The catalytic cycle of CYP enzymes is shown in Figure 8.8.

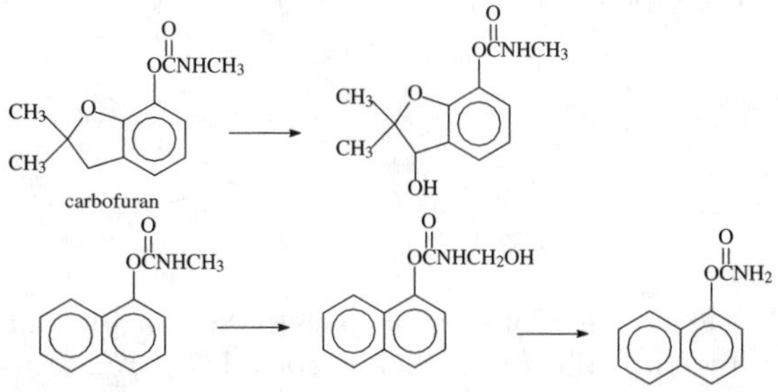

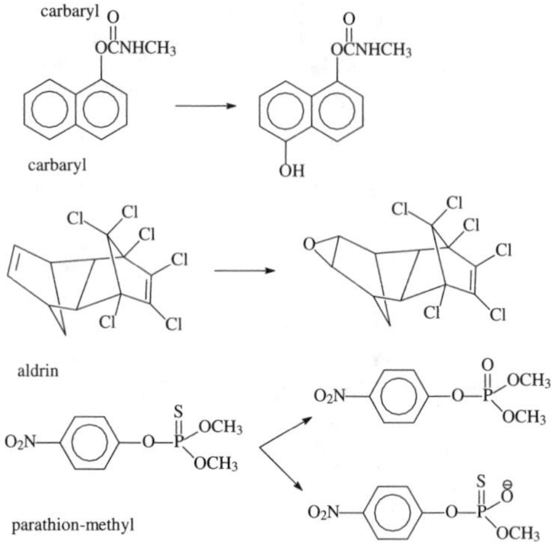

The reactions show some typical oxidations of insecticides catalyzed by CYP enzymes. Carbofuran is hydroxylated to another active compound; carbaryl is demethylated or hydroxylated; and aldrin is epoxidated to dieldrin, which also has insecticidal properties. Phosphorothionates must be oxidized to the phosphates by CYP enzymes in order to become inhibitors of acetyl cholinesterase. Parathion-methyl is transformed to the oxon analogue, paraoxon-methyl, which is the toxic compound. It also can be demethylated to the inactive desmethyl-parathion-methyl.

8.5.2   Epoxide hydrolase

The epoxide hydrolases are very important enzymes that render the epoxides formed by CYP enzymes harmless. Mammals have three distinct epoxide hydrolases. One microsomal form is specific for cholesterol-5,6α-oxide. It is induced by the same substances that induce the CYP enzymes. Another, less specific epoxide hydrolase is located in the endoplasmatic reticulum close to the CYP enzymes and is more important for xenobiotic substances. A third type, of some importance, is located in the cytosol. A typical substrate is trans-stilbene oxide, which also is an inducer. Some epoxides, such as dieldrin, are not detoxicated by these enzymes due to steric hindrance, but stilbene epoxide is a good substrate that is frequently used in experimental studies of these enzymes.

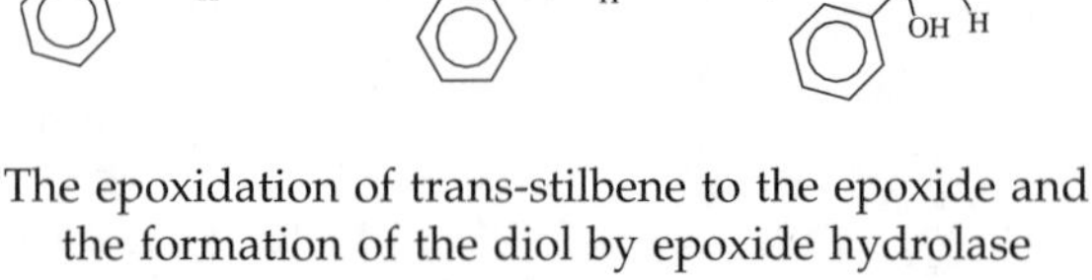

The epoxidation of trans-stilbene to the epoxide and
the formation of the diol by epoxide hydrolase

8.5.3 Glutathione transferase

All aerobic cells have a small tripeptide of glutamic acid, cysteine, and glycine. The glutamic acid is combined to the amino group of cysteine with its γ-carboxyl group:

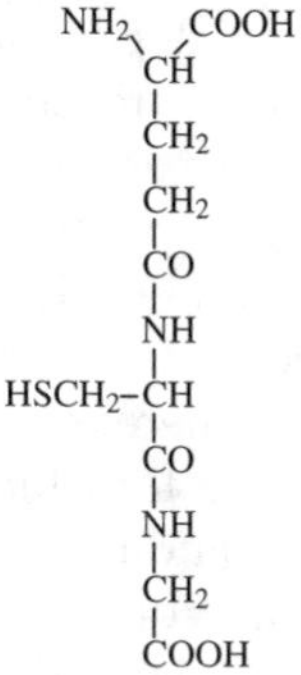

$$
\begin{array}{c}
NH_2 \quad COOH \\
\diagdown \diagup \\
CH \\
| \\
CH_2 \\
| \\
CH_2 \\
| \\
CO \\
| \\
NH \\
| \\
HSCH_2-CH \\
| \\
CO \\
| \\
NH \\
| \\
CH_2 \\
| \\
COOH
\end{array}
$$

The tripeptide is called glutathione or GSH. It is the main antioxidant in the cell, and aerobic life would have been impossible without it. The SH group of the cysteine combines very easily with other SH groups after oxidation. It easily forms free radicals by giving away an electron to other free radicals, but eventually it combines with another glutathione free radical to make the oxidized glutathione, often written as GSSG, which is fast reduced to 2GSH by glutathione reductase, at the expense of NADPH. GSH not only plays an important role in keeping the levels of free radicals low, but also reacts with electrophilic substances, which would otherwise react with some of the many nucleophilic atoms in proteins, nucleic acids, and lipids. Some of the basic reactions of glutathione are shown below:

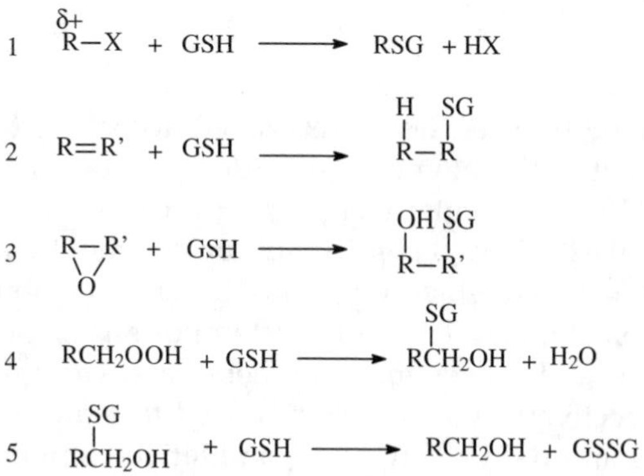

1. $\overset{\delta+}{R-X} + GSH \longrightarrow RSG + HX$

2. $R=R' + GSH \longrightarrow \underset{R-R}{\overset{\overset{H \quad SG}{|\quad|}}{}}$

3. $\underset{\underset{O}{\diagdown\diagup}}{R-R'} + GSH \longrightarrow \underset{R-R'}{\overset{\overset{OH\ SG}{|\quad|}}{}}$

4. $RCH_2OOH + GSH \longrightarrow \overset{\overset{SG}{|}}{RCH_2OH} + H_2O$

5. $\overset{\overset{SG}{|}}{RCH_2OH} + GSH \longrightarrow RCH_2OH + GSSG$

6    $GSSG + NADPH + H^+ \longrightarrow 2GSH + NADP^+$

7    $GSH + R^\bullet \longrightarrow RH + GS^\bullet$

8    $GS^\bullet + GS^\bullet \longrightarrow GSSG$

Reactions 1 to 5 are catalyzed by an important family of enzymes called glutathione transferases (GSTs). Halogen or nitro groups (–X) tied to an electrophilic atom are replaced by GS (reaction 1). GSH may add to a double bond if one atom is electrophilic (reaction 2). The epoxide bridge can be opened (reaction 3). A peroxy group can be reduced through two steps (reactions 4 and 5). Free radicals can be eliminated through two steps (reactions 7 and 8). Reactions 4 and 5 can also be catalyzed by a group of enzymes called glutathione peroxidases, whereas reaction 6 is catalyzed by glutathione reductase. The glutathione transferases constitute two families of enzymes: one is present in cytosol and one is embedded in the endoplasmic reticulum. The cytosolic GST may constitute as much as 10% of the total cytosolic protein in the rat liver. Many different pesticides that have an electrophilic atom are detoxicated by conjugation with GSH. Examples are lindane, dimethyl phosphorothionates, some other phosphorothionates such as EPN, atrazine, alachlor, and DDT in many resistant insect strains. The chemical structures of some of these compounds are shown, and the group replaced by GS is shown by a circle.

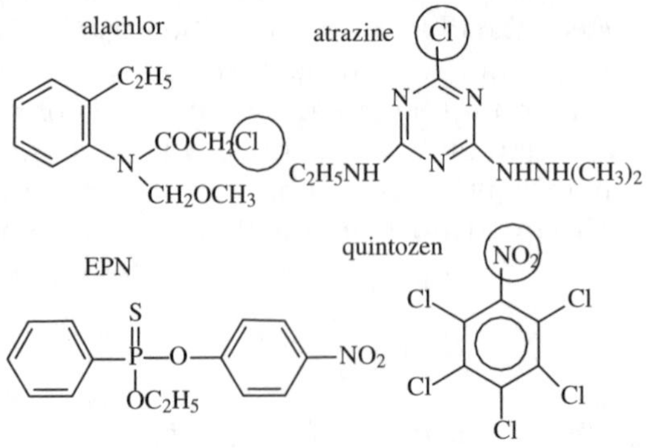

The glutathione transferases are classified according to their sequence similarities. The cytosolic enzymes are made up of two similar or identical peptide chains with a molecular weight of approximately 25,000 daltons. If the sequence similarity of two peptide subunits is more than 50%, they are placed in the same class, which is given a letter. The subunits themselves are given an Arabic number. The four most important classes of mammalian GSTs are called the alpha, beta, mu, and theta classes, or given the letters A, B, M, and T, respectively. A small letter is used to indicate the species. For instance, rGSTA1 and rGSTA2 are two quantitatively important enzymes in the A class that have been isolated and described from rats. A type called hGSTM1 is important for protecting smokers against lung cancer and may be one of the reasons why not all smokers get cancer. Anticancer compounds such as sulforafan, found in broccoli, induces GST. A diet with broccoli may therefore render insects less sensitive to certain insecticides and protect humans from carcinogens, because insecticides and carcinogens are then metabolized faster. Sulforafan is an isothiocyanate:

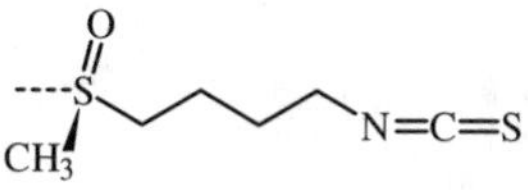

Other ways of indicating the relationships and similarities of glutathione transferases have been used earlier. The first method, to classify them according to their activity, i.e., what they *do*, would have been much more useful for the toxicologist, but their substrate specificity is overlapping and there is a myriad of known and unknown possible substrates. A structural basis for classification is therefore more rational than a functional basis, such as the earlier terms *aryl transferases, methyl transferases, epoxide transferases, DDT dehydrochlorinases*, etc.

8.5.4   Hydrolases

Many pesticides are esters or amides that can be activated or inactivated by hydrolysis. The enzymes that catalyze the hydrolysis of pesticides that are esters or amides are esterases and amidases. These enzymes have the amino acid serine or cysteine in the active site. The catalytic process involves a transient acylation of the OH or SH group in serin or cystein. The organophosphorus and carbamate insecticides acylate OH groups irreversibly and thus inhibit a number of hydrolases, although many phosphorylated or carbamoylated esterases are deacylated very quickly, and so serve as hydrolytic enzymes for these compounds. An enzyme called arylesterase splits paraoxon into 4-nitrophenol and diethyl-phosphate. This enzyme has cysteine in the active site and is inhibited by mercury(II) salts. Arylesterase is present in human plasma and is important to reduce the toxicity of paraoxon that nevertheless is very toxic. A paraoxon-splitting enzyme is also abundant in earthworms and probably contributes to paraoxon's low earthworm toxicity. Malathion has low mammalian toxicity because a carboxyl esterase that can use malathion as a substrate is abundant in the mammalian liver. It is not present in insects, and this is the reason for the favorable selectivity index of this pesticide.

8.5.5   Glucoronosyltransferase and sulfotransferase

Phenols, but also amines and carboxylic acids, are detoxicated by conjugation with sulfate and glucuronic acid. The amounts of and properties of these enzymes are different in various animals. Insects do not have much glucuronosyltransferase or sulfotransferase and conjugate phenols with glucose. Annelids have no or a small amount of this type of conjugation enzymes, with the consequence that phenolic pesticides (dinoseb, ioxynil, pentachlorophenol) or phenolic metabolites of pesticides (4-nitrophenol) are quite toxic to them.

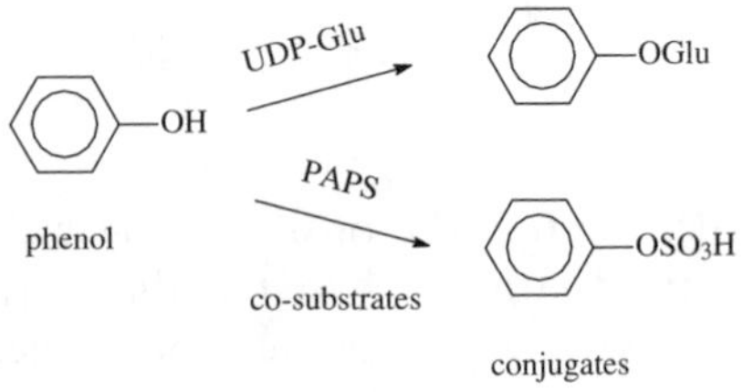

The co-substrates of the enzymes are uridine diphosphate glucuronic acid (UDP-Glu) and phosphoadenine-phosphosulfonate (PAPS).

8.5.6 Stereospecific biotransformation

Stereospecific biotransformation has been an often neglected but interesting phenomenon with some toxicological importance. Xenobiotics having a chiral center are usually produced as a 1:1 mixture of the two stereoisomers. The biological activity, rate, and route of the biotransformation of them will be different.

Asymmetric molecules can sometimes be formed from symmetric ones through specific biotransformation. We use the insecticide *methoxychlor* as an example to demonstrate this point (see Kishimoto and Kurihara, 1996 and Kishimoto et al., 1995 for experimental details and further references). Methoxychlor is very similar to DDT but has two methoxy groups (CH_3O-) instead of two chlorides in the para positions of the phenyl rings. The toxicity to insects resembles that of DDT, but some enzymes in the cytochrome P450 system, as well as many bacteria, may remove one or both of the methyl groups, and thus make the molecule susceptible for further degradation or elimination as conjugates. The estrogenic effect reported for methoxychlor is probably caused by the desmethyl methoxychlor.

Although the two methoxyphenyl groups look quite similar, they are not. One of them is pointing up to the right, the other to the left. If, for instance, the one to the right is tagged, it is not possible to turn the molecule in such a way that the tagged ring comes to the left without interchanging the position of the hydrogen and the trichloromethyl group. The oxidative demethylation is carried out by at least four different CYP enzymes in the rat liver (CYP2C6, CYP2A1, CYP2B1, and CYP2B2). In one experiment, 60 to 78% of the S-enantiomeric metabolite was formed when methoxychlor was incubated with a preparation of enzymes from the rat liver. By doing experiments with specific inhibition of different CYP enzymes and with specific antibodies (anti-CYP2B1, anti-CYP2C6), it was concluded that CYP2A1 was more stereoselective than the other CYP enzymes involved. The rate of metabolite formation becomes different after changing the hydrogen with deuterium in one of the methyl groups (CH_3O- changed to CD_3O-). The R-CD_3O-methoxychlor is metabolized more slowly than the S-CD_3)-enantiomer. There is reason to believe that the estrogenic effect of methoxychlor is mainly due to either the S-enantiomer or the R-enantiomeric metabolite. The stereospecific demethylation of methoxychlor is shown below:

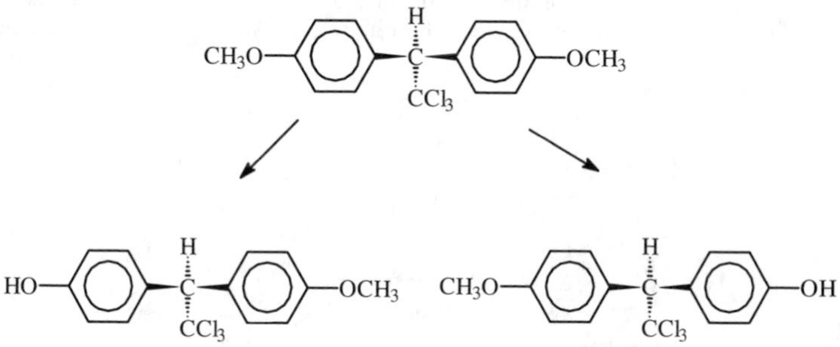

Stereospecific metabolism of methoxychlor

8.6 Designing pesticides that have low mammalian toxicity

The examples are picked from *The Pesticide Manual*, but the reader should also be familiar with the work of Krueger and O'Brien (1959) on malathion.

Carbofuran is a carbamate with high mammalian toxicity (LD50 (lethal dose in 50% of the population) oral toxicity for rats is 8 mg/kg). It is metabolized to 3-hydroxy- and 3-keto-carbofuran, which are also active cholinesterase inhibitors. Carbofuran is soluble in water (0.32 g/l) and is therefore systemic in plants. The high toxicity for mammals, birds (LD50 for quails is 2.5 mg/l), and earthworms makes it a less safe insecticide. Substituting the H in the carbamate moiety of carbofuran by a thiosulfenyl group led to the discovery of carbosulfan. This compound is metabolized back to the active carbofuran in insects, whereas the major detoxication pathway in mammals involves cleavage of the O–C bond, leading to the formation of nontoxic fragments. Carbosulfan thus has a much reduced mammalian toxicity (LD50 for rats in oral dose is 250 mg/kg). It is, however, rather toxic to fish and birds. In soil, it is also slowly transformed to carbofuran. The metabolism of carbosulfan and carbofuran in insects and mammals is shown in the reaction below:

Table 8.5 Toxicity of Some Organophosphorus Insecticides Where Metabolism Is Crucial for the Toxicity

Insecticide	ADI (mg/kg of body weight)	Toxicity Class (WHO)	LD50 (Rat) mg/kg, oral	Housefly mg/kg, topical
Acephate	0.03	III	1447	—
Metamidophos	0.004	1b	15.6	—
Malathion	0.3	III	1375–2800	17.4
Dimethoate	0.002	II	387	0.2
Demeton-S-methyl	0.0003	1b	30	—

Note: The toxicity is given as acceptable daily intake (ADI), the World Health Organization's (WHO) toxicity class, and LD50 for rats and houseflies, when known.

Source: Data are taken from Tomlin, C., Ed. 2000. *The Pesticide Manual: A World Compendium*. British Crop Protection Council, Farnham, Surrey. 1250 pp. Other sources.

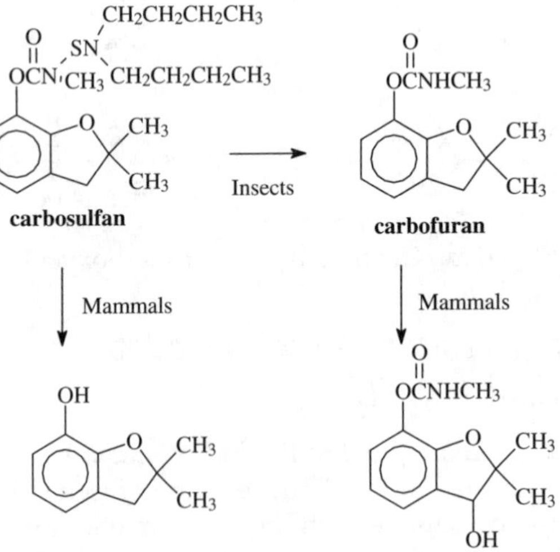

carbosulfan → Insects → carbofuran

Mammals

Mammals

8.6.1   Acephate

Acephate is an organophosphate with high activity toward insects, but it seems to have low toxicity to other animals. It is systemic and is metabolized in plants to metamidophos, which has a much higher toxicity. But this activation does not seem to be so important in mammals. Methamidophos is also used as an insecticide/nematicide. Table 8.5 shows the high difference in toxicity toward acephate and its metabolite.

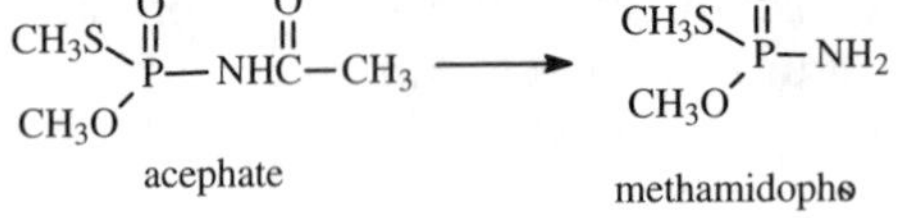

acephate → methamidophos

8.6.2   Malathion and dimethoate

Malathion and dimethoate were developed and marketed in 1952 and 1951, respectively. They may be regarded as first-generation organophosphates, but in spite of this, their mammalian toxicity is rather low because of different metabolisms in insects and mammals. R.D. O'Brien and co-workers detected the reason for this (Krueger and O'Brien, 1959). Compared to the more toxic alternative, demeton-S-methyl, dimethoate was often preferred for use in aphid and mite control as a result of the lower human toxicity. The ethyl analogue of demeton-S-methyl (demeton) is much more toxic and is now superseded. Malathion has such a low mammalian toxicity that it may be used for body lice control. Note that these pesticides at high temperature or long storage periods become more toxic because of isomerization reactions. Some basic toxicity data are given in Table 8.5.

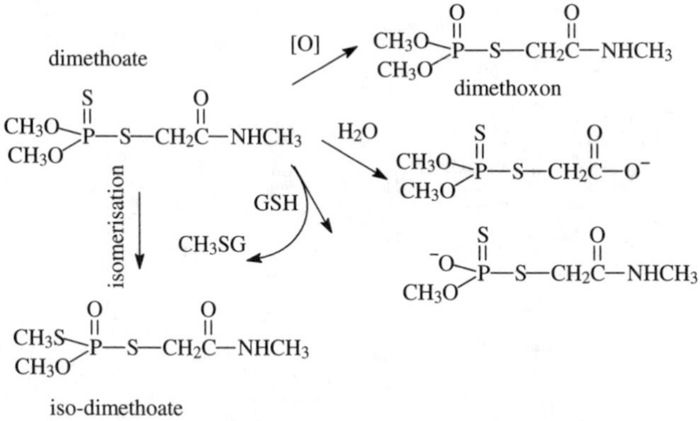

Various metabolic transformations of dimethoate in mammals

Dimethoate is split by an amidase in the mammalian liver, whereas it is activated to oxon in insects. Dimethoate may also be detoxicated by glutathione transferases or isomerized to the more toxic derivatives by heating. Demeton-S-methyl is metabolized to highly toxic compounds, such as demeton-S-methyl sulfoxide and sulfone in plants and animals:

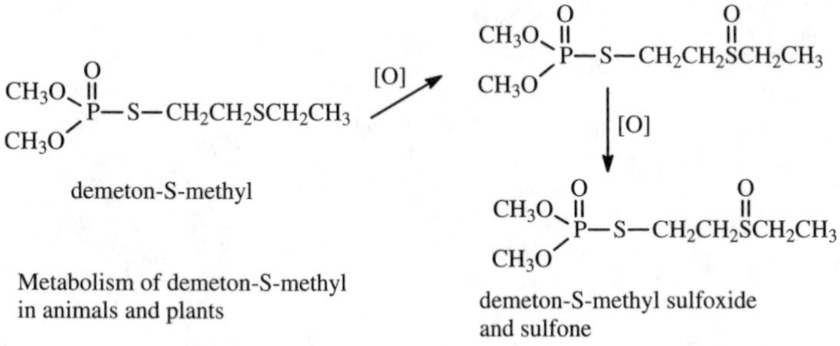

Metabolism of demeton-S-methyl
in animals and plants

demeton-S-methyl sulfoxide
and sulfone

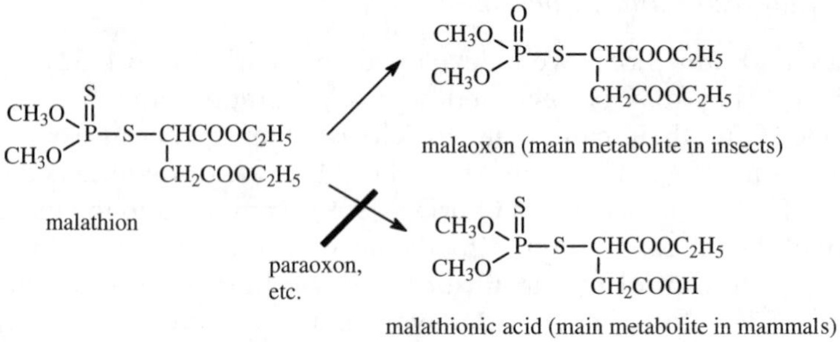

$$
\begin{array}{c}
\underset{CH_3O}{\overset{CH_3O}{\diagdown}} \overset{O}{\underset{\parallel}{P}} - S - CHCOOC_2H_5 \\
\qquad\qquad\qquad\quad | \\
\qquad\qquad\qquad CH_2COOC_2H_5
\end{array}
$$

malaoxon (main metabolite in insects)

$$
\begin{array}{c}
\underset{CH_3O}{\overset{CH_3O}{\diagdown}} \overset{S}{\underset{\parallel}{P}} - S - CHCOOC_2H_5 \\
\qquad\qquad\qquad\quad | \\
\qquad\qquad\qquad CH_2COOC_2H_5
\end{array}
$$

malathion

paraoxon, etc.

$$
\begin{array}{c}
\underset{CH_3O}{\overset{CH_3O}{\diagdown}} \overset{S}{\underset{\parallel}{P}} - S - CHCOOC_2H_5 \\
\qquad\qquad\qquad\quad | \\
\qquad\qquad\qquad CH_2COOH
\end{array}
$$

malathionic acid (main metabolite in mammals)

8.6.3   Nereistoxin

Nereistoxin is a toxin produced by a marine polychaete. It is too toxic to be a safe pesticide, but when built into another molecule, it is very useful.

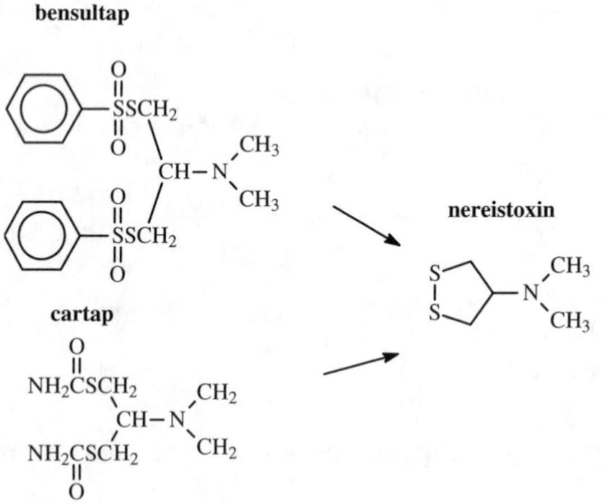

bensultap

cartap

nereistoxin

Nereistoxin is a cholinergic agonist.

chapter nine

Resistance to pesticides

The development of resistance to pesticides is generally considered to be one of the most serious obstacles to effective pest control today.

The first case was recognized in 1908 by Melander (1914), who noted an unusual degree of survival of San Jose scale (*Quadraspidiotus perniciosus* (Comstock)) after treatment with lime sulfur in Clarkston Valley of Washington. Oppenoorth (1965) has written a comprehensive review of the earlier studies of biochemical genetics of insecticide resistance. Newer issues in the "Annual Reviews" series regularly have articles about resistance in plants, insects, and pathogens (e.g., Hemingway and Ranson, 2000; Huang et al., 1999; Wilson, 2001). Anber's Ph.D. thesis (1989) gives a short and well-written introduction to the resistance problem, and the book *The Future Role of Pesticides in U.S. Agriculture* (Board on Agriculture and Natural Resources and Board on Environmental Studies and Toxicology, 2000) also describes the problem.

9.1 Definitions

Resistance to pesticides is the development of an ability in a population of a pest to tolerate doses of toxicants that would prove lethal to the majority of individuals within the same species. The term *behavioristic resistance* describes the development of the ability to avoid a dose that would prove lethal. Resistance is distinct from the *natural tolerance* shown by some species of pests. Here a biochemical or physiological property renders the pesticide ineffective against the majority of normal individuals.

Cross-resistance is a phenomenon whereby a pest population becomes resistant to two or more pesticides as a result of selection by one pesticide only. It must not be confused with *multiple resistance*, which is readily induced in some species with simultaneous or successive exposure to two or more pesticides. Cross-resistance is caused by a common mechanism (Winteringham and Hewlett, 1964).

9.2   Resistance is an inevitable result of evolution

Populations are polymorphous and show genetic variability between individuals in the same population. Even if they have been inbred for some time, the genetic difference of the individuals may be considerable. Every gene can occur as different versions, and these are known as alleles. New techniques in molecular genetics make it possible to study these differences with great precision. One insect specimen, the fruit fly, has 13,601 genes (Adams et al., 2000), and each of them can have hundreds of alleles. New alleles can be formed by mutations, and genes may also be duplicated to increase the total gene pool of the species. Most alleles are very rare, but if conditions change so that an allele becomes advantageous for survival and reproduction, it will in a few generations become the main allele in the population. One enzyme family, the CYP enzymes, often referred to as cytochrome P450 or mixed-function oxidases, is often involved in resistance because they are able to catalyze oxidation and detoxication of a wide variety of substances. The fruit fly has 90 different genes that code for these enzymes. Just one may be a rare allele of one of these genes, with a code for just one different amino acid, and may make an enzyme that is more active in degrading a specific pesticide (e.g., for a pyrethroid or a carbamate). This rare variant makes it easier to survive and reproduce in a pyrethroid- or carbamate-sprayed field. Other enzymes, such as the glutathione transferases, are important for detoxication of xenobiotics. They are also coded for by numerous genes that have many alleles.

In plants, nematodes, and microorganisms, the situation is similar, although resistance development to herbicides, fungicides, and nematicides appeared later. The small plant *Arabinopsis thaliana* has, for instance, 25,498 genes, and the free-living nematode *Chenorabditis elegans* has 19,099 genes. It is not very surprising that an allele of one or other of all these genes may make the organism less sensitive to an herbicide or nematicide.

Lethal toxicants in the environment will, of course, have a dramatic effect on the population. Only those individuals that for some reason survive are able to reproduce. An individual with alleles or gene duplications that make it less sensitive to the toxic environment will have much better opportunities to reproduce. The next generation of the pest will therefore have a higher frequency of these alleles. If the pest organism cannot be completely wiped out by the pesticide or by other means, resistance will appear sooner or later. Pesticides can therefore be regarded as consumable with a restricted time of usefulness. After having been used some years, the development of resistance may render them useless.

9.2.1   Time for resistance development

Because resistance is an inevitable result of evolution, it should have been predicted before becoming a problem. How fast resistance develops and in what species, as well as the biochemical mechanisms behind it and how

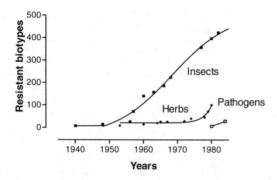

Figure 9.1 The appearance of resistant biotypes from 1940, when resistance started to become a problem, up to 1987. Herbicide resistance started to become a problem much later than insecticide resistance. Resistance in pathogenic fungi is newer and is more or less associated with the newer, more selective and systemic fungicides. (Data from Anber, H. 1989. *Studies on Pesticide Resistance. The Biochemical Genetics of Resistance to Organophosphates and Carbamates in Predacious Mite*, Amblyseius poten-tillae *(Garman)*. Faculty of Biology, Department of Pure and Applied Ecology, University of Amsterdam, Amsterdam. p. 90.)

many genes are involved, are matters of research. Insects often evolve resistance about a decade after the introduction of a new pesticide. Weeds evolve resistance within 10 to 25 years.

Resistance to insecticides was recorded as a problem just a few years after the introduction of the newer persistent insecticides, whereas resistance to herbicides and fungicides developed much later. Figure 9.1 shows an approximate graph of the development of resistance.

Fruit rot (*Botrytis cinera*) showed resistance against benomyl in 1971 or earlier, and a few years later resistance was observed in brown rot (*Monilinia laxa* and *Monilinia fructicola*). The benomyl-resistant biotypes had cross-resistance to thiophanate-methyl, fuberidazole, and thiabendazole (e.g., Dekker, 1972; Georgopoulos and Zaracovitis, 1967).

Resistance to triazine herbicides was recorded early (Ryan, 1970). Atrazine has been widely used in monocultures of maize, in orchards, and with vine crops, causing resistance in the weeds of the genuses *Amaranthus* and *Chenopodium*. Over 55 weed species had evolved triazine resistance before 1990 in the U.S. (Holt and Lebaron, 1990). Resistance in weeds to acetolactate synthase inhibitors includes *Kochia scoparia* and *Stellaria media* after 5 years of extensive use in cereals and maize. Acetyl-coenzyme A carboxylase inhibitors are widely used for annual grass control in soya, cotton, sugar beets, and cereals. Resistant weeds include *Avena fatua* and *Alopecurus myosuroides*. An approximate summary for first detection of resistance is found in Table 9.1. Palumbi (2001) has recently presented an overview with the expressive title "Evolution: Humans as the World's Greatest Evolutionary Force" that makes for highly recommended reading. The current edition of *The Pesticide Manual* also gives a very authoritative description and the current status (Tomlin, 2000).

Table 9.1 Approximate Date of Detection of Resistance
for a Number of Herbicides

Herbicide	Year Deployed	Resistance Observed	Type
2,4-D	1945	1954	Auxin agonist
Dalapon	1953	1962	Chlorinated fatty acid
Atrazine	1958	1968	Photosynthesis inhibitor
Picloram	1963	1988	Photosynthesis inhibitor
Trifluralin	1963	1988	Disrupts cell division
Triallate	1964	1987	Cell division inhibitor
Diclofop	1980	1987	Disrupts cell membrane

9.2.2   Questions about resistance

Since the 1950s, when resistance started to become a problem, several scientific questions had to be answered. Some of them are listed here and will be considered in this chapter.

- Are resistant insects more robust than sensitive ones?
- Is resistance caused by one allele in one gene locus, or is resistance acquired when sufficient resistance alleles from many genes have been accumulated in several individuals in a population?
- Do the pesticides cause resistance, or are the resistant individuals already there, before exposure to the pesticides?
- Is knowledge of the biochemical mode of action of resistance useful in the effort to find remedies against resistance?
- Why is resistance against the old biocidal pesticides not so common?
- Can resistance develop against the third- or fourth-generation insecticides based on pheromone-like or hormone-like action?
- Will resistant populations be susceptible if the use of the pesticide is terminated for a period?
- How can we use pesticides so that development of resistance is delayed or will not occur?

9.2.2.1   Are resistant insects more robust than sensitive ones?
The answer is no.

If the pests were hardier, natural selection would already have made them resistant before exposure to the pesticide. It is therefore unlikely that the resistant pests are stronger in other respects than in tolerance to the pesticides or other toxicants. It is, in fact, more likely that they are slightly less fit. Dr. Keiding (1967) at the Danish Pest Infestation Laboratory addressed this problem in the late 1960s.

9.2.2.2   Is resistance caused by one allele in one gene locus?
At a mild selection pressure, many alleles that increase survival and ability to reproduce in the toxic environment will accumulate in a population. Very

often we find several genetic differences between resistant and susceptible populations or strains. For instance, reduced uptake through the cuticle can be combined with increased detoxication caused by a higher titer of an esterase, a glutathione transferase, or a CYP enzyme, or a more efficient isoenzyme of a detoxication enzyme. However, quite often one gene dominates and is most important for resistance. The R-allele is very often partially recessive. The F_1 hybrids therefore have a susceptibility level somewhere between the susceptible and resistant strains.

9.2.2.3   Do pesticides cause resistance?

The answer is no.

The resistant individuals are present in the population before the pesticide is introduced. By careful testing for resistance in wild populations that have never been exposed to insecticides, it has been possible to detect some resistant individuals.

The question of heritability of acquired characteristics was a little controversial in the 1950s and 1960s because the Lamarckian ideas were still supported by the Soviet scientist Lysenko, who was very influential. It had been shown earlier that development of resistance in genetically homogeneous populations of insects did not occur even after selection pressure for many generations. But there were a few interesting exceptions that seemed to support Lysenko's ideas. The peach aphid (*Myzus persica*) forms clones because they are parthenogenetic. Nevertheless, in such clones it was possible to get a high degree of resistance after a few generations with selection pressure to parathion. However, this was nicely explained by the fact that gene amplifications were not very uncommon. The susceptible mother aphids had one gene for an esterase that degraded parathion, but this was not sufficient to make them parathion tolerant. But not uncommon individuals with gene duplications — up to 64 times — were resistant. This trait was partly inheritable and was transferred to the daughter aphids (see Figure 9.2).

An indirect proof that insecticides do not cause resistance is shown with sister selections. It was demonstrated that it is possible to make resistant strains by breeding pairs of insects separately. Some of the offspring were used for resistance testing, and the brothers and sisters of insects with low susceptibility were taken for breeding. By this method it was possible to produce resistant strains from insects that had never themselves, nor their ancestors, been exposed to insecticides.

9.3   Biochemical mechanisms

During the 1950s, an era when biochemical knowledge developed very fast, there was a very strong belief that by finding the biochemical mechanism for resistance, it should be easy to find some substance that counteracted it, for instance, inhibitors of enzymes that detoxicate the pesticide or a new pesticide that shows higher activity toward the resistant insects.

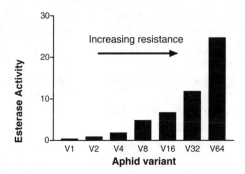

Figure 9.2 Devonshire and Sawicki (1979) at Rothamstead Experimental Station, Hertfordshire, U.K., found seven variants of the aphid *Myzus persicae* with different resistances to parathion. Excess production of a carboxylesterase as a result of one or several gene duplications was found to be the resistance mechanism. High levels of carboxylesterases take paraoxon away from acetylcholinesterase so that aphids become resistant.

A priori, i.e., without doing any empirical research, the following mechanisms have been postulated for insect resistance to insecticides. Most of the points also apply to weeds and fungi:

1. Behavior: Insects may have modified their behavior so that they avoid the areas sprayed with the insecticide. Such behavior may be genetically determined.
2. Reduced penetration of the pesticide through the cuticle or the intestine.
3. Lower transport into the target sites.
4. Lowered bioactivation: Some pesticides such as the sulfur-containing organophosphates may often be bioactivated.
5. Increased storage in fat depots or other inert organs.
6. Increased excretion of active ingredients.
7. Increased detoxication or decreased bioactivation.
8. Less sensitive receptors or enzymes that are inactivated or hyperactivated by the pesticides.
9. The development of alternative physiological pathways so that those disturbed by the pesticides are not so important.
10. More robust or bigger organisms so that they can tolerate bigger doses.

Extensive research has shown that points 7 and 8 are almost always involved, but with the other factors playing a modifying or additional role. Enhanced detoxication of the insecticide is often found in the resistant insect, or a modification of the biomolecule that is its target. The following are restricted to examples of these two mechanisms.

9.3.1 Increased detoxication

9.3.1.1 DDT dehydrochlorinase

Soon after DDT resistance had been observed in houseflies, it was shown that the resistant strains were able to metabolize DDT to 4,4'-dichlorodiphenyl-dichloroethylene (DDE) at a much faster rate than the susceptible ones. It was not certain that the enhanced DDT metabolism was the mechanism of resistance. A possibility was that resistant flies had a better opportunity to metabolize DDT. But very soon an enzyme named DDT dehydrochlorinase was detected in resistant flies (Sternburg et al., 1954). It was a big surprise to find an enzyme that had a completely artificial substrate, and a search for natural substrates was unsuccessful. The enzyme was dependent on the tripeptide glutathione and was difficult to measure because DDT has very low water solubility. Much effort was made in order to determine its activity, to purify and characterize it, and to measure the difference in activity between strains or individuals. Much later, after the enzyme family called *glutathione transferases* was detected and described, it was found that one or more of these enzymes were able to remove HCl from DDT (Clark and Shamaan, 1984). Other enzymes in this family were found to be able to degrade many other pesticides (Zhou and Syvanen, 1997). In the housefly, strains resistant to such different insecticides as lindane and the organo-phospate triester dimethyl parathion have enhanced levels of glutathione transferase. In this case it is cross-resistance between widely different pesti-cides. In DDT-resistant anophenline mosquitoes, resistance is mainly due to an increase in DDTase activity; i.e., the resistant mosquitoes have a high amount of a glutathione transferase that dehydrochlorinates DDT. The DDTase activity of the various isoenzymes of Anopheles isolated correlated with the activity toward 1,2-dichloro-4-nitrobenzene (DCNB), one of the substrates used in standard glutathione transferase assays. The resistant biotype had much more of the DCNB- and DDT-degrading isoenzymes (Prapanthadara et al., 1995a and b). The glutathione transferase constitutes a big enzyme family, with many variants, with different activity toward different electrophilic compounds. Natural populations may have individu-als with glutathione transferase having the unusual ability to degrade one or more pesticides. Two genetic mechanisms for the emergence of resistance have been considered. The first involves mutations in regulatory genes caus-ing overproduction of one or more glutathione transferases that are present in all flies. The second may involve the appearance of qualitatively different glutathione transferases with an exceptionally high ability to detoxify one or more pesticides. Increased amounts of glutathione transferase caused by amplification of *gst* genes are closely correlated to resistance.

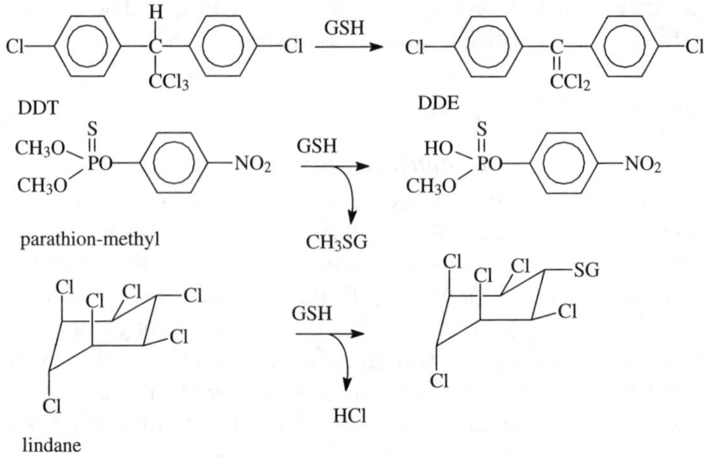

DDT

parathion-methyl

lindane

The diagram shows some reactions catalyzed by glutathione transferase that may give resistance to DDT or DDT analogues, methyl parathion but not the ethyl analogue, and lindane.

9.3.1.2 Hydrolases

Gene duplication and not point mutation may lead to more of a particular gene product.

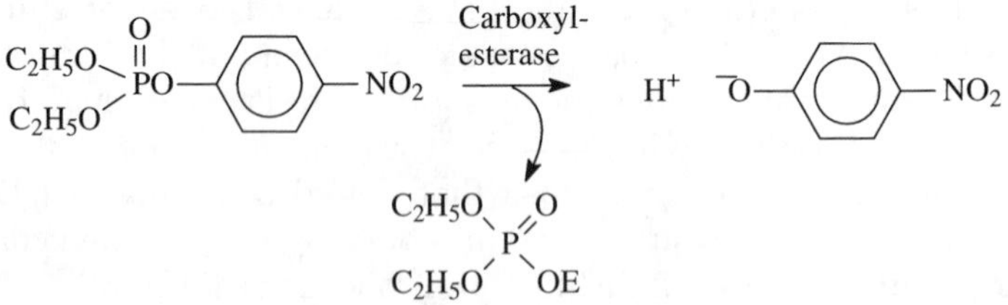

The reaction between paraoxon and carboxylesterase, rendering paraoxon unavailable for acetylcholinesterase inhibition

9.3.1.3 CYP enzymes in insects

The importance of high activity of the microsomal monooxygenases as a mechanism of insecticide resistance in insects was recognized some time ago. As mentioned, *Drosophila melanogaster* has 90 different *CYP* genes. It is therefore necessary to do extensive genetic and biochemical analysis to find out whether higher oxidative detoxication is due to higher expression of a particular *CYP* gene, or whether a different allele has evolved. Quite often the total amount of CYP enzymes are not so different in resistant and sensitive insects, but the catalytic properties are different. This may be caused by a different relative amount of the various CYP enzymes, and not due to new

Table 9.2 The Difference in Specific Activity of
Aldrin Epoxidase (Now Classified as One or
More CYP Enzymes) between Two Strains of
Housefly: Rutgers and NAIDM

	Aldrin Epoxidase Activity	
Fraction	Rutgers	NAIDM
B_1	297.9	1.15
C_1	160.0	41.6

Note: The activity is measured as pmol aldrin epoxidated to dieldrin per nmol cytochrome P450
and minute. It is very clear that the specific
activity is much higher in the Rutgers strain.

enzymes. The following example uses data from Yu and Terriere (1979) that
showed that a diazinon-resistant strain of housefly (Rutgers) had a 10 times
higher level of microsomal oxidase activity than a sensitive strain (NAIDM)
by using epoxidation. They were able to separate the CYP enzymes into six
different fractions (A_1, A_2, B_1, B_2, C_1, C_2) by chromatographic methods (ion
exchange and hydroxylapatite) and to compare the activity in two of them
(B_1, C_1), as seen in Table 9.2.

9.3.1.4 *CYP enzymes in plants*
In contrast to triazine resistance, there have been few reports of resistance
to phenyl urea herbicides, and cross-resistance is not usual. Blackgrass
(*Alopecurus myosuroides*), cleavers (*Galium aparine*), and annual ryegrass
(*Lolium rigidum*) have been reported to be resistant to chlortoluron due to
enhanced activity of CYP enzymes that demethylate chlortoluron by oxidation and hydroxylate the methyl group in the ring (see Burnet et al., 1993a,
1993b, 1994a, 1994b; Maneechote et al., 1994).

9.3.2 *Insensitive target enzyme or target receptor site*

9.3.2.1 *Acetylcholinesterase*
Acetylcholinesterase (AChE), the target enzyme for organophosphorus and
carbamate insecticides, may have variants with reduced sensitivity to inhibition. Such variants, which may be rare in the wild, unexposed population,
may be selected and cause resistance in insects and ticks. The insensitive
AChE in resistant strains sometimes, but not always, shows a reduced activity to substrates. An extreme case is the Ridgelands strain of the cattle tick
(*Boophilus*), with only 7% of the original activity toward the substrate acetylcholine. The number of species with this type of mechanism is growing,
and important mosquitoes species in France (*Culex pipiens*) and Japan (*Culex
tritaeniorhynchys*) have now acquired this resistance. Because this type of
resistance mechanism is caused by a slower rate of reaction with cholinesterase,
its effect can be greatly increased by a concomitant augmented detoxication.

Table 9.3 LD50 and Characteristics of Acetylcholinesterase
from Insecticide-Susceptible and -Resistant *H. armigera* Larvae

Pesticide	Susceptible Larvae	Resistant Larvae
Toxicity: LD50 (μg/larva)		
Profenofos	0.14	13.0
Methyl parathion	0.39	30.3
Chlorpyrifos	3.70	4.50
Biochemical Characteristics		
Km with acetylthiocholin	15 μM	45 μM
Vm	14.2 (arbitrary unit)	8.2 (arbitrary unit)
Bimolecular inhibition constant with methyl paraoxon	3.8×10^5 $M^{-1}mg^{-1}$	3.6×10^3 $M^{-1}mg^{-1}$

Note: The enzyme of resistant insects is a little less efficient by having a slightly higher
    Km value. This indicates a somewhat less efficient enzyme, but the difference
    is so slight that it does not cause any reduced fitness for insects because the
    amount of and activity of acetylcholinesterase are almost always much higher
    than strictly necessary. LD50 = lethal dose in 50% of the population. Vm =
    maximum enzyme velocity.

Source: Data from Gunning, R.V., Moores, G.D., and Devonshire, A.L. 1998. *Pest. Sci.*,
    54, 319–320.

The cotton bollworm, *Helicoverpa armigera*, in Australia is of great economic importance and is cross-resistant to parathion-methyl and profenofos,
but not to chlorpyrifos. It has an acetylcholinesterase with low sensitivity to
paraoxon-methyl and profenofos, but the sensitivity to chlorpyrifos is unal-
tered. As Table 9.3 shows, the enzyme of the resistant insects is a little less
efficient by having a slightly higher Km. (Km is the substrate concentration
at which an enzyme-catalyzed reaction proceeds at one-half its maximum
velocity.) This indicates a somewhat less efficient enzyme, but the difference
is so slight that it does not cause any reduced fitness for the insects. The
amount of and activity of acetylcholinesterase are almost always much
higher than strictly necessary.

9.3.2.2 *kdr resistance*

Characteristically, DDT resistance in flies does not extend to prolan; however,
strains that are resistant due to receptor-site modification are also resistant
to prolan and pyrethroids. This was observed early in houseflies and stable
flies (Busvine, 1953; Stenersen, 1966). The DDT resistance in stable flies did
not depend on metabolism (Stenersen, 1965).

The target biomolecules for DDT and the pyrethroids are the sodium
channels in the axon. One very common type of resistance is the so-called
knockdown resistance, or kdr resistance. In this case one or more amino
acids have been changed due to point mutation so that DDT or pyrethroids
do not bind. Whereas houseflies that are resistant due to the presence of the
DDT dehydrochlorinase type of glutathione transferase will be paralyzed by
DDT, it is found that when DDT has been detoxicated, the flies wake up and

fly away. When resistance is due to the altered sodium channels, the flies seem to tolerate DDT without being affected at all.

Prolan is an insecticidal DDT analogue that cannot be dehydrochlorinated and is toxic for DDTase-resistant insects.

prolan

Genetic analysis and biochemical studies have shown that kdr resistance is linked to one of the building blocks of the sodium channels in the axon (the gene for the so-called α-subunit). The gene has been sequenced in R- and S-flies. The substitution of a leucine residue with a phenylalanine residue makes the difference. Equivalent mutations have been found in pyrethroid-resistant strains of several other insect species. Two point mutations in the α-subunit of the sodium channels are associated with super-kdr resistance. Such mutations have been detected in the housefly, horn fly, diamond moth, and body louse. It may cause several hundred-fold resistance. It is interesting to note that if knockdown, and not death, is used as the endpoint, the resistance level is much higher.

9.3.3   Resistance in fungi

9.3.3.1   Benzimidazole

Resistance to benomyl and the other benzimidazole fungicides is widespread in pathogenic *Botrytis* spp. The mechanism of resistance is lower binding ability of the fungicides to the active site on β-tubulin due to point mutations that change one amino acid (e.g., valine with alanine, or phenylalanine with tyrosine). Resistant fungi may in some cases show higher sensitivity to other pesticides or be less tolerant to the cold. The phenylcarbamate diethofencarb is active toward *Botrytis* spp., showing resistance toward benzamidazoles, and may be used together with carbendazim or other benzimidazole preparations.

The amino acid sequence of tubulin has been highly conserved throughout evolution, but small and important changes can be made without seriously disturbing its function. It has been possible to change one amino acid residue (phenylalanine-167 to tyrosin-167) in the β-tubulin of baker's yeast (*Saccharomyces cerevisiea*) without altering its function seriously. However, the yeast shows a three- to fourfold resistance to carbendazim and nocodazole. The tyrosine mutant is cold sensitive and would therefore probably have a much lower fitness under natural conditions. More surprising, the mutant has an eightfold increased sensitivity to benomyl. Super sensitivity for some benzimidazole derivatives, coupled with resistance to others, has also been observed in natural mutants of *Aspergillus nidulans*. In this case alanine has been substituted with the very similar amino acid valine.

β-Tubulin from the naturally occurring ascomycete plant pathogen *Cochliobolus miyabeanus* has 447 amino acids. Other nonresistant ascomycetes had a phenylalanine at position 167, whereas this resistant biotype has tyrosine in this position, similar to the resistant baker's yeast. It therefore seems reasonable that this is the molecular basis of resistance (Gafur et al., 1998). Other substitutions may also cause resistance. In a strain of *Neurospora crassa* that was resistant to methylbenzimidazol-2-ylcarbamate (MBC) but sensitive to diethofencarb, glycine was substituted for glutamic acid at position 198. This was caused by a single base change in the β-tubulin gene and resulted in MBC resistance and diethofencarb sensitivity.

The sheep nematode *Haemonchus contortus* also has resistant varieties, and it is shown that its β-tubulin has less binding affinity to benzimidazoles.

9.3.3.2 Sterol biosynthesis inhibitors

The use of the fungicides acting by inhibition of ergosterol synthesis does not very quickly lead to resistance. Unlike the benzimidazoles, for which resistance can develop rapidly and is a very clear phenomenon, resistance to the sterol biosynthetic inhibitors shows a gradual reduction in sensitivity of the target fungal population to the particular fungicide. There is no cross-resistance between the DMI fungicides, squalene epoxidase inhibitors, and morpholine fungicides that inhibit reductases and isomerases.

Terbinafin resistance in *Necria haematocacca* may be caused by a less sensitive squalene epoxidase. The resistant biotypes have at least 10 times more squalene than the normal, and it seems that the epoxidase has less affinity to both squalene and terbinafine in the resistant mutants. Fenpropimorph causes certain sterols to accumulate (Δ8,14-sterols) and some to decrease (the Δ5,7 analogue). Resistance seems to be caused by tolerance to these changes, rather than to less sensitive enzymes.

9.3.4 Atrazine resistance and plants made resistant by genetic engineering

The target site for atrazine and other photosynthesis inhibitors is the so-called D1 protein of photosystem II. Herbicide binding causes a "bunging up" in the passage of electrons from chlorophyll to plastoquinone. Chloroplasts, the photosynthetic organelles, have their own DNA, and a plasmidic gene (called *psbA*) encodes the D1 protein. Atrazin resistance is very often caused by an allele of this gene. Serine instead of alanine in position 264 is enough to give resistance. The resulting D1 protein has a much lower binding affinity for atrazine. Atrazine-resistant biotypes show only minor effects on phenylureas, which bind to the same protein. Cross-resistance between the two groups of photosynthesis inhibitors is therefore rare. Specific mutations have been made in the *psbA* gene of the alga (*Chlamydomonas reinhardii*). When tested with various triazine herbicides, some mutants were resistant, whereas other mutants were more sensitive than the wild type.

Table 9.4 Specific Activity of Glutathione Transferase
Isolated from Atrazine-Resistant (R) and -Susceptible
(S) Biotypes of Velvetleaf (*Abutilon theophrasti*)

Biotype	Specific Activity with	
	Atrazine[a]	CDNB[b]
R	858.2	57.6
S	243.7	48.3

[a] Measured as nmol GS-atrazine produced per hour and mg
of purified glutathione transferase.

[b] Chloro-2,4-dinitrobenzene (CDNB) is a substrate often used
to monitor glutathione transferase activity because it is easy
to measure.

Another common resistance mechanism is enhanced detoxication in
plants. Atrazine has an electrophilic carbon that may be attacked by glutathione, and biotypes of various weeds may have glutathione transferases
that have a better ability to catalyze this reaction:

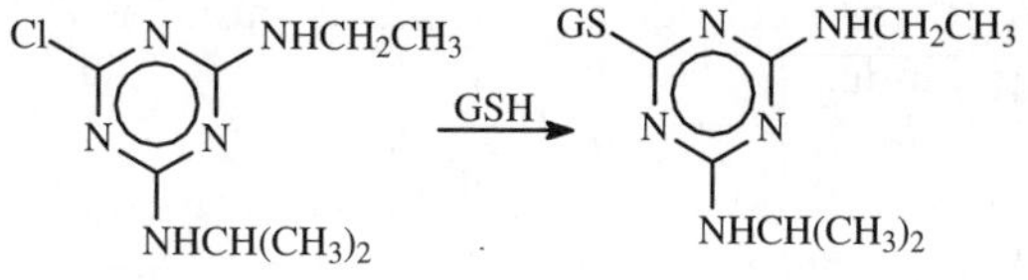

atrazine   conjugate of atrazine and glutathione

The data in Table 9.4 are extracted from data of Plaisance and Gronwald
(1999) and show the difference in the activity of glutathione transferase. In
this case there was no difference in the total amount of enzyme protein or
activity toward the standard substrate. The inheritance of this resistance
mechanism is in a nuclear gene.

9.3.5   Resistance to glyphosate

Resistance to glyphosate is not very common. In spite of 25 years of intensive
use, acquired resistance is seldom. By means of genetic engineering it is
possible to introduce genes taken from bacteria into plants, and engineered
glyphosate resistance has been tried quite often. Glyphosate has a very broad
botanical spectrum of herbicidal action. It has a very low vertebrate toxicity
and is degraded easily in soil. If crop plants had been resistant, glyphosate
would have been a perfect herbicide. Resistance may be made by either
increased detoxication or decreased sensitivity.

To illustrate engineered resistance, we may use some data from Arnaud
et al. (1998), summarized in Table 9.5. A gene from *Salmonella* (*aroA*) encodes
a glyphosate-tolerant EPSP synthase. EPSP synthase is the target enzyme for

Table 9.5 Summary of the Results of the
Work of Arnaud et al. (1998) on the Content
of EPSP Synthase and Its Properties

	Leaves		Roots	
Biotype	nkat	I_{50}	nkat	I_{50}
S	37.9	8.5	13	8
HR	475.5	980	32	100
WR	128.6	5600	12	4

Note: The activity (nkat) is in nmol EPSP produced per second in leaves or roots. Inhibition (I_{50}) is in μmol glyphosate.

glyphosate. The bacterial variant of the gene has been introduced into the tobacco plant by genetic engineering. The experiment was very successful, and the resulting plants tolerated much higher doses of glyphosate than the controls. Two biotypes, a highly resistant type (HR) and a less resistant type (WR), were studied and compared with controls (S) in detail. HR showed a growth inhibition of approximately 7.5% after 3 months by 800 g of glypho-sate/ha, while WR was inhibited 55% and S as much as 70%. The following conclusions were made:

1. The levels of EPSP synthase activity were much higher in HR plants than in WR plants and S plants.
2. The I_{50} (the glyphosate concentration necessary to inhibit 50% of the enzyme activity) of EPSP synthase was much higher in leaves of WR plants than in HR plants, in contrast to the situation inside the roots.
3. High levels of EPS synthase were maintained in the HR plants, from seedlings to 3-month-old plants, but the overproduction of EPSP synthase decreased in WR plants when they grew older.
4. No expression of resistant enzyme occurred in roots of WR plants.

It seems that HR was resistant due to overexpression of EPSP synthase in leaves, and only partly because of less sensitive enzyme, whereas WR was resistant mainly due to less sensitive enzyme.

Excellent glyphosate-resistant plants have also been made by introduc-ing a gene for glyphosate degradation (glyphosate oxidase (GOX)) into sev-eral crop plants. High and efficient detoxication is necessary because gly-phosate rapidly translocates from treated leaves to the meristematic points where the herbicide exerts its activity. The gene was taken from *Escherichia coli*, modified, and introduced into crop plants.

Glyphosate oxidase catalyzes the following reaction:

$$(HO)_2(O)P-CH_2NHCOOH \xrightarrow{\text{GOX}} (HO)_2(O)P-CH_2NH_2 + OHC-COOH$$

Table 9.6 Development of Resistance in Crop Plants by Genetic Engineering

Manipulated Target Enzyme	Herbicide	Reference
5-Enolpyruvylshikimate-3-phosphate synthase	Glyphosate	Klee et al., 1987; Shah et al., 1986
Acetolactate synthase	Chlorsulfuron	Haughn et al., 1988; Li et al., 1992
Manipulated Detoxication Enzyme	**Herbicide**	**Reference**
Phosphinothricin acetyltransferase	Glufosinate	Deblock et al., 1987
Nitrilase	Bromoxynil	Stalker et al., 1988
Dehalogenase	Dalapon	Buchananwollaston et al., 1992
Monooxygenase	2,4-D	Streber and Willmitzer, 1989
Carbamate hydroxylase	Phenmedipham	Streber et al., 1994
Glutathione transferase	Metolachlor	Andrews et al., 1997
CYP enzymes	Chlortoluron	Shiota et al., 1994; Inui et al., 2000

9.3.5.1 Summary

Herbicide-resistant crops to other inhibitors of amino acid biosynthesis include soybean, maize, cotton, canola, sugar beet, and wheat, and the herbicides include glyphosate, glufosinate, and acetolactate synthase inhibitors. Inui et al. (1999, 2000) have described how human detoxication enzymes (CYP enzymes) can be transferred to potatoes to make them herbicide resistant. They summarize the state of the art with regard to the making of genetically manipulated plants. In their introductions they give examples of manipulated target enzymes, which are summarized in Table 9.6.

9.3.6 Resistance to older biocides used as pesticides

Forgash (1984) has reviewed the early development of resistance in insects. Resistance toward the older general biocides used as insecticides is not so common, but in addition to the lime sulfur resistance in San Jose scale, we have seen several other examples.

California red scale has become resistant to hydrogen cyanide (1916), and codling moth to lead arsenate (1917). Captan is a multifunctional fungicide that reacts with SH groups in enzymes. Nevertheless, resistance has developed in *Botrytis cinera* (fruit rot). By exposure to captan the resistant fungus starts to produce more glutathione. This reacts with and inactivates captan. There are also examples of fungi that have become resistant to mercurials (*Drechslera gramini*) (Magnus, 1970). Chickweed has developed resistance to 2,4-D ((2,4-dichlorophenoxy) acetic acid). The snail that is the vector of the trematode that causes schistosomiasis may be resistant to pentachlorophenol, an uncoupler of oxidative phosphorylation. Rats have become resistant to warfarin, an anticoagulant, and the mechanism has been studied (see Searcey et al., 1977).

9.3.7   Resistance to third- and fourth-generation pesticides

The naive notion that insects would not be resistant to third- and fourth-generation insecticides (which rely on interference of hormone or pheromone function) was challenged first by Dyte (1972), who described populations of *Tribolium castaneum* that were cross-resistant to juvenile hormones.

The biochemistry and genetics of insect resistance to *Bacillus thuringiensis* have been recently reviewed (Ferre and Van Rie, 2002). Resistance to the toxins from *B. thuringiensis*, either used as such or engineered into plants, has also developed, e.g., in the pink bollworm moth (*Pectinophora gossypiella*). This is a serious problem and may make such plants less useful. The diamond moth (*Plodia xylostella*) is the only insect species that has evolved high levels of resistance in the field. The first case of field resistance was reported from Hawaii in 1985. Laboratory selection may produce several thousand-fold resistance, while field resistance is usually several hundred-fold. Sometimes resistance may be obtained for just one of the toxins, e.g., Cry1Ab, while leaving the moths susceptible to other toxins. Some mosquitoes (*Culex quinquefasciatus*) may be resistant to some of the toxins (Cry4 and Cry11), whereas other toxins (Cyt1A) counteract this by acting synergistically. It is possible that the presence of Cyt1A facilitates the binding of the other toxins in resistant larvae. The mode of action of *B. thuringiensis* toxins includes:

1. Solubilization of the toxic crystals
2. Cleavage of the protoxin by peptidases in the insect gut in order to produce the active toxin
3. Passage through the chitinous tube within the midgut, which separates the bulk food from the lining epithelium of the gut (the peritrophic membrane)
4. Binding to the receptors and insertion into the membrane
5. Formation of pores in the membrane
6. Osmotic disruption of the gut cell

Three different biochemical mechanisms of resistance have been observed so far. Interference with the binding of the toxin to the receptor and its insertion into the membrane is most common, although cleavage of the protoxin by peptidases has also been observed. Fast repair of damaged midgut cells has also been proposed as a mechanism in some cases. The different toxins have the same or different binding sites, making complicated and sometimes surprising patterns of cross-resistance. Detailed knowledge of this may be of value in strategies for resistance management.

9.4   How to delay development of resistance

Many scientists and administrators have addressed this problem, but there is no clear-cut simple method for preventing resistance (see Georghiou and

Taylor, 1977; Georghiou, 1990). Recent review articles about insecticide resistance are Hemingway and Ranson (2000) and Wilson (2001). Ferre and Van Rie (2002) have written a review about resistance to *B. thuringiensis*.

The only method to delay development of resistance is to slow down the evolutionary process, which means providing better reproduction possibilities to susceptible individuals. P. Richter (1992) argued that the aim of pesticide treatment is to reduce the target population substantially. The population recovers from three sources of founder individuals: immigrants, survivors in refuges, and resistant individuals. There is a short period after treatment during which the survivors are solitary. The population density may be too small for successful breeding. A rapid influx of (susceptible) immigrants will greatly enhance the chance of the resistant trait being conserved. Once a certain quantity of resistant insects escape extinction, the R-allele can quickly increase in frequency during following treatments. If this argument holds, resistance development from a very low frequency of R-alleles is dependent on survivors of (susceptible) insects in refuges or fast immigration. Otherwise, the few surviving R-individuals are not able to breed.

9.4.1 Refuge strategy

The strategy implies that some areas are not treated with pesticides, but are kept as a refuge for susceptible individuals. These insects can then mate without any selection pressure from pesticides. The (few) surviving, more or less resistant, individuals in the nearby treated fields will mate with the individuals from the refuge area and the resultant offspring will be heterozygotes and susceptible to the pesticide. The refuge strategy has two critical assumptions: that inheritance of resistance is recessive, and that random mating occurs between resistant and susceptible individuals. If resistance is recessive, hybrid first-generation (F_1) offspring produced by mating between S- and R-insects are killed by the pesticide. If mating is random, initially rare homozygous R-individuals are likely to mate with the more abundant homozygous susceptible insects that have not been exposed to the insecticide. In most cases, where the genetics of resistance have been studied, the resistance alleles are recessive or semidominant.

The primary strategy for delaying insect resistance to transgenic *B. thuringiensis* (*Bt*) cotton plants is to provide such refuges of cotton plants that do not produce *Bt* toxins. Pink bollworm resistance to *Bt* cotton is recessive. Survival of hybrid F_1 progeny is not higher than survival of the susceptible strain, but is markedly lower than the resistant strain. However, resistant larvae on *Bt* cotton require an average of 5.7 days longer to develop than larvae on non-*Bt* cotton. *Bt*-resistant moths from *Bt* plants will therefore probably mate more often with other R-moths, and the refuge method may not work. To achieve random mating, resistant adults from *Bt* plants and susceptible ones from refuge plants must emerge synchronously.

9.4.2   Mixing pesticides with different modes of action and different detoxication patterns

Two pesticides where the mechanism of resistance is likely to be caused by different genes may be mixed. It is highly unlikely that individuals with resistance to one will also be resistant to the other. Using such high doses of both pesticides such that they will kill heterozygote resistance will probably be prevented when the frequency of R-alleles for both is low, and none of the insects are homozygotes for mechanisms of resistance for both pesticides. Sometimes insecticides with widely different modes of action and structure may give resistance through a common mechanism. One example is parathion-methyl resistance in houseflies, which may be caused by a certain glutathione transferase. This same enzyme, however, is also able to detoxicate lindane.

Sequential or rotational use of pesticides will not prevent development of resistance, at least not if the survival of heterozygotes is better than that of normal susceptible ones. It may just make a delay because the selection pressure for each is lower.

9.4.3   Switching life-stage target

In many pests, for instance, in cabbage flies, the larvae, which are the real pest, will probably be much more robust and have a better repertoire of detoxifying enzymes (glutathione transferases) in order to detoxify the isothiocyanates and other natural pesticides present in the host plant (cabbage), while the adults may not eat anything during their short life span and hence do not have detoxication enzymes. The adults therefore will be more sensitive to pesticides and less likely to develop resistance. If pesticides are used against the adult stage and not against the real pest, the larvae, resistance development may be slowed down.

9.4.4   Increased sensitivity in resistant pests

If it were possible to develop two pesticides such that increased resistance to one of them led to increased susceptibility to the other, resistance development would at least be delayed. There is one example of this principle. The systemic fungicide *diethofencarb* is particularly effective against *Botrytis* spp., which are resistant against benzimidazole. Benzimidazoles like *carbendazim* and *thiophanate* bind to a site on the tubulin protein and inhibit mitosis. Resistant *Botrytis* has a tubulin that does not bind benzimidazoles, but may bind diethofencarb better.

9.4.5   Inhibition of detoxication enzymes

After DDT dehydrochlorinase was found as one of the causes of DDT resistance in flies, it was subsequently found that N,N-dibutyl-p-chlorobenzene sulfonamide inhibits the enzyme (Spiller, 1963). This substance, called WARF-Antiresistant, was tried, and although it increased the sensitivity of resistant insects, it was of no durable practical usefulness because either the flies developed other mechanisms of resistance or the amount of DDT dehydrochlorinase increased to overcome the action.

WARF-Antiresistant

9.5 Conclusions

- Resistance has been a problem in the whole pesticide era and is seen in animals, plants, fungi, and microorganisms.
- Resistance is caused by evolution — a fundamental biological process.
- The biochemical and physiological mechanisms may be many, but resistance is often due to an insensitive target for the pesticide or to increased detoxication.
- It is possible to curb or delay resistance development by making reproduction better for populations or subpopulations of pests that are not under selection pressure.
- Cross-resistance may occur for pesticides with related modes of action, but sometimes also between nonrelated pesticides if they can be detoxified by the same enzyme.
- It is possible to make plants resistant to herbicides by genetic engineering.

chapter ten

Pesticides as environmental hazards

We cannot describe all aspects of pesticides as environmental hazards, but we intend to address certain aspects of this problem; many books are available (Carson, 1962; Ecobichon, 2001; Edwards, 1973; Emden, 1996; Graham and Wienere, 1995; Mellanby, 1970; Ratcliff, 1967; Walker et al., 1996). It is also easy to find opinions and facts on the Internet. Our problem is that we meet people who believe that cigarette smoking causes cancer because the tobacco plants are sprayed with insecticides and who at the same time are convinced that food produced without pesticides is always safe. We are therefore often engaged in debates as defenders, and not as critics, to the unnecessary and unsafe use of pesticides, as we should be. But pesticides are poisonous to man, animals, plants, and sometimes to all types of life, as many of us have experienced.

10.1 *Pesticides are poisons*

There have been thousands of serious poisonings and fatalities caused by pesticides and pesticides are a constant threat to millions in the more illiterate parts of the rural populations in many countries. A quick search on the Internet, for instance, gave this information from Pesticide Action Network North America (http://panna.igc.org/resources/gpc/gpc_200008.10.2.24.dv.html):

> Growth of the agrochemical industry in China has been accompanied by problems related to quality control, unsafe application of chemicals, and pesticide residues. According to a report from the Chinese National Statistics Bureau, 48,377 pesticide poisoning cases, including 3,204 fatalities, were reported in 27 provinces in 1995. Another government estimate placed total farm worker fatalities due to pesticides at 7,000–10,000 annually.

Their sources seem trustworthy, but difficult to validate. Ecobichon (2001) writes:

> On the worldwide basis, intoxications attributed to pesticides have been estimated to be as high as 3 million cases of acute, severe poisoning annually, with as many or more cases and some 220,000 deaths. These estimates suffer from inadequate reporting of data for developing countries, but they may not be too far off the mark.

His source of information is a WHO report (World Health Organization, 1990). Ecobichon also states that as many as 25,000 cases of pesticide-related illnesses occur annually among agricultural workers in California. Those applying pesticides on the ground face the highest risk. Although California is at the top of the list of pesticide consumption, there are more poisonings in the developing nations. In the tropical countries more insecticides than herbicides are used. Insecticides are usually more toxic, and because of the hot weather, it is very unpleasant to wear protective clothes.

The epidemic poisoning from seed-dressed grain in Iraq is one of the well-described and well-documented tragedies that should not be forgotten. The production of pesticides is also a dangerous affair. Bhopal in India experienced one of the worst mass poisonings in history. During the production of methyl isocyanate something went wrong and a large amount of methyl isocyanate escaped. Methyl isocyanate is used to produce the more or less environmentally friendly carbamate carbaryl. The toxicity of the intermediate was not taken sufficiently into account and thousands of people were very seriously affected. Another accident happened in Seveso, Italy, where the production of pentachlorophenol, which was used as a pesticide and as an intermediate in the production of other pesticides, led to a severe dioxin contamination (see p. 238).

We should not forget that pesticides have been used as warfare agents, notably in Vietnam. The use of Agent Orange and other herbicides had an immediate and chronic effect on human health, and the vegetation has been seriously and permanently disturbed (Westing, 1975).

Pesticides are poisons and are made to be poisonous to one or several forms of life. The research necessary to develop pesticides may be dangerous, and the necessary use of animals to test toxicity and side effects may be ethically questionable. Dangers are involved in production, transport, and application. Contamination of the environment due to normal use, as well as by accidents, may occur and must be weighed against the advantages of using the pesticide. The residues in foods and the disposal of packing material and unused products need our attention. Is any alternative method more efficient, cheaper, and less dangerous?

The focus has often been on residues in food and environment as a consequence of normal and approved use. However, most accidents and harmful effects have been due to seriously incorrect use or carelessness.

Table 10.1 Types of Chemical Bonds in Biomolecules and in Compounds Often Made Synthetically

Biochemistry Textbooks		Organic Chemistry Textbooks	
Bond	ΔH (kcal/mol)	Bond	ΔH (kcal/mol)
C–H	120–70	C–F	103
C–C	128–60	C–Cl	97–83
C–O	127–50	C–Br	83–56
C–N	72–61	C–I	73–53
C–S	60–53	C–P	—
C–O–P	—	C–Si	—
P–O	—	C–Hg	68–41
N–H	110–75	C–Cd	54
S–H	—	N–O	73–40
S–S	—	P–S	—

Consequently, most countries have regulations and legislations for pesticide use. There are regulations about who is permitted to distribute, sell, and handle pesticides. In Norway, no organophosphorus insecticides and only two insecticidal carbamates (pirimicarb and ethiofencarb) are legal for use by nonprofessionals in plant protection. There are also regulations that state how much of a pesticide may be present in food. Extremely sensitive analytical methods have been developed to detect (and determine the concentration of) minute amounts of pesticides because it is difficult or impossible to be 100% certain that low-level residues do not harm anything. It is now impossible to state that by law no pesticide residues should be present, because good analytical methods mean they will be detected. Let us illustrate with an example: Hundreds of kilograms of DDT were once used in Norway to control spruce weevils and moths in the orchards. No local harm has ever been detected, but this use also adds to global contamination, although not by much. Suspend or dissolve just 1 kg in the waters of the Oslo Fjord. Will it cause any harm? Probably not. But let us dilute even more. Take 1 kg and distribute it evenly in all oceans of the world. All the oceans have approximately 1.37×10^{21} l of water. We can easily calculate that there will be 1240 molecules of DDT/l. (Multiply with Avogadro's number and divide by the molecular weight of DDT, and divide by 1.37×10^{21}.) The DDT will certainly not harm anything, but one day such low concentration will be possible to analyze.

10.1.1 Pesticides are xenobiotics

Pesticides are almost always xenobiotics. If you look at the chemical structures in organic chemistry and biochemistry textbooks, the difference is very striking (Table 10.1). The synthetic compounds have many so-called xenobiotic bonds, that is, bonds between carbons and halogens, direct bonds between carbons and phosphorus, and between carbons and metals or silicium.

Compounds with these bonds are rare in biomolecules, but they are very easy to make in chemical factories and laboratories.

Some of the enthalpies (ΔH) for homolytic bond cleavage of the chemical bonds are given in Table 10.1 to show that there are no major differences in stability between the more xenobiotic bonds and the more biological ones. Homolytic cleavage enthalpy gives a rough measure of the stability of the compound:

$$A–B \rightarrow A \cdot + \cdot B$$

It is important to know that nature can make a myriad of compounds with such xenobiotic bonds. See, for instance, Grimvall (1995) and Hjelm (1996). Good examples are the antibiotic called chloramphenicol and the hormone thyroxin. The former is an amide of dichloroacetic acid. Chloramphenicol has a nitrophenyl group, which is very rare in biological molecules. Thyroxin has an iodinated phenyl group.

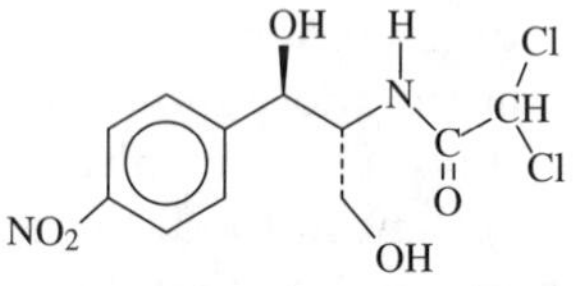

chloramphenicol

CH_3FCOOH
fluoroacetic acid

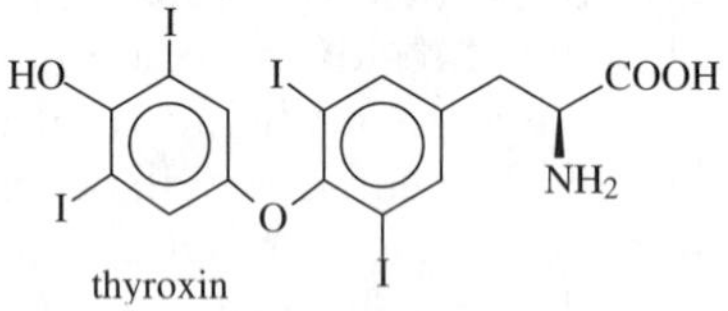

thyroxin

Fluoroacetic acid is an example of a pesticide produced by several species of plants in South Africa and Australia. Other organohalogens are both produced and mineralized during decomposition of organic matter in soil and are not the sole result of pesticide and polychlorinated biphenyl (PCB) fallout. Another interesting example of biologically produced xenobiotics is the ceramides in some invertebrates. The ceramides are analogues to phospholipids, but they contain C–P bonds. Methyl-mercury is another biological xenobiotic of greater relevance to us because it was once used quite extensively in seed dressings. However, as a side reaction in methionine synthesis, *Desulfobacter*, also produces methyl-mercury if inorganic mercury is present. Nevertheless, most pesticides have xenobiotic bonds and biomolecules do not. Pesticides also often have heterocyclic elements with structures that seldom are present in compounds of natural origin (triazines, triazoles, triazoles).

Another property of xenobiotics is that they are often mixtures of stereoisomers (racemates), whereas biological systems almost always choose one enantiomer. The different stereoisomers in a synthetic product will therefore behave differently in a biological system. They will be degraded differently and have different reactivity. Newer, more sophisticated methods for synthesis make it possible to produce the most active of the stereoisomers so that we can omit the contamination from the inactive form. Examples are the aryloxyalkanoic acid mecoprop and mecoprop-P. The former is the racemic mixture and the latter is the active enantiomer:

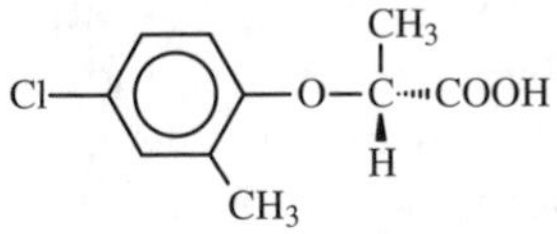

$$Cl-\bigcirc-O-\overset{CH_3}{\underset{H}{C}}-COOH$$

mecoprop-P

Substances with xenobiotic bonds are not necessarily more toxic than those without them. The aflatoxins (mycotoxins that are produced by *Aspergillus flavus*) are the most carcinogenic substances known but lack xenobiotic bonds, whereas the synthetic Freons are almost biologically inert but are also very stable. (They escape the lower atmosphere, reaching the stratosphere, where they are split into free radicals that react and destroy ozone, which helps to shield the Earth from too much ultraviolet light. These ozone-depleting reactions have dramatic environmental effects, but the direct toxicity of the Freons is modest.)

Although we can conclude that there is no absolute difference in structure or toxicity between biological substances and xenobiotics, the latter may often have structures that are very recalcitrant to degradation of biological metabolism.

10.1.2 Various types of bias

Judging environmental and health hazards objectively is a very difficult task because of the numerous ways bias can affect decisions. The misjudgments may be caused by insufficient scientific methods, by influence from media and organized groups fighting for a better environment, or from industry lobbyists. We shall summarize some of these problems in this chapter.

10.1.2.1 Publication bias

All readers of newspapers and viewers of television know that journalists like to choose the extraordinary. It is more important for them to get the attention of the reader than to inform. Newspapers write about murderers and rapists and pesticides that have extraordinary properties. We know this and are therefore very careful not to make important decisions based upon the messages from newspapers and commercial television programs. But we like to believe that scientific literature is without such bias. Nevertheless, bias in scientific publication is a real and serious problem to those who have to make risk assessments. We shall illustrate this problem by an example.

Suppose there is reason to believe that a pesticide is carcinogenic but the truth is that it is not. Two hundred laboratories carry out tests according to GLP (good laboratory practice). One hundred ninety-five laboratories could not reject the null hypothesis (H_0): *The frequency of tumors is identical in the exposed group and the control group.* Five laboratories had to reject H_0 and had to accept the alterative hypothesis (H_A): *There are more tumors in the exposed group.* This is exactly what we should expect when using the significance level $p < .05$ for rejecting H_0. If all 200 publish, then there is no publication bias, but if many of the 195 do not publish because they were too busy with other, more interesting problems or feel that negative results are uninteresting, and the 5 having positive results publish, the wrong conclusion (that the pesticide is carcinogenic) may be drawn.

Even high-rating scientific journals such as *Science* may have biased headings: "Pesticide Causes Parkinson's in Rats" (November 2000, p. 1068). The article is very interesting, and the conclusion is very clear:

> As to whether rotenone or other pesticides contribute to Parkinson's in humans, the researchers urge caution. So far, more than 15 epidemiological studies have linked Parkinson's to crude environmental risk factors, such as living in the countryside or working in the agricultural, chemical, or pharmaceutical industries. But no single chemical, including rotenone, has been reliably implicated as a risk factor. At this stage, Greenmyre [the author of the scientific article] suspects the risk of Parkinson's is a function of genetic predisposition — potentially related to how efficiently one metabolizes toxins — as well as environmental exposures.

The problem with such headings is that a busy journalist may miss the "rat" in the heading as well as the conclusion. When such unintentionally biased reports add to all those articles where the intention of the author *is* to misinform, the public and politicians will have difficulty deciding. Rational risk assessment will be impossible. A review by Begg and Berlin (1989) describes the problem.

10.1.2.2   Test bias

Examples of how difficult it is to trust even good tests are given below.

A toxicity test is carried out following standard protocols. The accuracy of the test is determined by its *sensitivity* and *specificity*. The test's sensitivity determines how many should-be positives are scored as positive, whereas the specificity tells how many inactive substances also score negative in the test. In the example below we suppose that the test has 95% sensitivity and 95% specificity. Suppose that the truth is that 2% of 1000 substances are carcinogenic. However, in spite of the good accuracy of the test, it may be too coarse for screening. Let us do some calculation to illustrate this:

Correct number of carcinogenic substances: 20 (2%)
Total number of innocuous substances: 980 (98%)

Positive scores:

Positives in tests that should have been negative (5% of 980): $980 \times 0.05 = 49$
Correctly scored positive results (95% of 20): $20 \times 0.95 = 19$

Total positive scores: 68

Negative scores:

Correctly scored as negative: $980 \times 0.95 = 931$
Erroneously scored as negative: $20 \times 0.05 = 1$

Total negative scores: 932

The test missed only one carcinogen, but as many as $(49 \times 100)/68 = 72\%$ of the substances positive in the test are not carcinogenic. This example shows that results from more than one test are necessary in order to classify a substance as a carcinogen or as safe.

10.1.2.3   Extrapolation bias

In risk assessment it is necessary to do extensive extrapolations. In human toxicology, we must extrapolate to man from experiments done on animals such as rats, guinea pigs, and hamsters. Even experiments performed on cells and bacteria are used in the assessment, which makes extrapolation even more extensive. In ecotoxicology we must extrapolate from one or few organisms, such as the most popular test organisms (e.g., the crayfish *Daphnia magna*, the earthworm *Eisenia fetida*, or the collembolan *Folsomia candida*) to all species in the environment. We must also extrapolate from laboratory experiments to the field situation.

We must almost always extrapolate from experiments carried out with high concentrations in a short time frame to real situations with low concentrations and long-term exposure.

In most cases it is impossible to calculate the risk (R) from an exposure level as an exact number (e.g., $R = 10^{-5}$). It may, however, be possible to agree on an exposure level that is so low that $R < 10^{-5}$. When performing risk extrapolations we bias *intentionally* in the worst-case direction to be sure that we do not *unintentionally* bias in the other (unwanted) direction. Figure 10.1 shows difficulties with extrapolation. We may find a good correlation between aflatoxins in the dose interval 1 to 100 ppb and response as frequency of tumors in mice between 0.05 and 1. However, we would like to know the doses that give a response of 10^{-5}. Is it possible to obtain this information from such data? Furthermore, we are not at all interested in mice. If we decide upon a safe aflatoxin exposure for mice, we must further extrapolate to humans.

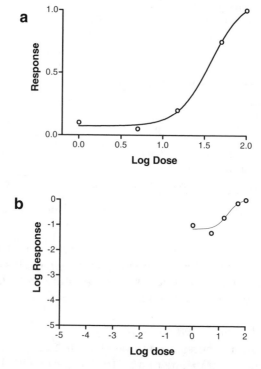

Figure 10.1 Tumor frequencies in mice as a function of dose of a carcinogen (e.g., aflatoxin) drawn in two scales. The same data set is used in (a) and (b) and shows the difficulties of extrapolating to low responses (e.g., 10^{-5}).

10.1.3   *Benchmark values*

Instead of doing more or less doubtful extrapolations down to responses of 10^{-5} or 10^{-6}, so-called benchmark values are sometimes determined. A dose–response regression line for the most appropriate endpoint is made. The dose producing 10% response and its 95% confidence interval are determined. The lower confidence limit is defined as the benchmark value (see Figure 10.2). This dose is then used to define a dose regarded as safe, by dividing it with an unsafety factor.

10.2   *Required toxicological tests for official approval of a pesticide*

The risk assessment of pesticides used for plant protection is similar but not identical in most countries. The European Union (EU), Organization for Economic Co-operation and Development (OECD), and other international organizations have specified legal requirements for toxicity documentations. The following tests should be carried out:

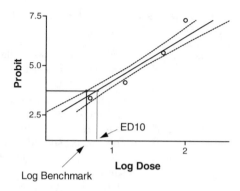

Figure 10.2 Determination of benchmark values.

- Acute oral, dermal, and inhalation studies on rats or other mammals
- Skin and eye irritation tests on rabbits
- Sensibilization tests on guinea pigs or rabbits to unveil the possibility of contact allergy
- Subacute studies on rats (14 days)
- Subchronic tests on rats and dogs (90 days)
- Chronic toxicity and carcinogenicity tests on rats (2 years) and mice (18 months)
- One-year toxicity tests on dogs
- Multigenerational reproduction tests on rats
- Teratology studies in rats, mice, or rabbits
- A battery of short-time *in vitro* and *in vivo* tests for mutagenicity and chromosomal effects
- Neurotoxicity
- Exposure by users
- Kinetics and biotransformation

Not all these tests may be required. If new data on toxicity are coming up, the approval may be withdrawn. Some countries (e.g., Norway) also require that the new pesticide be safer than or have other advantages compared with products already on the market. There may also be an approval period restricted to 5 years. Reapplication is necessary if marketing is to be continued. A fee must also be paid at the time of reapplication. This mechanism helps to keep the number of products on the market low.

The assessment of these compounds and the establishment of an ADI (acceptable daily intake), to be explained later, or MRL (maximum residue limit) are carried out by an international committee of the WHO (Joint Meeting on Pesticide Residues). The World Health Organization publishes toxicological and ecotoxicological evaluations of potential environmental toxicants in the *International Proceedings on Chemical Safety* in the series "Environmental Health Criteria."

Table 10.2 A Summary of the Necessary Steps in Residue Analysis

Stage	Step	Comment
Sampling	Collection	Avoidance of cross-contamination is as important as ensuring that a representative sample is taken
	Storage	No physical or chemical change in the residue should occur during storage
Sample preparation	Extraction	All residues, but as little as possible of other substances, should be extracted
	Concentration	A necessary step in low-concentration samples
	Isolation	Interfering substances must be removed
Analysis	Identification	The identity of the residues must be determined; expensive chromatographic and mass spectrographic methods may be needed
	Quantification	The amount or concentration of the substances must be determined, usually with a gas chromatographic method

10.3 Analysis of residues in food and the environment

10.3.1 Definitions

Pesticide residues are substances in food, forage, or the environment that are present because of the use of pesticides. Residues include derivatives such as degradation products, metabolites, reaction products, and impurities that may have toxicological significance.

Residue analysis is very expensive and requires much skill. Even more skill is needed to interpret the health aspect of minor residues of approved and well-tested pesticides. Analytical chemistry has become so advanced that almost anything can be found in minute concentrations. Social courage is needed to say explicitly that, even when the legal residue limits are exceeded, residues are not necessarily a health hazard. The residues may be regarded as "invisible rubbish" that should be kept as low as possible, even when no hazard is to be expected.

In most instances, pesticides from biological or environmental sources are too complex, too dilute, or incompatible with the chromatographic system to permit analysis by direct injection. Preliminary fraction, isolation, and concentration of the pesticides are needed before analysis can take place. A typical organic analysis can be divided into three stages: sampling, sample preparation or cleanup, and analysis (Table 10.2).

A pesticide will not be approved if an analytical method has not been developed and described, and if residue analysis is too expensive to be carried out.

10.3.2 Sampling

Because of the high cost of analysis, there must be good reason to do it, and the collection of samples must be well planned. Sometimes cross-contamination is extremely difficult to avoid. If high-residue samples are to be taken and analyzed together with low-residue materials, the low-residue samples should always be taken first. If, for instance, soil from inside and outside a fruit orchard is to be analyzed, different equipment should be used or the low-residue samples should be collected and stored before the high-residue samples are taken. It would be very easy to detect pesticide residues in Tutankamen's coffin if extreme caution was not taken. Some xenobiotics (phthalates) are present almost everywhere. A specific problem is taking below-surface soil or groundwater samples in order to look for leakage. The topsoil may be heavily contaminated and a minute amount of topsoil in the samples will spoil the results. The samples should be stored in an inert material to avoid possible sorption of the pesticides into the storage medium, or to prevent contamination of the sample from the storage medium. Often samples are kept in glass or Teflon® vessels. A good alternative is to wrap the samples in aluminum foil. Polyethylene or other plastics often contain phthalates or other softeners that may contaminate the samples with substances that interfere with the analysis. However, plastic is very versatile, lighter in weight, and cheaper than glass, and special freeze-tolerant vessels are commercially available. Laboratory equipment specialists or pesticide chemical analysts may offer advice about equipment. It is a good idea to try out various equipment before sampling is carried out. A thorough validation of the analytical procedure is always performed before analysis of real samples is done. The validation includes recovery experiments with known amounts of pesticides added to blank samples in order to have a measure of the accuracy and precision of the method.

Preferably, the samples should be frozen immediately using liquid nitrogen, dry ice, or a freezer at –80°C. Sometimes less rigid storage conditions are sufficient. Formaline, acetic acid, or some other bacteriostatic agent may be added.

10.3.3 Sample preparation

Samples must be weighed, whet, and dry. Before analysis, the samples have to be homogenized and extracted, and the pesticide must be concentrated and isolated. If the samples are too large, a subsample must be taken. Much effort must be carried out to make this representative. Sometimes the final steps of the analysis are too expensive, or the samples are too small to be individually analyzed. Pooled samples may then be made — in this case, representative individual samples are mixed and treated as one.

The extraction technique depends on the sample. Liquid samples may be extracted with an organic solvent or with a sorbent where compounds of interest are easily sorbed and desorbed. Solid-phase extraction, solid-phase microextraction, antibody–antigen interactions, or molecular imprinted polymers can be used for liquid samples. Different extraction techniques can be used for solid samples.

The residues must be soluble in the extraction liquid, whereas as few as possible of the interfering substances should be soluble. Possibilities to purify and reuse the solvents by distillation and chromatographic techniques in the laboratory are advantageous. If the extract has to be concentrated, a low boiling point is important.

Many pesticide residues are lipophilic, and a solvent not miscible with water may be used. Residues that have very low water solubility could be extracted with very unpolar solvents, like pentane, from samples that have been pulverized together with water-free sodium sulfate that take up the water.

All organic solvents represent a health hazard, but some are more harmful than others. Benzene is leukemic, hexane is neurotoxic, and tetrachloromethan is hepatotoxic and should be avoided. Diethyl ether is very inflammable and explosive and should be avoided wherever possible.

Pentane, cyclohexane, toluene, methylenechloride (dichloromethane), acetone or acetonitrile, ethanol, or methanol may be chosen, but these solvents are not without health hazards either. Pentane will only extract the most lipophilic substances, whereas methylenechloride, acetonitrile, and acetone will extract almost anything. If interfering substances are extracted together with the residues, further cleanup steps are necessary.

The most widely used adsorbents for sample cleanup are silica gel, alumina, magnesium oxide (Florisil), carbon, and diatomeous soil. If large volumes of organic solvents are used, the dilution may be too great for direct determination in the chromatographic system without preconcentration, which must be done in a way that minimizes loss of the residue components.

10.3.4 Analysis

10.3.4.1 Chromatographic methods

The final analysis consists of identification and quantitative determination. Very often these two analytical steps are done simultaneously. From 1960, gas–liquid chromatography (GLC) made it possible to analyze almost all (known) organic compounds that did not disintegrate at high temperatures and could evaporate. Chromatography is a negative analytical method. The analysis tells us that two substances are different because they have different retention times. However, two or more substances may have the same retention time. We can therefore not be sure about the identity of the substance. The identity must thus be verified with mass spectrometry after chemical derivatization. For instance, we know that DDT is converted to 4,4'-dichlorophenyldichloroethylene (DDE) by treatment with NaOH. If the sample contains a substance having the same retention time as DDT, but after treatment with NaOH has a substance with similar retention time as authentic DDE, there is reason to believe that DDT was present.

High-performance, or high-pressure, liquid chromatography (HPLC) is used when the substances are not volatile. It is possible to separate substances that are very similar, and by using a sensitive detector, accurate determinations can be made. Compounds containing chlorine, nitro groups, or a direct bond between phosphorus and carbon are very rare in nature, but quite common in pesticides. Specific detectors have been developed for such compounds. The electron capture detector is based on the principle that halogens take up electrons. The carrier gas from the separation column passes a source of soft beta radiation, and a device continuously monitors the radiation. When a chlorine-containing compound passes it adsorbs the radioactivity, and less radioactivity is then measured. Spectrophotometric methods are very useful in combination with HPLC. Different substances may absorb light in the visible or ultraviolet part of the spectrum. A spectrophotometer monitors the light absorbance of the eluate from the separation column at a wavelength suited for the specific compound. It is also possible to monitor the whole absorbance spectrum of the eluate continuously by means of a diode array detector. Taken together, the spectrum and the elution time from the separation column give the identity and amount of the compound. If the substances display fluorescence, very sensitive and specific detectors can be used.

All these analytical methods depend on comparison with known standard compounds. The presence of unknown compounds can be detected, but their identity must be sought by other *positive* methods. Mass spectrography is one such method and may be combined with gas–liquid chromatography. It is sensitive and specific, but it does require very expensive equipment. Dioxins are successfully analyzed by this method.

10.3.4.2 Biological methods

More elegant analytical methods exploit substances' biological or biochemical properties. This is simple for acetylcholinesterase-inhibiting pesticides. Acetylcholinesterase is easy to measure, and the enzyme may be bought from suppliers or extracted from flies, earthworms, or vertebrate nervous tissue. The enzyme may be measured with and without addition of the extract containing the insecticide. Some plant materials may contain natural cholinesterase inhibitors (e.g., solanine in potato) that will interfere with this analysis if not removed.

Growth-inhibiting fungicides may be analyzed by using a sensitive fungus as a test organism. A piece of filter paper with the extract is placed on an agar Petri dish inoculated with the fungus. If a zone without growth appears around the paper piece, the extract may contain a growth-inhibiting substance. Many algae are extremely sensitive to many herbicides and may be utilized in a similar way to detect the presence of herbicides. Immunological methods are also being developed. Antibodies against the pesticide, or sometimes against an adduct of the pesticide and a protein (e.g., bovine albumin), are produced by injection into rabbits. Plasma from the rabbit is then used to make a sensitive analytical method called ELISA (enzyme-linked immunosorbent assay).

In order to improve the credibility of the analysis, the laboratories follow certain standards and are certified as GLP laboratories. Ring tests, whereby many laboratories analyze identical samples, are carried out and the results compared.

10.4   Pesticide residues in food

Residues of pesticides are ranked very high as an important risk factor in society (see, for instance, Faustman and Omenn, 2001), although toxicologists do not think such residues are very significant for human health. Even in literature fundamentally critical to pesticides, the authors admit that pesticide residues in food are seldom a real toxicological problem (e.g., Emden, 1996). Gray and Graham (1995), referring to Zilberman et al. (1991), argue that too restrictive and careful practices may result in vegetables that are prohibitively expensive and therefore likely to lead to more illness due to malnutrition. Bruce Ames et al. (1990a, 1990b) present a similar view in two articles. Organic farming enthusiasts often challenge this view, and residue limits may play an important role in the trade war. Therefore, along with other reasons, pesticide residues in food are under strong legislation and regulation in most countries, and strict rules are a means of reducing the use of these poisonous chemicals. Usually the authorities set up a maximum residue limit (tolerance limit) based on toxicological data or on the expected residue level obtained when good agricultural practice (GAP) has been followed. GAP may be defined as the practice recommended by the agricultural authorities within a country. If an approved use of pesticide causes a certain residue level, a legal maximum residue limit should not be much higher than this level, even if much higher levels would be regarded as harmless. Acceptable daily intake is the key concept for determining tolerance levels of residues.

10.4.1   Toxicity classification of pesticides

WHO and other international and national organizations have developed classifications of pesticides according to their acute toxicities (Table 10.3). The terms are defined according to rat LD50 (lethal dose in 50% of the population), dermal and oral. Liquids are regarded as more dangerous than solids, dermal toxicity more dangerous than oral toxicity. Therefore, a higher dermal toxicity for a liquid is required in order to be classified as extremely hazardous.

A list of phrases that specify the hazard, called R-phrases or risk phrases, is also used. For example, R22 = harmful if swallowed; R28 = very toxic when swallowed; and R62 = possible risk of impaired fertility.

Table 10.3 Classification of Pesticides According to WHO

		LD50 for the Rat (mg/kg of body weight)			
		Oral Toxicity		Dermal Toxicity	
Class		Solids	Liquids	Solids	Liquids
Ia	Extremely hazardous	≤5	≤20	≤10	≤40
Ib	Highly hazardous	5–50	20–200	10–100	40–400
II	Moderately hazardous	50–500	200–2000	100–1000	400–4000
III	Slightly hazardous	≥501	≥2001	≥1001	≥4001
Table 5	Unlikely to present acute hazard in normal use	≥2000	≥3000	—	—
Table 6	Not classified; believed obsolete				
Table 7	Fumigants not classified under WHO				

10.4.1.1 Classification of carcinogenecity

Carcinogenecity is also classified according to well-defined terms, according to the strength of evidence, and not the potency, of carcinogens. A substance with extremely low but certain potency as a carcinogen, such as amitrole, will be classified together with very high potency substances (e.g., aflatoxin).

Below is the International Agencies for Research on Cancer (IARC) carcinogenicity evidence classification:

Group 1:   The agent is carcinogenic to humans.
Group 2A: The agent is probably carcinogenic to humans.
Group 2B: The agent is possibly carcinogenic to humans.
Group 3:   The agent is not classifiable as to carcinogenicity to humans.
Group 4:   The agent is probably not carcinogenic to humans.

10.4.2   Definitions of ADI and NOEL and tolerance limits

10.4.2.1   ADI

ADI is the amount of a chemical that can be ingested daily by humans for an entire lifetime without causing appreciable adverse effects and is expressed in mg/kg of body weight/day. It is obtained by dividing the no-observed-adverse-effect level (NOAEL, NOAEC, NEL by an unsafety factor (US), formerly called a safety factor (SF)). The factor is intended to make allowance for possible differences in sensitivity between the animal test species and humans, as well as the interindividual differences within the human population. Large safety factors are indicative of inadequate and uncertain data or unexpected toxic effects. The use of safety factors tends to be restricted to chemicals causing noncarcinogenic effects. The factor is by tradition often 100 = 10 × 2 × 5. Humans may be 10 times more sensitive than the most sensitive test organism. Some humans may be two times more sensitive due to differences in their toxicodynamics, and five times more sensitive due to differences in toxicokinetics. Unsafety factors of 10 or 1000 are also sometimes used. Table 10.4 shows that there is a correlation between LD50 values and ADI values.

The term *ADI* was first used by the joint Food and Agriculture Organizaton (FAO)/WHO Expert Committee on Food Additives in 1961 and subsequently was adopted by the FAO/WHO Expert Committee on Pesticide Residues in 1962. The ADI constitutes a useful benchmark that is employed by international and national agencies for establishing tolerances for pesticide residues in raw agricultural and other commodities and for developing health advisory guidelines for such residues in potable water. (The definition is taken, with a few modifications, from the *Dictionary of Toxicology* (Hodgson et al., 1998).)

10.4.2.2   NOEL

NOEL is defined as the highest dose level of a chemical that, in a given toxicity test, causes no observable adverse effect in the test organisms. The NOEL for the most sensitive test species, and therefore the most sensitive indicator of toxicity, is usually used for regulatory purposes. The value is calculated as mg/kg of body weight. Some experimental weaknesses are inherent in NOEL determinations. In an experiment, groups of test organisms are treated with different amounts of the pesticide, but the result from one group only is used to define NOEL. The results from the other groups are discarded. Another problem is that by using fewer organisms and endpoints the experimenter is "rewarded" with a higher NOEL value. Another possibility and scientifically more satisfactory method is to make a dose–response curve from the data and to calculate the ED10 (effective dose in 10% of the population), which is used instead of the NOEL. The lower confidence limit of the ED10 value has also been used as the benchmark value, as already explained. The L in NOEL is substituted with C if we specifically talk about concentrations in food or water. Sometimes the O is omitted. It is important because it reminds us that it is observed effects that are dealt with and not effects that have not been observed. If NOEL is calculated as mg/kg of body weight, it may be multiplied by 60 or 70. NOEL is divided by an unsafety factor (US), which may be 100 or selected after specific criteria:

$$ADI = \frac{NOEL}{US}$$

10.4.2.3   Residue tolerance limits

The maximum permitted residues in different crops are legal limits for specific compounds on and in specific foods. They are calculated from ADI values, taking into account the amount of the specific food that is likely to

Table 10.4 Typical ADI Values and Other Toxicity Data of Three Organophosphorus Insecticides, Two Insecticides Acting on the Sodium Channels, a Fungicide, and an Herbicide

Pesticide	Rat Oral LD50 (mg/kg)	Toxicity Class (WHO)	ADI (mg/kg of body weight)
Parathion	2	Ia	0.004
Azinphos-methyl	9	Ib	0.005
Dichlorvos	50	Ib	50
DDT	115	II	0.02
Deltamethrin	5000	II	0.01
Captan	9000	III	0.1
Glyphosate	5600	III	0.3

Source: The data are taken from Tomlin, C., Ed. 2000. *The Pesticide Manual: A World Compendium.* British Crop Protection Council, Farnham, Surrey. 1250 pp.

be consumed. The tolerance can therefore be higher for black pepper than for milk or potatoes. Decisions of tolerance limits may be based on residues obtained after practicing good agricultural practice and not on toxicological data, if the former is much lower than the latter. For practical purposes, a legal preharvest period is set up. This period must fit the agricultural practice and at the same time make sure that residue limits are not exceeded.

In many countries, very extensive analysis of food sampled from the market is carried out. Analysis of fruit and vegetables that have been treated with pesticides as recommended is also carried out in order to ensure that the recommendations are in fact safe. To control the amount of pesticides consumed by the general public, so-called Market Basket Programs are carried out. Samples of more than a hundred different items of food constituting a total human diet are taken from the market and analyzed for several different pesticides. Programs like this are probably good for assuring those who are concerned about the use of chemicals in food; it is very seldom that levels significant to life and health are found. If such programs are not too ambitious and too extensive, it is not necessarily a waste of money or skilled personnel because the results may mirror the general use of pesticides and may indicate irregularities and unjust use of these chemicals.

10.4.3 Comparing health hazards of pesticides with other toxicants present in the market basket

Bruce Ames and co-workers (1990a, 1990b) at the University of California were pioneers in developing *in vitro* methods to detect carcinogenic substances. Their method is often referred to as Ames' test. In some recent years, they have calculated that 99.99% (by weight) of the pesticides in the American diet are chemicals that plants produce to defend themselves. Americans eat about 1.5 g of natural pesticides per person and day, which is about

10.000 times more than they eat of synthetic pesticide residues. Fifty-two natural pesticides have been tested in high-dose animal cancer tests, and 27 of them are rodent carcinogens; these are shown to be present in many common foods. They conclude that natural and synthetic chemicals are equally likely to be positive in animal cancer tests. The comparative risk of synthetic pesticide residues is insignificant. In cabbage alone 49 natural pesticides from 6 different chemical groups are found. They are glucosinolates, indol glucosinolates, isothiocyanates, cyanides, terpenes, and phenols. The cooking of food is a major dietary source of potential rodent carcinogens. Cooking produces about 2 g per person per day of mostly untested burnt material that contains many rodent carcinogens — polycyclic hydrocarbons, heterocyclic amines, furfural, nitrosamines, and nitroaromatics. Roasted coffee is known to contain 826 of 990 volatile chemicals; 22 have been tested chronically and 16 are carcinogenic in rodents. A cup of coffee contains about 10 mg of carcinogens. However, coffee does not cause cancer in humans.

10.5 Elixirs of death

Rachel Carson (1962) used this expression to tag pesticides more or less in general. She starts Chapter 3 of her famous book *Silent Spring* by stating: "For the first time in the history of the world, every human being is now subjected to contact with dangerous chemicals, from the moment of conception to death." This statement, although far from true, still determines the agenda for much of the research and concern about pesticides. However, one substance, or more correctly, a group of substances deserve notoriously to be called an elixir of death.

The substance has the following properties: a very high acute lethal toxicity, but is slow acting. At very low doses, it may be carcinogenic, produce endometriosis and reduce semen quality, alter the immunoresponse, and reduce learning ability. The effects other than promoting cancer may be regarded as at least as important. At a somewhat higher dose it causes loss of appetite and reduces body weight. It is fetotoxic and causes enlarged spleen and skin lesions. The substance is known as dioxin, and it has all the properties an elixir of death should have. Some of its toxic effects, i.e., influencing of learning ability, immune response, a carcinogenic promoter, etc., are still under debate, and there are annual large international congresses devoted solely to dioxin and related substances.

The described symptoms may not always have relevance for human exposure. Nevertheless, the EU's Scientific Committee for Food (SCF) has recently published its risk assessment for dioxins and the PCBs related to the dioxins, and state that a weekly intake dose of 7 pg of dioxin/kg of body weight (or lower) is tolerable. The Environmental Protection Agency's Science Advisory Board in the U.S. also concluded that dioxins might give health effects at levels close to background exposures (see Kaiser, 2000). One of the problems is to decide if its toxicity has a threshold, and as yet, it has not been possible to agree on a safe dose.

The following concentrations may be found as typical average values in various foods:

Fatty fish	40.5 pg TEQ/g of fat
Egg	2.1 pg TEQ/g of fat
Milk	2.0 pg TEQ/g of fat
Meat	1.2 pg TEQ/g of fat

Note: The term TEQ is explained on p. 232.
The values are taken from Dybing
(2001). See also Lindström et al. (2000).

A weekly intake below 7 pg TEQ/kg of body weight is recommended. This is very difficult to achieve if fatty fish is consumed. On the other hand, it is recommended that we increase consumption of fatty fish at the expense of red meat, eggs, and milk. But the potency of dioxin is difficult to agree upon and is a matter of using different uncertainty factors and models for extrapolation from laboratory animals to humans. The toxicity between species varies extensively and makes extrapolations between species very problematic.

Substances often called dioxins have been present as contaminants in many pesticide formulations and in some bactericidal products. Dioxins are found in smoke from refuse incinerators and in effluents from the wood pulp industry, which uses chlorine as a bleaching agent. The magnesium industry used a production method that caused the formation of many "dioxins" (strictly speaking, dibenzofurans). Car exhaust fumes and cigarette smoke also have a low concentration. But dioxins may also be formed from more natural processes such as forest fires and cremations. Forest fires are suspected of producing 59 kg of dioxin per year in Canada alone. Humans have thus been exposed to dioxins long before the modern age.

10.5.1 Nomenclature and structure of dioxins

The term *dioxin* is used as a shortening of 2,3,7,8-tetrachlorodibenzo-p-dioxin and substances with similar modes of action, whereas the term *TCDD* (or better, 2,3,7,8-TCDD) is used for 2,3,7,8-tetrachlorodibenzo-p-dioxin only. Unsubstituted dibenzodioxin or chlorosubstituted dibenzodioxin with substitution patterns other than 2,3,7,8-TCDD may have low toxicity. Dioxin without dibenzo is very unstable.

1,4-dioxan   1,4-dioxin   2,3,7,8-tetrachlorodibenzo-1,4-dioxin

Many other substances have similar effects and modes of action. Common to them is a flat rectangular shape 3 × 10 Å in size and a charge distribution similar to that of TCDD. The most important ones are some polychlorinated biphenyls and chlorinated dibenzofurans.

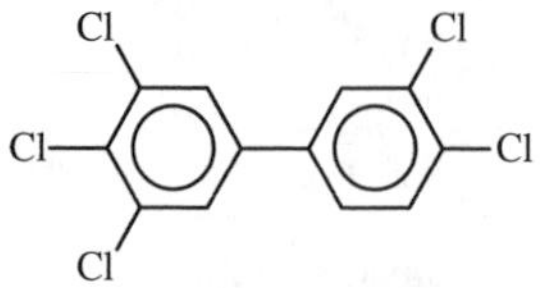

a non-*ortho* PCB with   tetrachlorodibenzofuran
dioxin-like toxicity

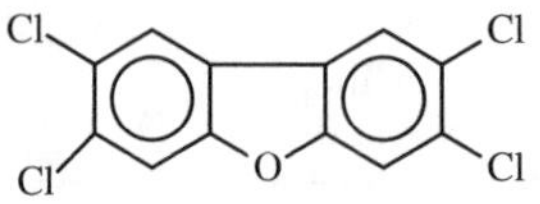

The PCBs and dibenzofurans have less relevance for pesticide science. The term *TEQ* is an abbreviation for TCDD equivalents. If the toxicity of 2,3,7,8-TCDD is 1, the toxicity of other congeners is reduced by a weighting factor. The extreme variation of potency among the TCDDs, TCDBs, and PCBs makes this necessary when risk assessments of residues are carried out. The residues are multiplied by these factors. The residue of octachlorodibenzodioxin is regarded as less potent and is multiplied by 0.001, whereas the concentration of 2,3,7,8-tetrachlorodibenzofuran is multiplied by 0.1, and so forth.

Dibenzodioxin may be substituted with chlorine in 75 different ways, and dibenzofurans in 135 different ways. The toxicity, as well as the environmental stability of the 210 derivatives, varies extremely. There are 209 possible chlorinated biphenyls, but only those with one or no chlorines in the ortho position have a mode of action similar to that of TCDD. (The ortho-substituted PCBs may also be toxic, but then through other mechanisms.)

Chemical properties important in their toxicology are very low water solubility (approximately 20 ng/l for TCDD) and low vapor pressure (1.7 × 10^{-7} mmHg for TCDD). TCDD is not stable in daylight because it absorbs light at a wavelength of 290 nm and is degraded. The half-life in surface water is about 6 days in the Nordic area. Soil adsorbs the dioxins strongly and quickly. The flat molecules with low electron density are bound very strongly to the clay and humus matrix through so-called charge transfer bonds. This property, together with the very low water solubility, reduces the significance of TCDD as the number one environmental xenobiotic. Inside the body, the situation is not so favorable. The biotransformation enzymes in the liver or other organs do not degrade TCDD and other dioxins lacking vicinal hydrogen atoms. Combined with the high lipid solubility, the body half-life becomes very high (e.g., 8 years).

10.5.2   Dioxins in pesticides

10.5.2.1   Vietnam

During the Vietnam War the U.S. used a considerable amount of herbicides in order to remove vegetation (defoliation) used as hiding areas by the enemy. In South Vietnam, from 1961 to 1971, the U.S. Air Force used more than 72 million liters of various combinations of 2,4-D, 2,4,5-T, picloram, and cacodylic acid (52 million kg of active ingredients). Agent Orange was a 50:50 mixture of n-butyl esters of 2,4-D and 2,4,5-T. More than 2.1 million ha (10% of the territory) was sprayed once. This intense use of herbicides brought considerable environmental impact, and TCDD contamination was just one of them. The 2,4,5-T was contaminated with dioxins and was blamed for the increase in the number of birth defects. The military use of herbicides very much influenced civilian use. Civilian use of 2,4,5-T (in the U.S.) dropped nearly 50% from 1964 to 1966. This decrease accompanied price increases and shortages of supply associated with the demand for 2,4,5-T as a defoliant and tactical weapon in Vietnam. During the last part of the 1960s and the first half of the 1970s there was strong scientific and political debate about the use of these pesticides. U.S. scientists were among some of the more vocal critics of their country's military politics. Their opposition, however, had little effect upon U.S. military activity. However, when the U.S. Department of Health, Education and Welfare published a report in 1969 announcing that 2,4,5-T was teratogenic in rats and mice, the herbicide program diminished. Sample sticks of Agent Orange retrieved from Vietnam for analysis by the manufacturer, the Dow Chemical Company, revealed dioxin to be present in concentrations of 0.05 to 47 ppm. It was estimated that at least 100 kg of dioxin was deposited in Vietnam (Bovey and Young, 1980; Hay, 1978a and b; Westing, 1975).

10.5.2.2 Presence of dioxins in pesticides in general
Woolson et al. (1972) examined 129 samples of 17 different pesticides derived from chlorophenols for polychlorinated dibenzo-p-dioxins. The tetra derivatives were found primarily in 2,4,5-T samples. Twenty of 42 samples contained more than 0.5 ppm. The presence of TCDD in some pesticides is easy to understand due to the chemical reactions used for its synthesis. The herbicide 2,4,5-T is produced by condensing 2,4,5-trichlorophenol with chloroacetic acid in the presence of NaOH at 105°C to yield the sodium salt of 2,4,5-T. Trichlorophenol may be made by the action of alkali on 1,2,4,5-tetrachlorobenzene produced by chlorination of the trichlorobenzenes resulting from dehydrochlorination of benzene hexachloride (BHC). Today most countries will not approve a pesticide without data on dioxin content.

Some of the reactions involved in 2,4,5-T production and the side reaction leading to TCDD are as follows:

tetrachlorobenzene   trichlorophenol

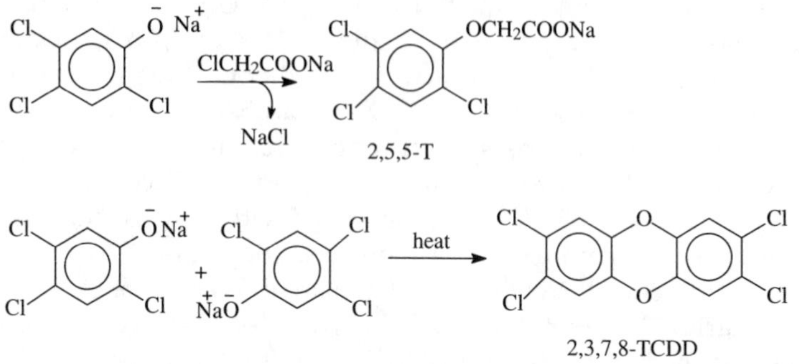

The amount of TCDD formed is higher at high temperature in the reaction vessel. Because the reaction between tetrachlorobenzene and sodium hydroxide is exothermic, cooling is necessary to minimize TCDD formation. But even at optimal conditions a small amount of TCDD is formed.

10.5.3   Toxicology

The toxicity of the dioxins was first observed in 1895 in workers in the chemical industry in Germany. They developed a skin complaint called chloracne, believing it was caused by chlorine. In 1957, millions of chickens in the eastern and midwestern U.S. died of a disease characterized by excessive fluid in the pericardial sac. The name chick edema factor was therefore applied to the component(s) in the feed responsible for this symptom. Liver and kidney damage was also observed. Similar outbreaks of the disease occurred in 1960 and 1969. A contaminant in trichlorophenol, or products produced from trichlorophenol, caused the chick edema disease and chloracne. Twelve years later its structure was published (Cantrell et al., 1969). The Dow Chemical Company had been aware of a highly toxic impurity formed in small amounts from the production of 2,4,5-T as early as 1950.

It is now established that dioxins are strong immunotoxicants, teratogens, carcinogens (as promotors), and make chloracne (Kohn et al., 1996; Melnick et al., 1996; Portier et al., 1996). Note, as already mentioned, that the toxicities for different animals are very different. Humans may be less sensitive than many other mammals. Table 10.5 shows LD50 values of 2,3,7,8-TCDD for various animals.

10.5.4   The target

Although the toxicity of dioxin and the possible threat to our health from these types of compounds are not yet settled, we know much about the primary target molecules and the reason for the difference in potency between substances and the difference in sensitivity between species.

Table 10.5 LD50 of 2,3,7,8-TCDD
in Various Animals

Organism	Single-Dose LD50 ($\mu g/kg$)
Guinea pig	0.6–2.0
Rabbit	115
Monkey	~70
Chicken	25–50
Rat	22–45
Dog	~100–200
Mouse	114–284
Frog	>1000
Hamster	1157–5051
Earthworm	Low toxicity
Plants	No effect

Note: The data in the table are collected from various sources, so other values may be found.

The target molecule is a so-called ligand-activated transcription factor belonging to the basic helix–loop–helix/Per-Arnt-Sim (bHLH-PAS) family of transcriptional regulatory proteins. (It is recommended that for further information about transcriptional regulatory proteins, a molecular biology textbook be consulted.) A transcription factor is a protein that binds to certain regions of DNA and induces transcription, i.e., production of mRNA that is the template for specific proteins. Hormones and other low-molecular-weight substances sometimes activate the transcription factor. As already mentioned, TCDD and other molecules with similar shape bind to the same transcription factor and have similar kinds of toxicity. There is a good correlation between these substances' binding ability and their toxicity. The transcription factor is usually called the Ah receptor because the most distinct result of this binding is an increased production of enzymes called CYP1A1 and CYP1A2. Ah is an abbreviation of aryl hydroxylase, which is another name of the CYP enzymes induced. The Ah receptor is a large protein with different regions — some of them are strongly conserved (Hahn et al., 1997). All vertebrates seem to have the enzymes CYP1A1 and CYP1A2 that increase in amounts soon after exposure to TCDD or similar toxicants. But as evolutionary distinct organisms, the nematode *Caenorabditis elegans* has genes with strong structural homology to the Ah genes in mouse, humans, and rats. *Drosophila* has a gene coding for a transcription factor (the *ss* gene) that is required for development of the antennae, establishment of the tarsal regions of the legs, and normal bristle growth. This gene has a base sequence with great homology to that of the Ah receptor in vertebrates. *Drosophila* also has a gene, called *Tgo*, that is similar to the so-called *Arnt* in vertebrates. *Arnt* (the aryl hydrocarbon receptor nuclear translocator) is important to translocate the Ah receptor when bound to dioxin into the nucleus and support its binding to the DNA. These findings suggest that dioxins, or similar substances, will still have undetected toxic manifestations in vertebrates or in various invertebrates (Emmons et al., 1999). It is interesting to note that the insect juvenile hormone also has a receptor belonging to the bHLH-PAS family of transcriptional regulators. The insecticide methoprene binds to this receptor, and insects resistant to this insecticide have a modified receptor with less affinity to methoprene (Ashok et al., 1998).

The first detected effect on the molecular level, the increase of the CYP1A enzymes, may lead to activation of carcinogenic substances that need metabolic activation. Other substances, however, are inactivated by the same enzymes. Steroid hormones, polyaromatic hydrocarbons, and many other endogenous and exogenous substances are also metabolized by the CYP enzymes. The production of a protein called plasminogen-activator-inhibitor-2 is also induced by dioxin. Plasminogen-activator-inhibitor-2 inhibits a specific protease, a plasminogen-activator that splits a protein (plasminogen) in order to make a product called plasmin, which by itself is a protease. Plasmin helps cells to dissolve their attachments to other cells. Overproduction of the plasminogen-activator may aid cancer cells in penetrating tissue. Inhibition may disturb embryonic development, wound healing, and inflammation. Another protein with the name cytokin-interleukin-1-β is also enhanced by dioxin. Cytokines are small proteins secreted from cells that bind to cell-surface receptors of certain cells and trigger their differentiation or proliferation. Cytokin-interleukin-1-β is important for inflammation reactions and the immunoresponse. More detailed explanations of the functions of plasminogen-activators and cytokines are given in textbooks of molecular biology. The interference with the normal regulation of the production of plasmin and cytokin-interleukin-1-β may explain many of the toxic properties of dioxin (Sutter et al., 1991).

10.5.4.1   Dioxin and metabolism of caffeine

The metabolism of caffeine is dependent on the types of degradation enzymes attacking it. Dioxin induces an increase of CYP1A2, which degrades caffeine differently than do the other caffeine-degrading enzymes that are always present. One of the metabolites of caffeine, paraxanthine, that are produced by CYP1A2 is excreted in urine where it may be analyzed. Halperin and co-workers (1995) analyzed the urine of factory workers that 15 years earlier had been exposed to dioxin in a 2,4,5-T factory. The result was negative. There were no differences between control groups and the workers. However, smoking in both groups had a dramatic effect on the caffeine metabolism.

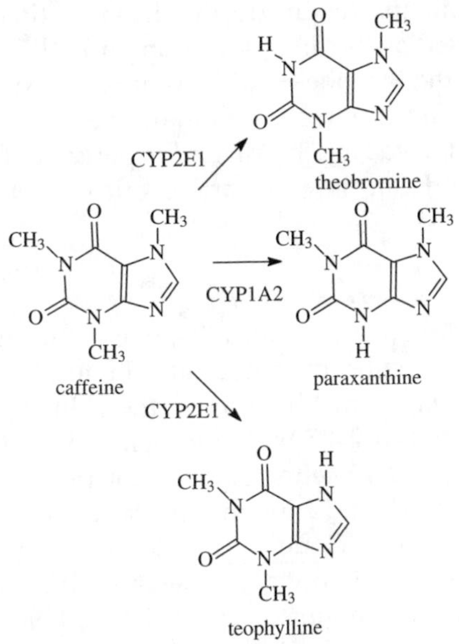

theobromine

caffeine

paraxanthine

teophylline

10.5.5 Analysis

The high toxicity and the diversity of the molecular species of the dioxins have been a challenge for analytical chemists. Very low concentrations have to be detected and quantified. As usual, the analytical protocol has four components: extraction, cleanup, identification, and determination. The material may be homogenized together with anhydrous sodium sulfate in order to bind the water and make residues of dioxin available for extraction. Extraction is then carried out with hexane, which is then passed through chromatographic columns of silica, potassium silicate, and sodium sulfate to trap undesired components. The extract is subsequently passed through a small amount of charcoal that adsorbs most of the dioxins, but also substances that disturb the final analysis. A mixture of solvents elutes these before the dioxins are gathered in toluene. The extract has to be purified further by various chromatographic methods before the final analysis can take place. The final purified extract is then analyzed by gas chromatography using a mass spectrograph as the detector (GC/MS). The problem is, of course, loss of dioxins during the complicated extraction and cleanup procedure and the final identification of the various peaks obtained on the mass spectrograph. This problem is partially solved by adding known amounts of dioxin isomers with peaks similar, but not identical, to those of the parent compounds on the mass spectrum. This is done by adding dioxins of known amounts and identities with some of their ^{12}C atoms replaced by ^{13}C isotopes. They can be detected and quantified, and by suggesting that they behave identically as the ordinary dioxins during the extraction and cleanup procedure, they can be used as the analytical standard. The GC/MS instruments are very expensive; the ^{13}C-labeled dioxins are also expensive. The solvents and column materials must be extremely pure, and the procedure is time-consuming and needs much skill and experience. Thus, dioxin analysis is very expensive and can only be carried out for very good reasons (see Oehme et al., 1989).

10.5.5.1 *Saturday, 12:30, July 10, 1976*

The accident that happened July 10, 1976, was a turning point, at least for Europe as far as production of dangerous chemicals was concerned. The town of Medan, not far from Milan in northern Italy, had a factory called Icmesa that had produced 2,4,5-trichlorophenol since 1969. An exothermic reaction in a tank with 1,2,4,5-tetrachlorophenol and NaOH had started. The cooling water was turned off during the weekend and a safety valve did not operate as it should have. A cloud of chemicals more than 6 km long and 1 km wide spread out over a highly populated area. About 40,000 people were exposed to an unknown amount of TCDD and other chemicals. Estimates of total TCDD content ranged from 0.3 to 150 kg. The explosion and emission of toxic chemicals came as a shocking surprise to the population. Three days after the accident, rabbits and birds were found dead, and later on children developed chloracne. Information from the factory management was scarce, and not until 10 days after the accident was it admitted that dioxin was present in the emission. Over the next 14 days people were prohibited from eating fruit and vegetables from the area. Much later it was found to be necessary to evacuate 736 persons from the most contaminated zone. Eventually 80,000 cattle had to be killed, while 4000 were already dead from poisoning. The surface soil from the contaminated area was removed, and many years later barrels with the soil were burned by Ciba-Geigy AG in Basel. However, there is some controversy as to the fate of the soil; a television program in 1993 suggested that the soil might have been placed in a rubbish landfill site in Germany.

Very soon after the Seveso accident people got skin lesions, though no increase in mortality was recorded. Total cancer incidents have not increased, but when looking at different types of cancer, mortality associated with leukemia in men and bone marrow cancer in women showed an increase. An increase in mortality due to heart disease in men has been recorded, but this may be attributed to stress. A further tragic result of the accident was the death of the factory director, who was shot dead while he accompanied his children to school. The so-called Seveso directive, valid in the EU since 1982, states that workers and the public shall know about the hazards from industry, and that security measures in case of accidents should be worked out and implemented.

10.5.6 Summary

In this short treatise, the following points should be noted:

1. The term *dioxin* is used here for a wide group of substances that have a structure similar to 2,3,7,8-tetrachlorodibenzo-p-dioxin. This latter term is abbreviated TCDD or 2,3,7,8-TCDD when it is necessary to distinguish it from other chlorinated dibenzodioxins.
2. TCDD is a supertoxicant for most mammals, but its potency varies widely.
3. Its primary reaction in the body is well understood, but the many toxic manifestations (e.g., promoter carcinogen, teratogen, immunotoxicant, skin lesions, loss of appetite, etc.) that it causes are not so easily explained and are a result of disturbance of regulatory processes at the DNA level.
4. It has low toxicity toward earthworms and other invertebrates.
5. It is strongly adsorbed to soil and has extremely low water solubility and a high degree of bioaccumulation potency in fat deposits in the body.
6. It is degraded by sunlight.
7. It was present in many pesticide formulations, notably 2,4,5-T, but is today mostly produced by natural processes, garbage incineration, gasoline combustion, cigarette smoking, forest fires, and so forth.

10.6 Angry bird-watchers, youth criminals, and impotent rats

This heading refers to eggshell thinning in peregrine falcon eggs caused by DDE; the proposal in the 1970s that DDT, as a nerve poison, led to an imbalance in youths by disturbing the nervous system, leading to criminal behavior; and later works indicating that rats may get erectile dysfunction because of DDE — a metabolite of DDT — which may also apply to humans.

DDT has resulted in a lot of environmental problems, some real and some rather speculative.

The first real threat was the development of resistance among flies and mosquitoes described in Chapter 9 (Brown, 1958). Rachel Carson (1962, p. 23) wrote: "No one yet knows what the ultimate consequences [of DDT usage] may be." She admits that the Food and Drug Administration chemists declared as early as 1950 that it is "extremely likely the potential hazard of DDT has been underestimated." She was not the first, but definitely the most vocal, in warning against the use of this substance. From that time, many felt that DDT would be harmful for something, but it was very difficult to find a real hazard caused by this compound. Its high persistence was well known, and a possible accumulation in the soil was believed to harm its fertility. It was suggested that in districts with rainy weather there would be no accumulation and no loss of soil fertility due to leakage and runoff. It was known that DDT was transferred from dairy milk to consumers, but so what? Good toxicological testing had been carried out, and when the Italian city Naples was liberated in the Second World War, its entire population had been dusted with DDT without any signs of harmful effects. Because of its transfer from hay or the cows' general environment to the milk, there were some restrictions on the use in cowsheds and in fodder production. The first real harmful effect attributed to a DDT analogue was probably the Clear Lake affair. The following description of the Clear Lake affair is taken (with permission) directly from K. Mellanby (1970, *Pesticides and Pollution*, Harper Collins, London, pp. 125–128).

10.6.1   Clear Lake

The greatest danger from the use of DDT over wide areas arises if water is contaminated. Damage to many forms of life may also occur when aquatic pests are controlled by chlorinated hydrocarbons. This is due to several causes. First, fish are particularly susceptible to DDT poisoning, so that fish deaths due to direct poisoning may occur immediately or soon after the insecticide has been applied. They may extract DDT, present in a low concentration, from the immense amount of water, which is passed through the gills for purposes of respiration. Second, insects and other invertebrates, which are the main food of some fish, may be exterminated, so the fish are starved. Third, the invertebrates may take up amounts of DDT, which are not immediately lethal; when fish eats these poisoned animals, the DDT may be retained in the bodies of the fish, which over a period obtain a toxic dose. To find clear-cut cases of the possible effects of chlorinated hydrocarbons we have again to turn to North American experience.

One classic case, which has been fully investigated, is that of Clear Lake in California. This has often been quoted, but it illustrates the case so well that I feel bound to deal with it again. This lake is used for recreation and fishing, but there have always been many serious complaints due to the clouds of a small gnat, a *Chaoborus*, whose larvae live in the water. This insect does not actually bite, but it occurs in such numbers that it is a serious pest. Holidaymakers also complain about the smaller numbers of biting mosquitoes and midges, which are a feature of many American inland holiday resorts. It is important to stress that there really was an insect problem; some accounts of Clear Lake give the impression that the authorities deliberately poisoned the area for no good reason. The next point to remember is that attempts were

made to use insecticides that would have the least possible harmful side effects. After considering all the chemicals available, it
was decided to use DDD (the same substance is also known as
TDE), an insecticide related to DDT but shown under experimental conditions to be less lethal to fish. It was known, as mentioned
above, that fish are very susceptible to DDT, and water containing
0.5 parts per million may be lethal. DDD concentrations of this
level did not appear to harm most fish in preliminary tests. In the
first instance, in 1949, DDD was applied at Clear Lake in an
amount which, had the whole amount been dispersed through
the total volume of the lake, concentrations in the region of 0.015
parts per million would have been obtained. As it was expected
that most of the DDD would sink to the bottom, where the midge
larvae were found, the exposure of the fish was assumed to be
slight. In 1949 the operation was a spectacular success. The midges
were almost completely eliminated, other invertebrates were certainly harmed, but not seriously, and the populations seemed to
build up rapidly. The fish, and other wildlife, including fish-eating
birds, seemed unharmed. In 1950, although no further control
measures were introduced, the midges did not return in sufficient
numbers to cause serious complaints. In the next two years the
midges did increase in numbers up to approximately their original level, but control from one application of an insecticide had
lasted for several years, and it was assumed that a safe and eco-
nomical method of midge control had been found.

Further applications of DDD were made, and gave reasonable,
though not always 100 per cent, midge control. However, by 1954
serious side effects were suspected. In fact, the first application
of DDD appears to have seriously reduced the breeding of the
Western Grebe in the lake area, but it was not until the winter of
1954 that large numbers of dead grebes were found and a public
outcry occurred. The exact cause of these deaths took several
years to find. It now seems that this was a striking case of the
possible concentration of a poison in a food-chain, with death
caused only to the animals at the end of the chain. As mentioned
above, the amount of DDD originally added was only sufficient
to give a very low concentration if equally dispersed on the water
or the bottom mud. Even after several applications this amount
is still low. However, all forms of life have concentrated the DDD.
Plankton is found to contain about five parts per million. Small
fish, which feed on plankton, contain about twice as much. The
predatory fish which eat these small fish contain much higher
amounts, but the greatest concentrations of DDD and other substances produced by its metabolism have been found in the
grebes, which have as much as 1,600 parts per million in their

visceral fat. We shall discuss the significance of small insecticide residues in birds and other animals later, but there is no doubt that large amounts like this are sufficient to cause fatal poisoning, even if lower residues may not always be harmful. The interesting fact to emerge at Clear Lake was that this concentration in food chains could occur, and could kill the *fish-eating birds* at the end of the chain. This has caused ecologists to worry in case similar effects should occur in many other ecosystems.

There have been many other reports from America of invertebrates, fish and fish-eating birds being killed when streams and lakes are treated, accidentally or on purpose, with insecticides.

10.6.2   *Peregrine falcons and other birds of prey*

DDT may have serious effects on reproduction in birds and other vertebrates. DDT's countdown started when Ratcliff (1967) at Monks Wood Experimental Station, Abbots Ripton, U.K., found that a decrease in eggshell weight in peregrine and sparrow hawk had been synchronous, rapid, and widespread from 1945. He wrote in his "Decrease in Eggshell Weight in Certain Birds of Prey" (Ratcliff, 1967):

> There was a notable boom in organic insecticides, fungicides and herbicides containing chlorine, mercury, phosphorus and sulfur after 1945, and environmental contamination by persistent residues have been widespread. British peregrines, sparrow hawks and golden eagles have shown widespread contamination by pp'-DDT, pp'-DDE, γ-BHC, dieldrin and heptachlor epoxide. The introduction of DDT into general use (about 1945–46) coincided closely with the onset of the eggshell change [Figure 10.3]. Dieldrin, aldrin and heptachlor appeared 10 years later and have been used extensively ever since.

Ratcliff's suggestion that DDT was responsible was correct. During the following years many reports substantiated his findings and suggestions. For instance, Stickel and Rhodes (1970) found a good negative correlation between DDT in the food given to coturnix quail and the hatchability of their eggs, the shell thickness, and egg numbers produced per bird. Cade's group (1971) found a correlation between DDE residues and eggshell thickness in the Alaskan falcon and hawks. The biochemical mechanism behind the eggshell thinning is now known. DDE is a specific inhibitor of an enzyme, prostaglandin synthase, thus inhibiting synthesis of a type of prostaglandin that regulates transport of calcium into the eggshell gland. A review explaining the details is written by Lundholm (1997).

As early as 1966 Sladen and co-workers (1966) reported that DDT residues were present in adipose tissue of the Antarctic Adelie penguin (*Pygoscelis adeliae*) and crabeater seal (*Lobodon carcinophagus*). The animals were taken

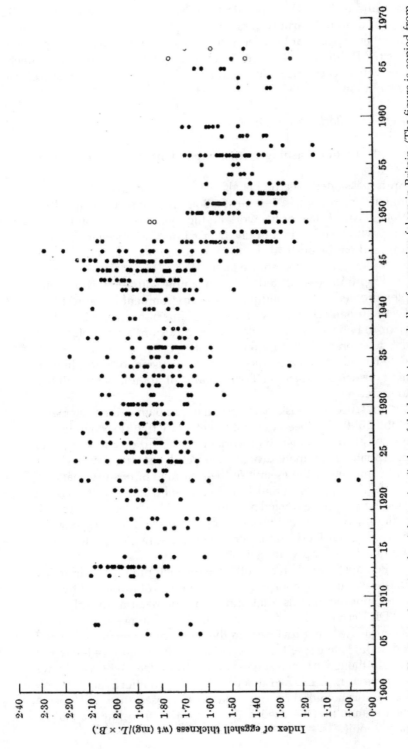

Figure 10.3 Change in the ratio of weight to size (index of thickness) in eggshells of peregrine falcon in Britain. (The figure is copied from Ratcliff, D. 1967. *Nature*, 215, 208–210. With permission.)

to Ross Island in 1964. The concentrations were low — between 16 and 152 µg/kg — and do not harm the animals.

Not all accepted that bird reproduction toxicity was a serious problem and that DDT should be banned. In October 1971 *Nature* had a comment from the Nobel Peace Price winner Norman E. Borlaug that symbolized the green revolution.

10.6.2.1 Borlaug's warning

By our Washington Correspondent (Anonymous, 1971)

An impassioned plea on behalf of DDT was delivered in Washington last week by Norman E. Borlaug, recipient of the 1970 Nobel Peace Prize for his work on high yield wheat strains. Borlaug, who testified in a public hearing on the cancellation order imposed on the pesticide by the Environmental Protection Agency, said at a press conference later:

"Environmentalists today seek a simple solution to very complex problems. The pollution of the environment is the result of every human activity as well as the whims of nature. It is a tragic error to believe that agricultural chemicals are a prime factor in the deterioration of our environment.

"The indiscriminate cancellation, suspension, or outright banning of such *pesticides as* DDT is a game of dominoes we will live to regret.

"DDT, because it is a name popularly known to most segments of the public, has been the first target. Once that is accomplished, the so-called ecologists will work on hydrocarbons, then organophosphates, carbamates, weed killers, and, perhaps, *even fertilizers* will come under the assault of their barrage of misinformation.

"If this happens — and I predict it will if most DDT uses are cancelled — I have wasted my life's work. I have dedicated myself to finding better methods of feeding the world's starving populations. Without DDT and other important agricultural chemicals, our goals are simply unattainable.

"Perhaps more than any other single factor in the world today DDT has a unique contribution to the relief of human suffering. I need not reiterate its vital importance in malaria control.

"DDT critics will say, of course, that only domestic uses of the chemical are being reviewed in the hearings at which I appeared today. But I have spent my life working in the nations of the world to help them feed themselves. I know how they will react if we *terminate uses* of DDT in this country and, in effect, label it 'poison'. If it is not good enough for your purposes, they will reason, then it shouldn't be used in our countries. The impact will be catastrophic."

In spite of Borlaug's warning, DDT has been phased out in almost all countries. Borlaug's view still gets support, but now more from the health workers. Smith (2000) wrote:

> In many regions of the world, especially Europe and the USA, people have forgotten what it is like to have endemic malaria. One of the most important reasons why these regions are no longer endemic for malaria is the use of DDT.
>
> There is little evidence that chronic low-level exposure to DDT produces serious deleterious effects, the current debate on potential "endocrine disruptors" has brought up the possibility of other toxicological effects. DDE has been found to be an antiandrogen, and in addition to its proposed link to breast cancer, DDT is commonly cited as having estrogenic effects. In one study of the most heavily exposed workers in a DDT factory, there seemed to be no effect on their ability to father children. In interpreting possible toxic hormonal effects of DDT, it should be noted that in-vitro studies often employ the o,p-isomer of DDT, which have weak estrogenicity in vivo but has constituted only a tiny percentage of the total DDT used.

o,p'-DDT constitutes up to 30% of the technical product, and is usually not removed. However, it is not so persistent as DDT and DDE, and constitutes a very low part of the total DDT residues in the environment. Its importance for low-level exposure through food from environmental exposure is therefore negligible.

Indoor spraying with DDT is still the safest and most cost-effective approach to malaria control.

DDT has been used for 55 years. Any negative effects on humans have not been confirmed. "There are probably few other chemicals that have been studied in as much depth as has DDT, experimentally or in human beings" (cited from Smith, 2000).

A recent search in the database ScienceFinder Scholar using the word *DDT* alone or combined with other words gave the following hits:

DDT alone	33,479 hits
plus *World Trade Center*	1
plus *intelligence*	6
plus *cancer*	1143
plus *toxic*	7580
plus *fatigue syndrome*	1
plus *ADHD*	2
plus *impotence*	1

Note: ScienceFinder searched several databases and therefore many hits are duplicates. Nevertheless, the resulting search is impressive.

10.6.2.2   DDT and impotence?

According to Susan Brien et al. (2000, the single hit obtained by combining *DDT* and *impotence*), 33 million American men have erection problems, which may be caused by environmental contaminants. This problem inspired her and her co-workers to see if p,p-DDE disturbed erection in rats. They found that high doses (500 mg/kg) of DDE indeed had such an effect and concluded: "The endocrine disruptor p,p-DDE can markedly interfere with erectile function and demonstrates persistence after a single dose. This supports our novel concept that environmental hormones may cause erectile dysfunction."

The knowledge of the estrogenicity of o,p-DDT is old, whereas the effect of p,p-DDE as a potent androgen receptor antagonist (an antiandrogen) is from more recent data (Kelce et al., 1995a, 1995b). The significance of this finding to human health is not quite clear. The epidemiological evidence is not very convincing, and the results from studies on rodents are not easy to interpret due to doubtful extrapolations from high doses to low doses, and from one species to another. A few years ago environmental contaminants were always blamed for being carcinogenic and other possible effects were almost totally neglected. The focus on xenoestrogens and other hormone agonists or antagonists has therefore been important. The search for toxic effects other than carcinogenecity is important. Such effects of DDT are not restricted to birds and alligators, although they definitely have a more serious effect on these animals than on humans and other mammals.

One effect of DDT at the molecular level is a change of metabolism of other substances. DDT and similar compounds increase the level of various CYP enzymes. In a study by Sierra-Santoyo and co-workers (2000), it was found that female rats dosed with DDT had an 18-fold increase of CYP3A2, whereas CYP2B1 and 2B2 increased 19-fold. Males did not react in this manner. The toxicological relevance of this finding for humans is uncertain.

Beard and co-workers (2000) found a correlation between bone mineral density and DDE level in plasma of 68 women who had an adequate intake of calcium. Other factors, such as age and years of hormone replacement therapy, were also negatively correlated to bone mineral density. However, only 21% of variance of the mineral density was explained by age, DDE, and hormonal replacement therapy.

The lesson we have learned from the many laboratory and field studies of DDT is that real-world observations, as well as sophisticated laboratory experiments, are important in order to unveil possible side effects of pesticides.

10.7   Conclusions

Probably there is no synthetic substance that has saved more human lives than DDT, and no substance that has been scrutinized for side effects so rigorously.

In spite of this, the possible negative effects on human health are speculative and must be subtle. They include a weak positive correlation to breast cancer, reduced bone mineral density, erection problems, and other disturbances of male sexual functions.

Dramatic ecotoxicological effects caused by DDT have been observed and are explained by the hormone-like effect of DDE and DDE's inhibition of prostaglandin synthase, combined with its high stability and bioconcentration ability. Disturbance of male sexual development in alligators has recently been observed and explained by the antiandrogenic activity of DDE. Eggshell thinning and decline of the population of peregrine falcons far outside the areas where DDT has been applied are the most serious problems observed so far, but DDT has also caused local mass death of birds due to poisoning through the food web.

Literature

Adams, M.D. et al. 2000. The genome sequence of *Drosophila melanogaster*. *Science*, 287, 2185–2195.

Alberts, B., Bray, D., Johnson, A., Lewis, J., Raff, M., Roberts, K., and Walter, P. 1998. *Essential Cell Biology: An Introduction to the Molecular Biology of the Cell*. Garland Pub., New York. 630 pp.

Alberts, B., Johnson, A., Lewis, J., Raff, M., Roberts, K., and Walter, P. 2002. *Molecular Biology of the Cell*. Garland Science, Taylor & Francis Group, London. 1463 pp.

Aldridge, W.N. and Davison, A.N. 1952a. The inhibition of erythrocyte cholinesterase by tri-esters of phosphoric acid. 1. Diethyl p-nitrophenyl phosphate (E600) and analogues. *Biochem. J.*, 51, 62–70.

Aldridge, W.N. and Davison, A.N. 1952b. The inhibition of erythrocyte cholinesterase by tri-esters of phosphoric acid. 2. Diethyl p-nitrophenyl thionphosphate (E605) and analogues. *Biochem. J.*, 51, 663–671.

Aldridge, W.N. and Reiner, E. 1972. *Enzyme Inhibitors as Substrates: Interactions of Esterases with Esters of Organophosphorus and Carbamic Acids*, Vol. XVI. North-Holland Pub. Co., Amsterdam. 328 pp.

Altenburger, R., Bödeker, W., Faust, M., and Grimme, L. 1990. Evaluation of the isobologram method for the assessment of mixtures of chemicals. Combination effect studies with pesticides in algal biotests. *Ecotoxicol. Environ. Saf.*, 20, 98–114.

Ames, B.N., Profet, M., and Gold, L.S. 1990a. Dietary pesticides (99.99-percent all natural). *Proc. Natl. Acad. Sci. U.S.A.*, 87, 7777–7781.

Ames, B.N., Profet, M., and Gold, L.S. 1990b. Natures chemicals and synthetic chemicals: comparative toxicology. 3. *Proc. Natl. Acad. Sci. U.S.A.*, 87, 7782–7786.

Amrhein, N., Deus, B., Gehrke, P., and Steinruecken, H.C. 1980. The site of the inhibition of the shikimate pathway by glyphosate. II. Interference of glyphosate with chorismate formation *in vivo* and *in vitro*. *Plant Physiol.*, 66, 830–834.

Anber, H. 1989. *Studies on Pesticide Resistance. The Biochemical Genetics of Resistance to Organophosphates and Carbamates in Predacious Mite*, Amblyseius potentillae *(Garman)*. Faculty of Biology, Department of Pure and Applied Ecology, University of Amsterdam, Amsterdam. p. 90.

Andrews, C.J., Skipsey, M., Townson, J.K., Morris, C., Jepson, I., and Edwards, R. 1997. Glutathione transferase activities toward herbicides used selectively in soybean. *Pest. Sci.*, 51, 213–222.

Anon. 1971. Borlaug's warning. *Nature*, 233, 444.

Anthony, N.M., Harrison, J.B., and Sattelle, D.B. 1993. *GABA Receptor Molecules of Insects*. Birkhäuser Verlag, Basel, Switzerland. 172–209 pp.

Arnaud, L., Sailland, A., Lebrun, M., Pallett, K., Ravanel, P., Nurit, F., and Tissut, M. 1998. Physiological behavior of two tobacco lines expressing EPSP synthase resistant to glyphosate. *Pest. Biochem. Physiol.*, 62, 27–39.

Arntzen, C.J., Ditto, C.L., and Brewer, P.E. 1979. Chloroplast membrane alterations in triazine-resistant *Amaranthus* retroflexusbiotypes. *Proc. Natl. Acad. Sci. U.S.A.*, 76, 278–282.

Ashok, M., Turner, C., and Wilson, T.G. 1998. Insect juvenile hormone resistance gene homology with the bHLH-PAS family of transcriptional regulators. *Proc. Natl. Acad. Sci. U.S.A.*, 95, 2761–2766. (See comments.)

Axelsen, P.H., Harel, M., Silman, I., and Sussman, J.L. 1994. Structure and dynamics of the active-site gorge of acetylcholinesterase: synergistic use of molecular-dynamics simulation and x-ray crystallography. *Protein Sci.*, 3, 188–197.

Bakke, A. 1977. Field response to a new pheromonal compound isolated from *Ips typographus*. *Naturwissenschaften*, 64, 98.

Bakke, A. 1978. Barkbillenes kjemiske språk. *Naturen*, 1, 31–37.

BarretBee, K. and G. Dixon. 1995. Ergosterol biosynthesis inhibition: a target for antifungal agents. *Acta Biochim. Polonica*, 42, 465–479.

Beard, J., Marshall, S., Jong, K., Newton, R., Triplett-McBride, T., Humphries, B., and Bronks, R. 2000. 1,1,1-Trichloro-2,2-bis (p-chlorophenyl)-ethane (DDT) and reduced bone mineral density. *Arch. Environ. Health*, 55, 177–180.

Begg, C.P. and Berlin, J.A. 1989. Publication bias and dissemination of clinical research. *J. Natl. Cancer Inst.*, 81, 107–115.

Belden, J. and Lydy, M. 2000. Impact of atrazine on organophosphate insecticide toxicity. *Environ. Toxicol. Chem.*, 19, 2266–2274.

Berlin, M. 1986. Mercury. In *Specific Metals*, 2nd Ed., 386–445. Vol. 2, Friberg, L., Nordberg, G., and Nordman, C., Eds. Elsevier, Amsterdam.

Bhupinder, P.S., Khambay, B.B. and Bromilow, R.H. 2000. Discovery of pesticides. In *The New Chemistry*, Hall, N., Ed. Cambridge University Press, Cambridge, U.K. pp. 232–258.

Bloomquist, J.R. 1996. Ion channels as targets for insecticides. *Annu. Rev. Entomol.*, 41, 163–190.

Board on Agriculture and Natural Resources and Board on Environmental Studies and Toxicology, C.o.L.S. 2000. *The Future Role of Pesticides in U.S. Agriculture/ Committee on the Future Role of Pesticides in U.S. Agriculture*. 301 pp.

Borg, K., Wanntorp, H., Erne, K., and Hanko, L. 1969. Alkyl mercury poisoning in terrestrial Swedish wildlife. *Viltrevy*, 6, 301–379.

Bovey, R.W. and Young, A.L. 1980. *The Science of 2,4,5-T and Associated Phenoxy Herbicides*. John Wiley & Sons, New York. 462 pp.

Breidbach, O. and Kutsch, W. 1995. *The Nervous Systems of Invertebrates: An Evolutionary and Comparative Approach*. Birkhäuser Verlag, Basel, Switzerland. 448 pp.

Brien, S.E., Heaton, J.P., Racz, W.J., and Adams, M.A. 2000. Effects of an environmental anti-androgen on erectile function in an animal penile erection model. *J. Urol.*, 163, 1315–1321.

Brown, A.W.A. 1958. *Insecticide Resistance in Arthropods*. World Health Organization, Geneva.

Brown, V. 1988. *Acute and Sub-Acute Toxicology*. Hodder & Stoghton. London. 125 pp.

Buchananwollaston, V., Snape, A., and Cannon, F. 1992. A plant selectable marker gene based on the detoxification of the herbicide dalapon. *Plant Cell Rep.*, 11, 627–631.

Burnet, W.M.W., Christopher, J.T., Holtum, J.A.M., and Powles, S.B. 1994a. Identification of 2 mechanisms of sulfonylurea resistance within one population of rigid ryegrass (*Lolium-rigidum*) using a selective germination medium. *Weed Sci.*, 42, 468–473.

Burnet, M.W.M., Hart, Q., Holtum, J.A.M., and Powles, S.B. 1994b. Resistance to 9 herbicide classes in a population of rigid ryegrass (*Lolium-rigidum*). *Weed Sci.*, 42, 369–377.

Burnet, M.W.M., Loveys, B.R., Holtum, J.A.M., and Powles, S.B. 1993a. Increased detoxification is a mechanism of simazine resistance in *Lolium-rigidum*. *Pest. Biochem. Physiol.*, 46, 207–218.

Burnet, M.W.M., Loveys, B.R., Holtum, J.A.M., and Powles, S.B. 1993b. A mechanism of chlorotoluron resistance in *Lolium-rigidum*. *Planta*, 190, 182–189.

Busvine, J.R. 1953. Forms of insecticide-resistance in houseflies and body lice. *Nature*, 171, 118–119.

Butenandt, A., Beckmann, R., Stamm, D., and Hecker, E. 1959. Über den Sexual-Lockstoff des Seidenspinners *Bombyx mori*. Reindarstellung und Konstitutionsermittlung. *Z. Naturforschung*, 14b, 283–284.

Cade, T.J., Lincer, J.L., White, C.M., Roeseneau, D.G., and Swartz, L.G. 1971. DDE residues and eggshell changes in Alaskan falcons and hawks. *Science*, 172, 955–957.

Cantrell, J., Webb, N., and Mabis, A. 1969. *Acta Crystallogr. Sect. B.*, 25, 150.

Carde, R.M. and Minks, A.L. 1997. *Insect Pheromone Research*. Chapman & Hall, London. 684 pp.

Carson, R. 1962. *Silent Spring*. The Riverside Press, Boston, MA. 368 pp.

Carter, R.H. 1947. Estimation of DDT in milk by determination of organic chloride. *Ind. Eng. Chem. Anal.*, 1947, 54.

Casida, J.E. and Quistad, G.B. 1998. Golden age of insecticide research: past, present, or future? *Annu. Rev. Entomol.*, 43, 1–16.

Caux, P. and More, R. 1997. A spreadsheet program for estimating low toxic effects. *Environ. Toxicol. Chem.*, 16, 802–806.

Chambers, J. and Yardbrough, J. 1982. *Effects of Chronic Exposures to Pesticides*. Raven Press, New York. 250 pp.

Chatonnet, A., Hotelier, T., and Cousin, X. 1999. Kinetic parameters of cholinesterase interactions with organophosphates: retrieval and comparison tools available through ESTHER database: ESTerases, alpha/beta hydrolase enzymes and relatives. *Chem.-Biol. Interact.*, 119/120, 567–576.

Chatonnet, A. and Lockridge, O. 1989. Comparison of butyrylcholinesterase and acetylcholinesterase. *Biochem. J.*, 260, 625–634.

Christiansen, E. 1979. Chemical repellent prevents moose browsing. *Meddelelser Norsk Inst. Skogforskning*, 34, 241–248.

Clark, A.G. and Shamaan, N.A. 1984. Evidence that DDT-dehydrochlorinase from the housefly is a glutathione S-transferase. *Pest. Biochem. Physiol.*, 22, 249–261.

Crickmore, N., Zeigler, D.R., Feitelson, J., Schnepf, E., Van Rie, J., Lereclus, D., Baum, J., and Dean, D.H. 1998. Revision of the nomenclature for the *Bacillus thuringiensis* pesticidal crystal proteins. *Microbiol. Mol. Biol. Rev.*, 62, 807.

Daalen, v.J.J., Meltzer, J., Mulder, R., and Wellinga, K. 1972. A selective insecticide with a novel mode of action. *Naturwissenschaften*, 59, 312–313.

Dahm, P.A., Founaine, F.C., Panlaskie, J.E., Smith, R.C., and Atkeson, F.W. 1950. The effects of feeding parathion to dairy cows. *J. Dairy Sci.*, 33, 747–757.

Davidse, L.C. and Flach, W. 1978. Interaction of thiabendazole with fungal tubulin. *Biochim. Biophys. Acta*, 543, 82–90.

Deblock, M., Botterman, J., Vandewiele, M., Dockx, J., Thoen, C., Gossele, V., Movva, N.R., Thompson, C., Vanmontagu, M., and Leemans, J. 1987. Engineering herbicide resistance in plants by expression of a detoxifying enzyme. *EMBO J.*, 6, 2513–2518.

Dekker, J. 1972. Resistance. In *Systemic Fungicides*, Marsh, E., Ed. John Wiley & Sons, New York. pp. 156–174.

Devine, M., Duke, S.O., and Fedke, C. 1993. *Physiology of Herbicide Action*. Prentice Hall, New York. 441 pp.

Devonshire, A.L. and Sawicki, R.M. 1979. Insecticide-resistant *Myzus persicae* as an example of evolution by gene duplication. *Nature*, 280, 140–141.

deWaard, M.A. 1997. Significance of ABC transporters in fungicide sensitivity and resistance. *Pest. Sci.*, 51, 271–275.

Dewaard, M.A., Georgopoulos, S.G., Hollomon, D.W., Ishii, H., Leroux, P., Ragsdale, N.N., and Schwinn, F.J. 1993. Chemical control of plant-diseases: problems and prospects. *Annu. Rev. Phytopathol.*, 31, 403–421.

Dhadialla, T., Carlson, G., and Le, D. 1998. New insecticides with ecdysteroidal and juvenile hormone activity. *Annu. Rev. Entomol.*, 43, 545–569.

Duggan, R.E. and Weatherwax, J.R. 1967. Dietary intake of pesticide chemicals. *Science*, 157, 1006–1010.

Duran, I., Parrilla, A., Feixas, J., and Guerrero, A. 1993. Inhibition of antennal esterases of the Egyptian armyworm *Spodoptera-littoralis* by trifluoromethyl ketones. *Bioorg. Med. Chem. Lett.*, 3, 2593–2598.

Dybing, E. 2001. Dioksin. *Nytt Miljø Samfunnsmedisin*, 5, 1.

Dyte, C.E. 1972. Resistance to synthetic juvenile hormone in brain of flour beetle, *Tribolium castaneum*. *Nature*, 238, 48.

Ecobichon, D.J. 2001. Toxic effects of pesticides. In *Cassarett and Doull's Toxicology. The Basic Science of Poisons*, Klaassen, C., Ed. McGraw-Hill, New York. pp. 763–810.

Edwards, C. 1973. *Environmental Pollution by Pesticides*. Plenum Press, New York. 542 pp.

Elliott, M., Farnham, A.W., Janes, N.F., Needham, P.H., Pulman, D.A., and Stevenson, J.H. 1973. A photostable pyrethroid. *Nature*, 246, 169–170.

Elliott, M., Janes, N.F., and Potter, C. 1978. The future of pyrethroids in insect control. *Annu. Rev. Entomol.*, 23, 443–469.

Ellman, G.L., Courtney, K.O., Anders, A., and Featherstone, R.M. 1961. A new and rapid colorimetric determination of acetylcholinesterase activity. *Biochem. Pharmacol.*, 7, 88–95.

Emden, H. and Peaball, D.B. 1996. *Beyond Silent Spring, Integrated Pest Management and Chemical Safety*. Chapman & Hall. London. 322 pp.

Emmons, R.B., Duncan, D., Estes, P.A., Kiefel, P., Mosher, J.T., Sonnenfeld, M., Ward, M.P., Duncan, I., and Crews, S.T. 1999. The spineless-aristapedia and tango bHLH-PAS proteins interact to control antennal and tarsal development in *Drosophila*. *Development*, 126, 3937–3945.

Engels, A.J.G., Holub, E.F., Swart, K., and DeWaard, M.A. 1998. Genetic analysis of resistance to fenpropimorph in *Aspergillus niger*. *Curr. Genet.*, 33, 145–150.

Escartin, E. and Porte, C. 1997. The use of cholinesterase and carboxylesterase activities from *Mytilus galloprovincialis* in pollution monitoring. *Environ. Toxicol. Chem.*, 16, 2090–2095.

Eto, M., Casida, J.E., and Eto, T. 1962. Hydroxylation and cyclisation reaction involved in the metabolism of tri-o-cresyl phosphate. *Biochem. Pharmacol.*, 11, 337–352.

Eto, M. and Kuwano, K. 1992. Prenyl imidazoles and related compounds controlling hormonal development processes of insects. In *Insecticides: Mechanism of Action and Resistance*, Otto, D. and Weber, B., Eds. Intercept Ltd., Andover, U.K. pp. 155–165.

Faustman, E.M. and Omenn, G.S. 2001. Risk assessment. In *Cassarett and Doull's Toxicology. The Basic Science of Poisons*, Klaassen, C.D., Ed. McGraw-Hill, New York. pp. 83–104.

Fedke, C. 1982. *Biochemistry and Physiology of Herbicide Action.* Springer-Verlag, Heidelberg, Germany. 202 pp.

Fent, K. 1996. Ecotoxicology of organotin compounds. *CRC Crit. Rev. Toxicol.*, 26, 1–117.

Ferre, J. and Van Rie, J. 2002. Biochemistry and genetics of insect resistance to *Bacillus thuringiensis. Annu. Rev. Entomol.*, 47, 501–533.

Fimreite, N. 1970. Mercury uses in Canada and their possible hazards as sources of mercury contamination. *Environ. Pollut.*, 1, 119–131.

Fimreite, N. 1974. Mercury contamination of aquatic birds in northwest Ontario. *J. Wildl. Manage.*, 38, 120–131.

Finney, D. 1971. *Probit Analysis.* Cambridge University Press, Cambridge, U.K. 333 pp.

Fisher, H.K. and Williams, G. 1976. Paraquat is not bacteriostatic under anaerobic conditions. *Life Sci.*, 19, 421–426.

Flueckiger, C.R., Kristinsson, H., Senn, R., Rindlisbacher, A., Buholzer, H., and Voss, G. 1992. CGA 215944: a novel agent to control aphids and whiteflies. In *Brighton Crop Protection Conference: Pests and Diseases (1992).* British Crop Protection Council, Hampshire, U.K. pp. 43–50.

Fobert, P.R., Miki, B.L., and Iyer, V.N. 1991. Detection of gene regulatory signals in plants revealed by T-DNA-mediated fusions. *Plant Mol. Biol.*, 17, 837–851.

Fock, A., Kettner, J., and Boettger, M. 1992. The occurrence of 4-chlorotryptophan in *Vicia faba. Phytochemistry*, 31, 2327–2328.

Forgash, A.J. 1984. History, evolution and consequences of insecticide resistance. *Pest. Biochem. Physiol.*, 22, 178.

Friedman, H.B. 1992. DDT (Dichlorodiphenyltrichloroethane): a chemist's tale. *J. Chem. Educ.*, 69, 362–365.

Gafur, A., Tanaka, C., Shimizu, K., Ouchi, S., and Tsuda, M. 1998. Molecular analysis and characterization of the *Cochliobolus heterostrophus* beta-tubulin gene and its possible role in conferring resistance to benomyl. *J. Gen. Appl. Microbiol.*, 44, 217–223.

Gahan, L.J., Gould, F., and Heckel, D.G. 2001. Identification of a gene associated with bit resistance in *Heliothis virescens. Science*, 293, 857–860.

Georghiou, G.F. and Taylor, C.F. 1977. Genetic and biological influences in evolution of insecticide resistance. *J. Econ. Entomol.*, 70, 319–323.

Georghiou, G.P. 1990. Overview of Insecticide Resistance. ACS Symposium Series. American Chemical Society, Washington, D.C. 421, 18–41.

Georgopoulos, S.G. and Zaracovitis, P. 1967. Tolerance of fungi to organic fungicides. *Ann. Rev. Phytopathol.*, 5, 109.

Glare, R.T. and O'Callaghan, M. 2000. Bacillus thuringiensis: *Biology, Ecology and Safety.* John Wiley & Sons, Ltd., Chichester, U.K.

Glynn, P. 1999. Neuropathy target esterase. *Biochem. J.*, 344 (Pt. 3), 625–631.

Goodwin, M., Gooding, K.M., and Regnier, F. 1979. Sex pheromone in the dog. *Science*, 203, 559–561.

Goyer, A. and Clarkson, T.W. 2001. Toxic effects of metals. In *Cassarett and Doull's Toxicology. The Basic Science of Poisons*, Klaassen, C.D., Ed. McGraw-Hill, New York. pp. 811–868.

Graham, J. and Wienere, B. 1995. *Risk versus Risk.* Harvard University Press, Cambridge, MA. 337 pp.

Gray, G.M. and Graham, J.D. 1995. Regulating pesticides. In *Risk vs. Risk*, Vol. 1, Graham, J.D. and Wiener, J.B., Eds. Harvard University Press, Cambridge, MA. pp. 173–192.

Gregus, Z.K. and Klaasen C.D. 2001. Mechanisms of toxicity. In *Cassarett and Doull's Toxicology. The Basic Science of Poisons*, Klaassen, C., Ed. McGraw-Hill, New York. pp. 35–83.

Gressel, J. 2002. *Molecular Biology of Weed Control*, Vol. XVI. Taylor & Francis, London. 504 pp.

Grimvall, A. 1995. Evidence of naturally produced and man-made organohalogens in water and sediments. In *Naturally-Produced Organohalogens*, Grimvall, A. and deLeer, E., Eds. Kluwer Academic Publishers Inc, Dordrecht, Netherlands. pp. 1–20.

Gullan, P. and Cranston, P. 2000. *The Insects. An Outline of Entomology*. Blackwell Science Ltd., London. 470 pp.

Gunning, R.V., Moores, G.D., and Devonshire, A.L. 1997. Esterases and fenvalerate resistance in a field population of *Helicoverpa punctigera* (Lepidoptera: Noctuidae) in Australia. *Pest. Biochem. Physiol.*, 58, 155–162.

Gunning, R.V., Moores, G.D., and Devonshire, A.L. 1998. Insensitive acetylcholinesterase causes resistance to organophosphates in Australian *Helicoverpa armigera* (Hubner) (Lepidoptera: Noctuidae). *Pest. Sci.*, 54, 319–320.

Hahn, M.E., Karchner, S.I., Shapiro, M.A., and Perera, S.A. 1997. Molecular evolution of two vertebrate aryl hydrocarbon (dioxin) receptors (AHR1 and AHR2) and the PAS family. *Proc. Natl. Acad. Sci. U.S.A.*, 94, 13743–13748.

Hajjar, N.P. and Casida, J.E. 1978. Insecticidal benzoylphenyl ureas. Structure-activity relationships as chitin synthesis inhibitors. *Science*, 200, 1499–1500.

Halperin, W., Kalow, W., Sweeney, M.H., Tang, B.K., Fingerhut, M., Timpkins, B., and Wille, K. 1995. Induction of P-450 in workers exposed to dioxin. *Occup. Environ. Med.*, 52, 86–91.

Harborne, J.B. 1978. *Biochemical Aspects of Plant and Animal Coevolution: Proceedings of the Phytochemical Society Symposium, Reading, April 1977*, Vol. xvii. Academic Press, London. 435 pp.

Harborne, J.B. 1993. *Introduction to Ecological Biochemistry*. Academic Press, London. 318 pp.

Haughn, G.W., Smith, J., Mazur, B., and Somerville, C. 1988. Transformation with a mutant arabidopsis acetolactate synthase gene renders tobacco resistant to sulfonylurea herbicides. *Mol. Gen. Genet.*, 211, 266–271.

Hay, A. 1978a. Dioxin source is 'safe.' *Nature*, 274, 526.

Hay, A. 1978b. U.S. producers fear ban on 2,4,5-trichlorophenol. *Nature*, 275, 471.

Hayes, A.W. 2001. *Principles and Methods of Toxicology*, Vol. XIX. Taylor & Francis, Philadelphia. 1887 pp.

Hayes, W.J. and Laws, E.R. 1991a. *Handbook of Pesticide Toxicology. Vol. 1: General Principles*. Academic Press, San Diego. 1–496 pp.

Hayes, W.J. and Laws, E.R. 1991b. *Handbook of Pesticide Toxicology. Vols. 2 and 3: Classes of Pesticides*. Academic Press, San Diego. 497–1576 pp.

Hemingway, J. and Ranson, H. 2000. Insecticide resistance in insect vectors of human disease. *Annu. Rev. Entomol.*, 45, 371–391.

Hertel, R., Thomson, K.-S., and Russo, V.E.A. 1972. *In vitro* auxin binding to particulate cell fractions from corn coleoptiles. *Planta*, 107, 325–340.

Hewlett, P. and Plackett, R. 1979. *An Introduction to the Interpretation of Quantal Responses in Biology*. Edward Arnold, London. 85 pp.

Hill, I. and Wright, S. 1978. *Pesticide Microbiology*. Academic Press, London. 844 pp.

Hjelm, O. 1996. *Organohalogens in Forest Soil*. Department of Water and Environmental Studies, Linköping University, Linköping.

Hodgson, O., Mailman, R.B., Chambers, J.E., and Dow, R.E. 1998. *Dictionary of Toxicology.* MacMillan, New York. 504 pp.

Hoel, L. 1999. *Bruk av enzymer i meitemark som biomarkører for forurensning av tungmetaller og organofosfater.* Biological Institute, University of Oslo, Oslo. 80 pp.

Holan, G. 1969. New halocyclopropane insecticides and the mode of action of DDT. *Nature*, 221, 1025–1029.

Holt, J.S. and Lebaron, H.M. 1990. Significance and distribution of herbicide resistance. *Weed Technol.*, 4, 141–149.

Hoostal, M.J., Bullerjahn, G.S., and McKay, R.M.L. 2002. Molecular assessment of the potential for *in situ* bioremediation of PCBs from aquatic sediments. *Hydrobiologia*, 469, 59–65.

Huang, F., Buschman, L.L., Higgins, R.A., and McGaughey, W.H. 1999. Inheritance of resistance to *Bacillus thuringiensis* toxin (Dipel ES) in the European corn borer. *Science*, 284, 965–967.

Ingles, P.J., Adams, P.M., Knipple, D.C., and Soderlund, D.M. 1996. Characterization of voltage-sensitive sodium channel gene coding sequences from insecticide-susceptible and knockdown-resistant house fly strains. *Insect Biochem. Mol. Biol.*, 26, 319.

Inui, H., Kodama, T., Ohkawa, Y., and Ohkawa, H. 2000. Herbicide metabolism and cross-tolerance in transgenic potato plants co-expressing human CYP1A1, CYP2B6, and CYP2C19. *Pest. Biochem. Physiol.*, 66, 116–129.

Inui, H., Veyama, Y., Shiota, N., Ohkawa, Y., and Ohkawa, H. 1999. Herbicide metabolism and cross-tolerance in transgenic potato plants expressing human CYP1A1. *Pest. Biochem. Physiol.*, 64, 33–46.

Ishaaya, I. 1992. 13 selective insect control agents: mechanisms and agricultural importance. In *Insecticides: Mechanism of Action and Resistance*, Otto, D. and Weber, B., Eds. Intercepts Ltd., Andover, U.K. pp. 127–134.

Janicki, R.H. and Kinter, W.B. 1971. DDT: disrupted osmoregulatory events in the intestine of the eel *Anguilla restrata* adapted to seawater. *Science*, 173, 1146–1148.

Jaworski, E.G., 1972. Mode of action of N-phosphonomethylglycine. Inhibition of aromatic amino acid biosynthsis. *J. Agric. Food Chem.*, 20, 1195–1198.

Johnson, M.K. 1982. The target for initiation of delayed neurotoxicity by organophosphorus esters: biochemical studies and toxicological application. In *Reviews of Biochemical Toxicology*, Vol. 4, Hodgson, E., Bend, J., and Philpot, R., Eds. Elsevier, New York. pp. 141–212.

Johnson, M.K. and Lotti, M. 1980. Delayed neurotoxicity caused by chronic feeding of organophosphates requires a high-point of inhibition of neurotoxic esterase. *Toxicol. Lett.*, 5, 99–102.

Jones, D. 1998. *Piperonyl Butoxide.* Academic Press, London. 323 pp.

Jones, W.A. and Jacobson, M. 1968. Isolation of N,N-diethyl-m-toluamide (deet) from female pink bollworm moths. *Science*, 159, 99–100.

Kaiser, J. 2000. Toxicology: panel backs EPA dioxin assessment. *Science*, 290, 1071.

Karlson, P. and Butenandt, A. 1959. Pheromones (ectohormones) in insects. *Annu. Rev. Entomol.*, 4, 39–58.

Karlson, P. and Lüscher, M. 1959. "Pheromones": a new term for a class of biological active substances. *Nature*, 183, 55–56.

Kearney, P.C. and Kaufman, D.D. 1975. *Herbicides: Chemistry, Degradation, and Mode of Action.* Marcel Dekker, New York. 1036 pp.

Keiding, J. 1967. Persistance of resistant populations after the relaxation of selection pressure. *World Rev. Pest Control*, 6, 115–130.

Kelce, W.R., Stone, C.R., Laws, S.C., Earlgray, L., Kemppainen, J.A., and Wilson, E.M. 1995a. Ego boost for toxicology elsewhere: persistent DDT metabolite p,p'-DDE is potent androgen receptor antagonist. *Hum. Exp. Toxicol.*, 14, 850–850. (Comment.)

Kelce, W.R., Stone, C.R., Laws, S.C., Gray, L.E., Kemppainen, J.A., and Wilson, E.M. 1995b. Persistent DDT metabolite p,p'-DDE is a potent androgen receptor antagonist. *Nature*, 375, 581–585.

Keyserlingk, v.H.C. and Willis, R.J. 1992. The GABA-activiated Cl– chammel in insects as target for insecticide action: a physiological study. In *Insecticides: Mechanism of Action and Resistance*, Vol. 1, Otto, D. and Weber, B., Eds. Intercept, Andover, U.K. pp. 205–236.

Kim, Y.S., Kim, D.H., and Jung, J. 1998. Isolation of a novel auxin receptor from soluble fractions of rice (*Oryza sativa* L.) shoots. *FEBS Lett.*, 438, 241–244.

Kishimoto, D. and Kurihara, N. 1996. Effects of cytochrome P450 antibodies on the oxidative demethylation of methoxychlor catalyzed by rat liver microsomal cytochrome P450 isozymes: isozyme specificity and alteration of enantiotopic selectivity. *Pest. Biochem. Physiol.*, 56, 44–52.

Kishimoto, D., Oku, A., and Kkurihara, N. 1995. Enantiotopic selectivity of cytochrome P450-catalyzed oxidative demethylation of methoxychlor: alteration of selectivity depending on isozymes and substrate concentrations. *Pest. Biochem. Physiol.*, 51, 12–19.

Klaassen, C., Ed. 2001. *Cassarett and Doull's Toxicology. The Basic Science of Poisons*. McGraw-Hill, New York. 1236 pp.

Klaassen, C. and Eaton, D. 1991. Principles of toxicology. In *Cassarett and Doull's Toxicology. The Basic Science of Poisons*, Vol. 1, Amdur, M., Doull, D., and Klaassen, C., Eds. Pergamon Press, New York. pp. 12–49.

Klee, H.J., Muskopf, Y.M., and Gasser, C.S. 1987. Cloning of an arabidopsis-thaliana gene encoding 5-enolpyruvylshikimate-3-phosphate synthase: sequence-analysis and manipulation to obtain glyphosate-tolerant plants. *Mol. Gen. Genet.*, 210, 437–442.

Koch, R.B. 1969. Inhibition of animal tissue ATPase activies by chlorinated hydrocarbon pesticides. *Chem.-Biol. Interact.*, 1, 199–209.

Koellner, G., Kryger, G., Millard, C.B., Silman, I., Sussman, J.L., and Steiner, T. 2000. Active-site gorge and buried water molecules in crystal structures of acetylcholinesterase from *Torpedo californica*. *J. Mol. Biol.*, 296, 713–735.

Kohn, M.C., Sewall, C.H., Lucier, G.W., and Portier, C.J. 1996. A mechanistic model of effects of dioxin on thyroid hormones in the rat. *Toxicol. Appl. Pharmacol.*, 136, 29–48.

Köller, W. 1992. *Target Sites of Fungicide Action*. CRC Press, Boca Raton, FL. 328 pp.

Krämer, F. and Mencke, N. 2001. *Flea Biology and Control: The Biology of Cat Flea Control and Prevention with Imidacloprin in Small Animals*. Springer-Verlag, Berlin. 192 pp.

Krueger, H.R. and O'Brien, R.D. 1959. Relationship between metabolism and differencial toxicity of malathion in insects and mice. *J. Econ. Entomol.*, 52, 1063–1067.

Kuthiala, A., Gupta, R.K., and Davis, E.E. 1992. Effect of the repellent deet on the antennal chemoreceptors for oviposition in *Aedes-aegypti* (Diptera, Culicidae). *J. Med. Entomol.*, 29, 639–643.

Landers, J.P. and Bunce, N.J. 1991. The Ah receptor and the mechanism of dioxin toxicity. *Biochem. J.*, 276, 273–287.

Le Baron, H.M. and Gressel, J. 1982. *Herbicide Resistance in Plants*. John Wiley & Sons, New York. 401 pp.

Lee, C.Y., Hemingway, J., Yap, H.H., and Chong, N.L. 2000. Biochemical characterization of insecticide resistance in the German cockroach, *Blattella germanica*, from Malaysia. *Med. Vet. Entomol.*, 14, 11–18.

Levine, B. and Murphy, S.D. 1977. Effects of piperonyl butoxide on the metabolism of dimethyl and diethyl phosphorothionate insecticides. *Toxicol. Appl. Pharmacol.*, 40, 393–406.

Levitan, I.K. and Kaczmarek, L.K. 2002. *The Neuron Cell and Molecular Biology*. Oxford University Press, Oxford. 603 pp.

Leyser, O. 2002. Molecular genetics of auxin signaling. *Annu. Rev. Plant Biol.*, 53, 377–398.

Li, Z.J., Hayashimoto, A., and Murai, N. 1992. A sulfonylurea herbicide resistance gene from arabidopsis-thaliana as a new selectable marker for production of fertile transgenic rice plants. *Plant Physiol.*, 100, 662–668.

Lindström, G., Småstuen, H.L., and Nicolaysen, T. 2000. *International Intercalibration on Dioxin in Food 2000: Final Report*.

Longnecker, M.P., Klebanoff, M.A., Zhou, H.B., and Brock, J.W. 2001. Association between maternal serum concentration of the DDT metabolite DDE and preterm and small-for-gestational-age babies at birth. *Lancet*, 358, 110–114.

Lundholm, C.E. 1997. DDE-induced eggshell thinning in birds: effects of p,p'-DDE on the calcium and prostaglandin metabolism of the eggshell gland. *Comp. Biochem. Physiol. C Pharmacol. Toxicol. Endocrinol.*, 118, 113–128.

Lunt, G.G. and Olsen, R.W. 1988. *Comparative Invertebrate Neurochemistry*, Vol. viii. Croom Helm, London. 327 pp.

Magnus, A.H. 1970. Mercury-tolerance in *Pyrenophora avenae* in Norway. *Meldinger Norges Landbrukshøyskle*, 49, 1–8.

Maneechote, C., Holtum, J.A.M., Preston, C., and Powles, S.B. 1994. Resistant acetyl-Coa carboxylase is a mechanism of herbicide resistance in a biotype of *Avena-sterilis* ssp. Ludoviciana. *Plant Cell Physiol.*, 35, 627–635.

Mattes, C., Bradley, R., Slaughter, E., and Browne, S. 1996. Cocaine and butyrylcholinesterase (BChE): determination of enzymatic parameters. *Life Sci.*, 58, L257–L261.

McGready, R., Hamilton, K.A., Simpson, J.A., Cho, T., Luxemburger, C., Edwards, R., Looareesuwan, S., White, N.J., Nosten, F., and Lindsay, S.W. 2001. Safety of the insect repellent N,N-diethyl-M-toluamide (DEET) in pregnancy. *Am. J. Trop. Med. Hyg.*, 65, 285–289.

Mehl, A., Rolseth, V., Gordon, S., Bjoraas, M., Seeberg, E., and Fonnum, F. 2000. Brain hypoplasia caused by exposure to trichlorfon and dichlorvos during development can be ascribed to DNA alkylation damage and inhibition of DNA alkyltransferase repair. *Neurotoxicology*, 21, 165–173.

Melander, A.L. 1914. Can insects become resistant to sprays? *J. Econ. Entomol.*, 7, 167.

Mellanby, K. 1970. *Pesticides and Pollution*. Harper Collins, London. 221 pp.

Melnick, R.L., Kohn, M.C., and Portier, C.J. 1996. Implications for risk assessment of suggested nongenotoxic mechanisms of chemical carcinogenesis. *Environ. Health Perspectives*, 104, 123–134.

Michael, G. 1986. *Dioxin, Agent Orange: The Facts*. Plenum Press, New York. 289 pp.

Mineau, P., Ed. 1991. *Cholinesterase-Inhibiting Insecticides: Their Impact on Wildlife and the Environment*. Elsevier, Amsterdam. 348 pp.

Miyamato, J., Kaneko, H., Hutson, D.H., Esser, H.O., Gorbach, S., and Dorn, E. 1988. *Pesticide Metabolism: Extrapolation from Animals to Man.* Blackwell, Oxford. 120 pp.

Morcillo, Y., Borghi, V., and Porte, C. 1997. Survey of organotin compounds in the western Mediterranean using molluscs and fish as sentinel organisms. *Arch. Environ. Contam.Toxicol.*, 32, 198–203.

Morin, J.-P., Rochat, D., Malosse, C., Lettere, M., deChenon, R.D., Wibwo, H., and Descoins, C. 1996. Ethyl 4-methyloctanoate, major component of *Oryctes rhinoceros* (L.) (Coleoptera, Dynastidae) male pheromone. *C. R. Acad. Sci. Ser. III Sci. Vie-Life Sci.*, 319, 595–602.

Munthe, K. and Ørpen, H.M. 2001. *Plantevern. Kjemiske og biologiske midler 2001–2002.* Landbruksforlaget. Oslo. 262 pp.

Myers, E.W., Sutton, G.G., Delcher, A.L., Dew, I.M., Fasulo, D.P., Flanigan, M.J., Kravitz, S.A., Mobarry, C.M., Reinert, K.H.J., Remington, K.A., Anson, E.L., Bolanos, R.A., Chou, H.H, Jordan, C.M., Halpern, A.L., Lonardi, S., Beasley, E.M., Brandon, R.C., Chen, L., Dunn, P.J., Lai, Z.W., Liang, Y., Nusskern, D.R., Zhan, M., Zhang, Q., Zheng, X.Q., Rubin, G.M., Adams, M.D., and Venter, J.C. 2000. A whole-genome assembly of *Drosophila. Science*, 287, 2196–2204.

Nachmansohn, D. and Wilson, I.B. 1951. The enzymatic hydrolysis and synthesis of acetylcholine. *Adv. Enzymol.*, 12, 259–339.

Nakagawa, Y., Matsumura, F., and Hashino, Y. 1993. Effect of diflubenzuron on incorporation of [3H]-N-acetylglucosamine ([3H]NAGA) into chitin in the intact integument from the newly molted American cockroach *Periplaneta americana. Comp. Biochem. Physiol. C Comp. Pharmacol. Toxicol.*, 106, 711–715.

Nelson, D.L. and Cox, M.M. 2000. *Lehninger Principles of Biochemistry.* Worth Publisher, New York. 1150 pp.

Neumann, J. and Peter, H.H. 1987. Insecticidal organophosphates: nature made them first. *Experientia*, 43, 1235–1237.

Nicolas, G. and Sillans, D. 1989. Immediate and latent effects of carbon dioxide on insects. *Annu. Rev. Entomol.*, 34, 97–116.

Nishimura, K. and Okimoto, H. 1997. Quantitative structure-activity relationships of DDT-type compounds in a sodium tail-current in crayfish giant axons. *Pest. Sci.*, 50, 104–110.

Nolte, D.L. and Barnett, J.P. 2000. A repellent to reduce mouse damage to longleaf pine seed. *Int. Biodeterior. Biodegrad.*, 45, 169–174.

O'Brien, R.D. 1976. Acetylcholinesterase and its inhibition. In *Insecticide Biochemistry and Physiology*, Vol. XXII, Wilkinson, C.F., Ed. Plenum Press, New York. 768.

Oehme, M., Manø, S., Brevik, E., and Knutzen, J. 1989. Determination of polychlorinated dibenzofuran (PCDF) and dibenzo-p-dioxin (PCDD) levels and isomer patterns in fish, crustacea, mussel and sediment samples from a fjord region polluted by Mg-production. *Fresenius Z. Anal. Chem.*, 335, 987–997.

Oliva, J., Navarro, S., Barba, A., Navarro, G., and Salinas, M.R. 1999. Effect of pesticide residues on the aromatic composition of red wines. *J. Agric. Food Chem.*, 47, 2830–2836.

Oppenoorth, F.J. 1965. Biochemical genetics of insecticide resistance. *Annu. Rev. Entomol.*, 10, 185–206.

Otto, D. and Weber, B. 1992. *Insecticides: Mechanism of Action and Resistance.* Intercept, Andover, U.K. 499 pp.

Otto, G. 1838. *Haandbog i toxikologien.* Græbe & Søn, Copenhagen. 504 pp.

Ozoe, Y., Yagi, K., Nakamura, M., Akamatsu, M., Miyake, T., and Matsumura, F. 2000. Fipronil-related heterocyclic compounds: structure-activity relationships for interaction with gamma-aminobutyric acid- and voltage-gated ion channels and insecticidal action. *Pest. Biochem. Physiol.*, 66, 92–104.

Pajuelo, L., SanchezAlonso, J.A., delHoyo, N., Pulido, J.A., and PerezAlbarsanz, M.A. 1997. Non-muscarinic- and non-adrenergic-mediated effects of lindane on phosphoinositide hydrolysis in rat brain cortex slices. *Neurochem. Res.*, 22, 57–62.

Palumbi, S.R. 2001. Evolution: humans as the world's greatest evolutionary force. *Science*, 293, 1786–1790.

Parkinson, A. 2001. Biotransformation of xenobiotics. In *Cassarett and Doull's Toxicology. The Basic Science of Poisons*. Klaassen, C.D., Ed. McGraw-Hill, New York. pp. 133–225.

Parrilla, A. and Guerrero, A. 1994. Trifluoromethyl ketones as inhibitors of the processionary moth sex-pheromone. *Chem. Senses*, 19, 1–10.

Peterson, G.E. 1967. The discovery and development of 2,4-D. *Agric. Hist.*, 41, 243–253.

Plaisance, K.L. and Gronwald, J.W. 1999. Enhanced catalytic constant for glutathione S-transferase (atrazine) activity in an atrazine-resistant *Abutilon theophrasti* biotype. *Pest. Biochem. Physiol.*, 63, 34–49.

Pöch, G., Reiffenstein, R., and Unkelbach, H.D., 1990. Application of isobologram technique for the analysis of combined effects with respect to additivity as well as independence. *Can. J. Physiol. Pharmacol.*, 68, 682–688.

Pokorny, R. 1941. Some chlorophenoxyacetic acids. *J. Am. Chem. Soc.*, 63, 1768.

Portier, C.J., Sherman, C.D., Kohn, M., Edler, L., KoppSchneider, A., Maronpot, R.M., and Lucier, G. 1996. Modeling the number and size of hepatic focal lesions following exposure to 2,3,7,8-TCDD. *Toxicol. Appl. Pharmacol.*, 138, 20–30.

Prapanthadara, L., Hemingway, J., and Ketterman, A.J. 1995a. Ddt-resistance in *Anopheles-gambiae* (Diptera, Culicidae) from Zanzibar, Tanzania, based on increased Ddt-dehydrochlorinase activity of glutathione S-transferases. *Bull. Entomol. Res.*, 85, 267–274.

Prapanthadara, L.A., Kuttastep, S., Hemingway, J., and Ketterman, A.J. 1995b. Characterization of the major form of glutathione transferase in the mosquito *Anopheles-dirus-A. Biochem. Soc. Trans.*, 23, S81.

Quensen, J.F., Mousa, M.A., Boyd, S.A., Sanderson, J.T., Froese, K.L., and Giesy, J.P. 1998a. Reduction of aryl hydrocarbon receptor-mediated activity of polychlorinated biphenyl mixtures due to anaerobic microbial dechlorination. *Environ. Toxicol. Chem.*, 17, 806–813.

Quensen, J.F., Mueller, S.A., Jain, M.K., and Tiedje, J.M. 1998b. Reductive dechlorination of DDE to DDMU in marine sediment microcosms. *Science*, 280, 722–724.

Quero, C., Malo, E.A, Fabrias, G., Camps, F., Lucas, P., Renou, M., and Guerrero, A. 1997. Reinvestigation of female sex pheromone of processionary moth (*Thaumetopoea pityocampa*): no evidence for minor components. *J. Chem. Ecol.*, 23, 713–726.

Ramel, C. 1977. *Chlorinated Phenoxy Acids and Their Dioxins: Mode of Action, Health Risks, and Environmental Effects*. Swedish Natural Science Research Council, Stockholm.

Ratcliff, D. 1967. Decrease in eggshell weight in certain birds of prey. *Nature*, 215, 208–210.

Reeder, N.L., Ganz, P.J., Carlson, J.R., and Saunders, C.W. 2001. Isolation of a deet-in-sensitive mutant of *Drosophila melanogaster* (Diptera: Drosophilidae). *J. Econ. Entomol.*, 94, 1584–1588.

Richardson, M. 1996. *Environmental Xenobiotics.* Taylor & Francis, London. 492 pp.

Richter, P. 1992. Possible genetic start points and end points of insecticide resistance evolution. In *Mechanism of Action and Resistance*, Otto, D. and Weber, B., Eds. Intercept Ltd., Andover, U.K. pp. 355–363.

Robbins, W.E., Thompson, M.J., Yamamoto, R.T., and Shortino, T.J. 1965. Feeding stimulants for the female house fly, *Musca domestica* Linneaus. *Science*, 147, 628–630.

Roch, F. and Alexander, M. 1997. Inability of bacteria to degrade low concentrations of toluene in water. *Environ. Toxicol. Chem.*, 16, 1377–1383.

Rockstein, M. 1978. *Biochemistry of Insects.* Academic Press, New York. 649 pp.

Ruder, F.S., Benson, J.A., and Kayser, H. 1991. The mode of action of the insecticide/acaricide diafenthiuron. In *Insecticides: Mechanism of Action and Resistance*, Otto, D. and Weber, B., Eds. Intercept, Andover, U.K. pp. 263–276.

Ryan, G.F. 1970. Resistance of common groundstill to simazine and atrazine. *Weed Sci.*, 18, 614–616.

Schnepf, E., Crickmore, N., Van Rie, J., Lereclus, D., Baum, J., Feitelson, J., Zeigler, D.R., and Dean, D.H. 1998. *Bacillus thuringiensis* and its pesticidal crystal proteins. *Microbiol. Mol. Biol. Rev.*, 62, 775.

Schrader, G. 1951. *Die Entwicklung neuer insektizide auf Grundlage organischer Fluor- und Phosphor-Verbindungen.* Verlag Chemie, Weinheim, Germany. 92 pp.

Schrader, G. 1963. *Die Entwicklung neuer insektizider Phosphrsäure-Ester.* Verlag Chemie GMBH, Weinheim/Bergstr., Germany. 444 pp.

Searcey, M.T., Graves, C.G., and Olson, R. 1977. Isolation of a warfarin binding protein from liver endoplasmatic reticulum of Sprague-Dawley and warfarin-resis-tant rats. *J. Biol. Chem.*, 252, 6260–6267.

Seawright, A.A. and Eason, C.T. 1993. Proceeding of the science workshop on 1080. Miscellaneous Series 28. In *Workshop on 1080*, Zealand, The Royal Society of New Zealand, Ed. Christchurch, New Zealand. p. 173.

Secor, W.E., Freeman, G.L., and Wirtz, R.A. 1999. Short report: prevention of *Schisto-soma mansoni* infections in mice by the insect repellents AI3-37220 and N,N-di-ethyl-3-methylbenzamide. *Am. J. Trop. Med. Hyg.*, 60, 1061–1062.

Shah, D., Horsch, R.B., Klee, H.J., Kishore, G., Winther, J., Turner, N.E., Hironake, C.M., Sanders, P.R., Gasser, C.S., Aybert, S., Siegel, N.R., Rogers, S.G., and Fraley, R.T. 1986. Engineering herbicide tolerance in transgenic plants. *Science*, 233, 478–481.

Shiota, N., Nagasawa, A., Sakaki, T., Yabusaki, Y., and Ohkawa, H. 1994. Herbicide-re-sistant tobacco plants expressing the fused enzyme between rat cytochrome P4501a1 (Cyp1a1) and yeast NADPH-cytochrome P450 oxidoreductase. *Plant Physiol.*, 106, 17–23.

Shivanandappa, T., Joseph, P., and Krishnakumari, M.K. 1988. Response of blood and brain cholinesterase to dermal exposure of Bromophos in the rat. *Toxicology*, 48, 199–208.

Siegel, M. 1981. Sterol-binding fungicides: effect on sterol biosynthesis and site of action. *Plant Dis.*, 65, 986–989.

Sierra-Santoyo, A., Hernandez, M., Albores, A., and Cebrian, M.E. 2000. Sex-depen-dent regulation of hepatic cytochrome P-450 by DDT. *Toxicol. Sci.*, 54, 81–87.

Silver, A. 1974. *The Biology of Cholinesterases*, Vol. XIV. North-Holland Pub. Co., Amsterdam. 596 s. pp.

Silverstein, R.M. 1981. Pheromones: background and potential for use in insect pest control. *Science*, 213, 1326–1331.

Silverstein, R.M. 1984. Chemistry of insect communication. In *Insect Communication*, Vol. 1. Royal Entomological Society, London. pp. 105–120.

Singer, A.G., Agosta, W.C., O'Connell, R.J., Pfaffman, C., Bowen, D.W., and Field, F.H. 1976. Dimethyl disulphide: an attractant in pheromone in hamster vaginal secretion. *Science*, 191, 948–950.

Sladen, W., Menzie, C., and Reichel, W. 1966. DDT residues in Adelie penguins and crabeater seal from Antarctica: ecological implications. *Nature*, 210, 670–673.

Smith, A. 2000. How toxic is DDT? *Lancet*, 356, 267–268.

Smith, C.N., Gilbert, I.H., and Gouck, H.K. 1960. *Use of Insect Repellents*. U.S. Department of Agriculture, Agricultural Research Service. pp. 1–8.

Soderlund, D.M., Clark, J.M., Sheets, L.P., Mullin, L.S., Piccirillo, V.J., Sargent, D., Stevens, J.T., and Weiner, M.L. 2002. Mechanisms of pyrethroid neurotoxicity: implications for cumulative risk assessment. *Toxicology*, 171, 3–59.

Spencer, W.F. and Cliath, M.M. 1972. Volatility of DDT and related compounds. *Agric. Food Chem.*, 20, 645–649.

Spiller, D. 1963. Insecticide resistance: effects of WARF antiresistant on toxicity of DDT to adult houseflies. *Science*, 142, 585–586.

Stalker, D.M., McBride, K.E., and Malyj, L.D. 1988. Herbicide resistance in transgenic plants expressing a bacterial detoxification gene. *Science*, 242, 419–423.

Stenersen, J.H. 1965. DDT-metabolism in resistant and susceptible stable-flies and in bacteria. *Nature*, 207, 660–661.

Stenersen, J. 1966. Cross-resistance to Dilan and DDT plus Warf antiresistant in DDT-resistant stable flies. *Norsk Entomol. Tidsskrift.*, 12, 11–16.

Stenersen, J. 1972. Pesticides for plant protection in Norway: legislation, use, and residues. In *Residue Reviews*, Vol. 42, Gunther, F. and Gunther, D., Eds. Springer-Verlag, Heidelberg, Germany. pp. 91–102.

Stenersen, J. 1979. Action of pesticides on earthworms. Part I. The toxicity of cholinesterase inhibiting insecticides towards earthworms as evaluated by laboratory tests. *Pest. Sci.*, 10, 66–74.

Stenersen, J. 1981. *The Mode of Action of Cholinesterase-Inhibiting Pesticides on Earthworms: Toxicity, Inhibition of Cholinesterases and Degradation*. Norwegian Plant Protection Institute, UiO, Ås. p. 86.

Stenersen, J. 1992. Uptake and metabolism of xenobiotics by earthworms. In *Ecotoxicology of Earthworms*, Greigh-Smith, P.W., Becker, H., Edwards, P.J., and Heimbach, F., Eds. Intercept Ltd., London. pp. 118–127.

Stenersen, J. and Friestad, H. 1969. Residues of DDT and DDE in soil from Norwegian fruit orchards. *Acta Agric. Scand.*, 19, 240–243.

Stenersen, J., Gilman, A., and Vardanis, A. 1973. Carbofuran: its toxicity to and metabolism by earthworm (*Lumbricus terrestris*). *J. Agric. Food Chem.*, 21, 166–171.

Stenersen, J. and Sømme, L. 1963. Notes on cross-resistance and genetics of resistance to the DDT-group insecticides in the stable fly (*Stomoxys calcitrans*). *Norsk Entomol. Tidsskrift.*, 12, 113–117.

Sternburg, J., Kearns, C.W., and Moorefield, H. 1954. Resistance to DDT. DDT-dehydrochlorinase, an enzyme found in DDT-resistant flies. *J. Agric. Food Chem.*, 2, 1125–1130.

Stickel, L. and Rhodes, L. 1970. The thin eggshell problem. In *The Biological Impact of Pesticides in the Environment*, Gillett, J., Ed. Oregon State University, Corvallis. pp. 31–35.

Strathern, P. 2000. *Mendeleyev's Dream*. Hamish Hamilton, London. 309 pp.

Streber, W.R., Kutschka, U., Thomas, F., and Pohlenz, H.D. 1994. Expression of a bacterial gene in transgenic plants confers resistance to the herbicide phenmedipham. *Plant Mol. Biol.*, 25, 977–987.

Streber, W.R. and Willmitzer, L. 1989. Transgenic tobacco plants expressing a bacterial detoxifying enzyme are resistant to 2,4-D. *Biotechnology*, 7, 811–816.

Stringer, A. and Wright, M.A., 1973. Effect of benomyl and some related compounds on *Lumbricus terrestris* and other earthworms. *Pesticide Science*. 4, 165–170.

Stromstedt, M., Rozman, D., and Waterman, M.R. 1996. The ubiquitously expressed human CYP51 encodes lanosterol 14 alpha-demethylase, a cytochrome P450 whose expression is regulated by oxysterols. *Arch. Biochem. Biophys.*, 329, 73–81.

Sur, N., Pal, S., Banerjee, H., Adityachaudhury, N., and Bhattacharyya, A. 2000. Photodegradation of fenarimol. *Pest. Manage. Sci.*, 56, 289–292.

Sussman, J.L., Harel, M., Frolow, F., Oefner, C., Goldman, A., Toker, L., and Silman, I. 1991. Atomic structure of acetylcholinesterase from *Torpedo californica*: a prototypic acetylcholine-binding protein. *Science*, 253, 872–879.

Sutter, T.R., Guzman, K., Dold, K.M., and Greenlee, W.F. 1991. Targets for dioxin: genes for plasminogen activator inhibitor-2 and interleukin-1 beta. *Science*, 254, 415–418.

Taitz, L. and Zeiger, E. 1998. *Plant Physiology*. Sinauer Associates, Inc., Sunderland.

Thompson, H.M. 1991. Serum "B" esterases as indicators of exposure to pesticides. In *Cholinesterase-Inhibiting Insecticides. Their Impact on Wildlife and the Environment*, Vol. 2, Mineau, P., Ed. Elsevier, Amsterdam. pp. 109–126.

Timbrell, J. 2000. *Principles of Biochemical Toxicology*. Taylor & Francis, London. 394 pp.

Tomizawa, M., Otsuka, H., Miyamoto, T., Eldefrawi, M.E., and Yamamoto, I. 1995a. Pharmacological characteristics of insect nicotinic acetylcholine-receptor with its ion-channel and the comparison of the effect of nicotinoids and neonicotinoids. *J. Pest. Sci.*, 20, 57–64.

Tomizawa, M., Otsuka, K., Miyamoto, T., and Yamamoto, I. 1995b. Pharmacological effects of imidacloprid and its related-compounds on the nicotinic acetylcholine-receptor with its ion-channel from the torpedo electric organ. *J. Pest. Sci.*, 20, 49–56.

Tomlin, C., Ed. 1994. *The Pesticide Manual: Incorporating the Agrochemicals Handbook*. British Crop Protection Council, Farnham, Surrey.

Tomlin, C., Ed. 2000. *The Pesticide Manual: A World Compendium*. British Crop Protection Council, Farnham, Surrey. 1250 pp.

Torstenson, L., Stark, J., and Göransson, B. 1975. The effect of repeated applications of 2,4-D and MCPA on their breakdown in soil. *Weed Res.*, 15, 159–164.

Trevors, J. 1986. A BASIC program for estimating LD50 values using the IBM-PC. *Bull. Environ. Contam. Toxicol.*, 37, 18–26.

Tripathi, A.K., Prajapati, V., Khanuja, S.P.S., and Kumar, S. 2002. Chitin synthesis inhibitors as insect-pest control agents. *J. Med. Aromatic Plant Sci.*, 24, 104–122.

Vargas, R.I., Stark, J.D., Kido, M.H., Ketter, H.M., and Whitehand, L.C. 2000. Methyl eugenol and cue-lure traps for suppression of male oriental fruit flies and melon flies (Diptera: Tephritidae) in Hawaii: effects of lure mixtures and weathering. *J. Econ. Entomol.*, 93, 81–87.

Vinggaard, A.M., Breinholt, V., and Larsen, J.C. 1999. Screening of selected pesticides for oestrogen receptor activation *in vitro*. *Food Addit. Contam.*, 16, 533–542.

Walker, C.H., Hopkin, S.P., Sibly, R.M., and Peakall, D.B. 1996. *Principles of Ecotoxicology*. Taylor & Francis, London. 321 pp.

Walker, C.H. and Thompson, H.M. 1991. *Phylogenetic Distribution of Cholinesterases and Related Esterases* in *Cholinesterase-Inhibiting Insecticides: Their Impact on Wildlife and the Environment*, Mineau, P., Ed. Elsevier, Amsterdam. 1–19 pp.

Watanabe, T. and Sano, T. 1998. Neurological effects of glufosinate poisoning with a brief review. *Hum. Exp. Toxicol.*, 17, 35–39.

Wehtje, G., Walker, R.H., and Shaw, J.N. 2000. Pesticide retention by inorganic soil amendments. *Weed Sci.*, 48, 248–254.

West, T.F. and Campbell, G.A. 1950. *DDT and Newer Persistent Insecticides*. Chapman & Hall Ltd., London. 632 pp.

Westing, A.H. 1975. Environmental consequences of the second Indochina war: a case study. *AMBIO*, 4, 216–222.

White, R.A.J., Franklin, R.T., and Agosin, M. 1979. Conversion of a-pinene to a-pinene oxide by rat liver and the bark beetle *Dendroctonus terebrans* microsomal fractions. *Pest. Biochem. Physiol.*, 10, 233–242.

Wilkinson, C.F. 1976. *Insecticide Biochemistry and Physiology*, Vol. XXII. Plenum Press, New York. 768 pp.

Williams, C. 1967. The juvenile hormone. II. Its role in the endocrine control of molting, pupation, and adult development in the cecropia silkworm. *Biol. Bull. Woods Hole*, 121, 572–585.

Williams, D.R., Fisher, M.J., and Rees, H.H. 2000. Characterization of ecdysteroid 26-hydroxylase: an enzyme involved in molting hormone inactivation. *Arch. Biochem. Biophys.*, 376, 389–398.

Wilson, E.O. 1963. Pheromones. These substances are used for chemical communication by some animal species. *Sci. Am.*, 208, 100–114.

Wilson, T.G. 2001. Resistance of *Drosophila* to toxins. *Annu. Rev. Entomol.*, 46, 545–571.

Winder, B.S., Strandgaard, C.S., and Miller, M.G. 2001. The role of GTP binding and microtubule-associated proteins in the inhibition of microtubule assembly by carbendazim. *Toxicol. Sci.*, 59, 138–146.

Winteringham, F.P.W. and Hewlett, P.S. 1964. Insect cross-resistance phenomena: their practical and fundamental implications. *Chem. Ind.*, 35, 1512–1518.

Woodwell, G.M. 1967. Toxic substances and ecological cycles. *Sci. Am.*, 216, 24–31.

Woodwell, G.M., Craig, P.P., and Johnson, H.A. 1971. DDT in the biosphere: where does it go? *Science*, 174, 1101–1107.

Woolson, E., Thomas, R., and Ensor, P. 1972. Survey of polychlorodibenzo-p-dioxin content in selected pesticides. *Agric. Food Chem.*, 20, 351–353.

World Health Organization (WHO). 1974. *Conference on Intoxication due to Alkylmercury Treated Seed, Baghdad, Iraq*. p. 46.

World Health Organization (WHO). 1990. *Public Health Impact of Pesticide Use in Agriculture*. WHO, Geneva.

Worthing, C., Ed. 1979. *The Pesticide Manual: A World Compendium*. British Crop Protection Council, Croydon. 655 pp.

Wright, R.H. 1963. Chemical control of chosen insects. *New Sci.*, 5, 598–600.

Wright, R.H. 1964. After pesticides: what? *Nature*, 204, 121–125.

Yamamoto, I., Yabuta, G., Tomizawa, M., Saito, T., Miyamoto, T., and Kagabu, S. 1995. Molecular mechanism for selective toxicity of nicotinoids and neonicotinoids. *J. Pest. Sci.*, 20, 33–40.

Yu, S. and Terriere, L. 1974. Possible role of microsomal oxidases in metamorphosis and reproduction in the housefly. *J. Insect. Physiol.*, 20, 1901–1912.

Yu, S.J. and Terriere, L.C. 1979. Cytochrome P-450 in insects. 1. Differences in forms present in insecticide resistant and susceptible house flies. *Pest. Biochem. Physiol.*, 12, 239–248.

Zeidler, O. 1874. Verbindungen von Chloral mit Brom- und Chlorbenzol. *Berichte*, 7, 1180.

Zeleny, J., Havelka, J., and Slama, K. 1997. Hormonally mediated insect-plant relationships: arthropod populations associated with ecdysteroid-containing plant, *Leuzea carthamoides* (Asteraceae). *Eur. J. Entomol.*, 94, 183–198.

Zhou, Z.H. and Syvanen, M. 1997. A complex glutathione transferase gene family in the housefly *Musca domestica*. *Mol. Gen. Genet.*, 256, 187–194.

Zilberman, D., Schmitz, A., Casterline, G., Lichtenberg, E., and Siebert, J.B. 1991. The economics of pesticide use and regulation. *Science*, 253, 518–522.

Zlotkin, E. 1999. The insect voltage-gated sodium channel as target of insecticides. *Annu. Rev. Entomol.*, 44, 429–455.

Zlotkin, E., Devonshire, A.L., and Warmke, J.W. 1999. The pharmacological flexibility of the insect voltage gated sodium channel: toxicity of AaIT to knockdown resistant (kdr) flies. *Insect Biochem. Mol. Biol.*, 29, 849–853.

Zwiernik, M.J., Quensen, J.F., and Boyd, S.A. 1998. FeSO4 amendments stimulate extensive anaerobic PCB dechlorination. *Environ. Sci. Technol.*, 32, 3360–3365.

Index

Numbers